ALLE ZEIT WACH

SJ

1842

Alfred Grafe

A History of Experimental Virology

Translated by
Elvira Reckendorf

Springer-Verlag
Berlin Heidelberg New York
London Paris Tokyo
Hong Kong Barcelona
Budapest

Dr. Alfred Grafe
Deutsche Gesellschaft für Hygiene
und Mikrobiologie
Institut für Medizinische Virologie
der Universität Heidelberg
Im Neuenheimer Feld 324
W-6900 Heidelberg 1
Federal Republic of Germany

ISBN-13:978-3-642-75252-0 e-ISBN-13:978-3-642-75250-6
DOI: 10.1007/978-3-642-75250-6

Library of Congress Cataloging-in-Publication Data
Grafe, Alfred, 1922- A history of experimental virology / Alfred Grafe; translated by Elvira Reckendorf.
p. cm. Includes bibliographical references and indexes.
ISBN-13:978-3-642-75252-0

1. Virology--History. 2. Virology--Experiments--History. I. Title. [DNLM: 1. Virology--history. QW 11.1 G736h] QR359.G73 1991 576'.64--dc20 DNLM/DLC for Library of Congress 91-5003 CIP

Softcover reprint of the hardcover 1st edition 1991

Typesetting: Data conversion by Elsner & Behrens GmbH, Oftersheim
31/3145-543210 - Printed on acid-free paper

Preface

> By their powers of reason scientists will be able to extract from nature the answers to their questions.
> *From:* Critique of Pure Reason, 1781
> Immanuel Kant (1724–1804), German Philosopher

History is a composite of stories. The history of the biological disciplines has been written by all those who opened the gates of new knowledge by generating ideas and the experiments to support them. Previous authors have attempted various approaches to the history of virology, as is reflected in the numerous books and book-series issuing from the publishing houses. This volume is an attempt at a comprehensive yet compact survey of virology, which has meant penetrating the rigid limits of the separate disciplines of biology in which virologists have worked.

Writing this history of experimental virology was really a search for the origins and for vital signposts to portray the wide scope of the knowledge attained thus far. This was done in complete awareness of the fact that every presentation depends heavily upon the perspective of the observer, and of necessity communicates only a part of the whole. The present scientific story hopes to recount the most important knowledge achieved during this past century – the first century of the exciting developments in virology.

Though every history is influenced by its "Zeitgeist", it is essentially created by people. These are more readily recognizable in the earlier than in the later stages of virology, which becomes very apparent simply from the number of scientists participating in a single publication; this has swollen from 1–2 in the past to more than 20 at present. A publication by "Geheimrat Prof. Dr. Emil von Behring and Kitasato" belongs to the annals of history not merely for chronological reasons.

This trend has also had an impact on this book. To mention each of the most important publications in the history of experimental virology would mean a list of more than 2,000 titles. The limits of the book would have diverged from traditional printing norms. Thus, not a list of references but a bibliography has been prepared to serve only as an orientation and a help for those who desire more profound details of the works quoted.

This book has entailed many long years of study, during which I experienced the kind-hearted support of a circle of friends to whom I owe my sincere gratitude. The critical help afforded by my colleagues at the Institute, Angela Rösen-Wolff, Gholamreza Darai and Rolf Flügel deserves special acknowledgement. My indebtedness also extends to the libraries and their librarians for unmeasurable assistance. My young friend Timothy Rebbeck in the Department of Human Genetics at the University of Michigan, Ann Arbor helped fill in many a literature gap - a war remnant in the German libraries, and Robert J.Moore from the National Institute for Medical Research, Mill Hill, London helped to find some biographical data. Monika Lechner from the Hygiene Institute at the University of Heidelberg, and Ingeborg Helmy from the Paul Ehrlich Institute at Frankfurt on Main/Langen gave me access to valued treasures which mark the beginnings of parasitology, bacteriology and hygiene, immunology and serology, and, of course, virology. In trying to cope with the ever-growing avalanche of contemporary literature, I came to appreciate deeply the unceasing amiability and cooperation of the library assistants at the University of Heidelberg and Boehringer Mannheim GmbH: Bernd List, Jutta Klockner, Therese Stockmann and Peter Stadler. I would like to thank in particular Dr. Patrick King for his painstaking and very critical revision of the text.

Heidelberg, September 1991 Alfred Grafe

Contents

Introduction 1

1 2000 Years Preceded the Emergence of Experimental Virology 3

1.1 Antiquity and the Origin of the Plague 3
1.2 The Cornerstones of Generatio Spontanea Begin to Quake 5
1.3 The Germ Theory is Born and Proven 10
1.4 The Germ Theory Crystallizes into the Germ Doctrine 16
1.5 The Theoretical and Experimental Basis of the Germ Doctrine at the End of the 19th Century .. 22
1.5.1 Theoretical Basis 23
1.5.2 Experimental Basis 27

2 Microscopes and Culture Media Fail 30

2.1 The Virus and Koch's Postulates 30
2.1.1 The Use of the Term "Virus" 31
2.1.2 Koch's Postulates 31
2.1.2.1 "Koch's Postulates" Do Not Exist! 32
2.1.2.2 Reliable Recognition of Pathogens Using Koch's Proof 33
2.2 A New Type of Pathogenic Agent? 34
2.2.1 The Discussion Begins 34
2.2.1.1 Tobacco Mosaic Disease (TMD) 35
2.2.1.2 Foot-and-Mouth Disease (FMD) 37
2.2.1.3 Bovine Pleuropneumonia 39
2.2.1.4 Rabbit Myxomatosis 39
2.2.1.5 African Horse Sickness 40
2.2.1.6 Fowl Plague 40
2.2.1.7 Yellow Fever 41

2.2.2 The Pathogen Remains While the Definition of the Virus Changes ... 42
2.2.2.1 Invisible Microbes ... 42
2.2.2.2 Non-Culturable Microbes ... 43
2.2.2.3 Filterable Microbes ... 43
2.2.3 The Filterable Viruses are Recognized ... 44
2.2.4 Who was the Founder of Virology? ... 45

3 Are Filterable Viruses Miniscule Bacteria? ... 47

3.1 The Use of Physical Methods ... 48
3.1.1 Filtration ... 49
3.1.2 Centrifugation ... 51
3.1.3 Adsorption ... 54
3.1.4 Electrophoresis ... 55
3.1.5 Optical Methods ... 56
3.2 Biochemical Investigations with Filterable Viruses ... 57
3.3 Are Cell Inclusions Specific to Filterable Viruses? ... 61
3.4 Culturing Filterable Viruses In Vivo and In Vitro ... 63
3.4.1 The Search for Experimental Animals ... 64
3.4.2 Is "Pure Culture" of Filterable Viruses Possible? ... 65
3.5 Filterable Viruses as Antigens ... 70
3.5.1 Neutralization Tests ... 71
3.5.2 Aggregation Tests ... 74
3.5.2.1 Agglutination and Precipitation ... 75
3.5.2.2 Complement-Fixation Reaction (CFR) ... 76
3.5.3 Immunity and Immunization ... 76
3.5.3.1 Immunity ... 77
3.5.3.2 Passive Immunization ... 78
3.5.3.3 Active Immunization ... 79
3.6 Filterable Viruses as Pathogens ... 83
3.6.1 Infectious Diseases Caused by Viruses ... 84
3.6.2 Tumour Development ... 86
3.7 Viruses are not Just Very Small Bacteria ... 88
3.7.1 Linking Virus Reproduction to Living Cells ... 89
3.7.2 Viruses are Autonomous, Organized Particles ... 90

4 On the Way to a Definition of the Virus ... 93

4.1. Viruses as Pathogens ... 99
4.1.1 New Knowledge about Animal Viruses ... 101

4.1.1.1 First Contact With a Virus 106
4.1.1.2 Tissue and Organs In Which Cell Infection is Initiated . 107
4.1.1.3 Secondary or Generalized Symptomatic Infections 110
4.1.1.4 Uncommon Interactions Between the Virus and the Host Organisms 112
4.1.1.5 Tumour Virus Infections 113
4.1.1.6 Interference 114
4.1.1.7 Immunological Problems 115
4.1.1.8 Immunization 118
4.1.2 Disease-Producing Insect Viruses 127
4.1.2.1 Inclusion Diseases 127
4.1.2.2 Non-Inclusion Diseases 129
4.1.3 Virus Diseases in Higher Plants 130
4.1.3.1 First Contact With a Virus 130
4.1.3.2 Intracellular Virus Multiplication 132
4.1.3.3 The Spread of Viral Plant Diseases 132
4.2 A Biochemical Approach to the Virus 133
4.2.1 The Morphology and Size of Viruses 135
4.2.2 The Biochemistry of Viruses 137
4.2.2.1 Nucleic Acids 137
4.2.2.2 Proteins 140
4.2.2.3 Virus Enzymes and Other Substances 140
4.2.3 The Structure of Viruses 141
4.2.3.1 Plant Viruses 142
4.2.3.2 Enteroviruses 143
4.2.3.3 Complex, Medium-sized Animal Viruses 143
4.2.3.4 Complex, Larger Animal Viruses 144
4.2.3.5 Bacteriophages 144
4.3 Viruses as Genetic Macromolecules 146
4.3.1 The Genetics of Bacteriophages 146
4.3.2 Plant Virus Mutants 154
4.3.3 The Genetics of Animal Viruses 155
4.4 What Exactly is a Virus? 158

5 The Period of Biochemistry and Molecular Genetics ... 164

5.1 The Virion 172
5.1.1 Virion Morphology and Capsid Symmetry 172
5.1.2 Virion Structure 179
5.1.2.1 The Biochemical Structure of the Virion 180
5.1.2.2 The Molecular-Genetic Structure of Virions 186

5.1.3 The Taxonomy and Classification of Viruses 189
5.2 The Virus Multiplication Cycle 197
5.2.1 Adsorption 198
5.2.2 Uptake by the Cell 200
5.2.3 Uncoating 203
5.2.4 The Synthesis Phase 203
5.2.4.1 DNA Viruses 204
5.2.4.2 RNA Viruses 206
5.2.4.3 Retroviruses 207
5.2.4.4 Hepadnaviruses 208
5.2.5 Virus Assembly 209
5.2.6 Virus Maturation 211
5.2.7 Virus Release 211
5.3 Viruses as Pathogens 212
5.3.1 Acute Virus Infections 214
5.3.1.1 Animal Viruses 214
5.3.1.2 Plant Viruses 219
5.3.2 Inapparent Virus Infections 220
5.3.2.1 Persistent Infections 221
5.3.2.2 Latent Infections 222
5.3.3 Slow Virus Infections 223
5.3.3.1 Slow Infections Caused by Viruses 223
5.3.3.2 Slow Infections Caused by Unconventional Agents ... 225
5.3.4 Viruses and Tumours 229
5.3.4.1 In Vitro Tests 230
5.3.4.2 Virus-Specified Proteins in Tumours and Transformed Cells 235
5.3.4.3 Genes and Their Products in Cell Transformation 235
5.3.4.4 Results on Some DNA (Tumour) Viruses 238
5.3.4.5 Results with RNA Tumour Viruses 242
5.4 Viruses and Immunity 244
5.4.1 Humoral and Cell-Mediated Immunity as Parameters of Protection 251
5.4.2 The Diagnosis of Virus Infection 255
5.4.3 Immunoprophylaxis in Virus Infections 260
5.5 Viral Pathogenicity 269
5.5.1 Histological and Anatomical-Pathological Effects 269
5.5.1.1 Teratogenic Virus Effects 270
5.5.1.2 Mutagenic Virus Effects 271
5.5.1.3 Virus Diseases Dependent on Virus-Induced Defects of the Immune System 272

5.5.2 Molecular-Genetic Assessment of Virus Pathogenicity . 274
5.5.2.1 Cell and Tissue Tropism 274
5.5.2.2 Virulence and Attenuation 275
5.5.2.3 Latency 277
5.6 Virus Chemotherapy 277
5.6.1 The Antibiotic Phase 278
5.6.2 The Anti-metabolite Phase 279
5.6.3 Will the Molecular-Genetic Approach Help? 282

Epilogue 287

Glossary 291

Bibliography 295

Name Index 311

Subject Index 329

Introduction

The field of science presently designated "virology" is a branch of microbiology and is thus linked to biology. The classification of a virus is a challenge because it simply does not exercise the basic functions attributed to a living entity: a virus does not have a metabolic system with its own respiration and growth, it lacks the ability to respond to stimuli, and it reproduces in a unique way.

Microscopic perception of minute, living entities was accomplished by the Dutchman van Leeuwenhoek in the second half of the 17th century. He had improved upon the compound microscope, probably developed by his countryman Janssen in 1590, to the extent that a 270-fold magnification could be achieved. Since the French surgeon Sedillot first used the term microorganism in 1878, it has been employed generally in the literature.

The word virus had already been employed to mean a poisonous substance in ancient Greek and Latin. Ios in the ancient Greek was the visible poison which caused man and animal alike to become ill or even die. It was recognizable as the venomous secretion of a viper, the poisonous potion from a plant, or the saliva of a rabid dog. The Latin word virus denotes a slimy, poisonous, malodorous liquid and, hence, corresponds to the ancient Greek word. This visible, tangible poison was contrasted with the invisible, intangible, putrid secretion termed miasma, which was assumed emanated from inside the earth and polluted the surrounding air with poisonous vapours. Miasma was deemed to be the source of various infectious diseases which caused widespread death and were generally referred to as the plague, simply because it was impossible to differentiate effectively between diseases. It has always been, and will continue to be, a challenge to medical historians to classify the ancient types of plague according to our present-day knowledge of epidemic diseases.

Attic schools of medicine not only practiced but also further developed the healing arts adopted from the Orient: the belief that men are subject to protective as well as to vengeful gods, goddesses and demons. By means of natural disasters such as earthquakes, floods, hailstones etc., the gods afflicted men with contagious diseases or warned them of an imminent plague. The gods then accompanied their ill-fated victims throughout their miserable suffering to their deaths, or rewarded them with a cure.

Consequently, we find that the science of the virus begins in a mystical period of the medical arts and spans more than three millennia, up to and including the modern field of molecular genetics. It is now an intrinsic part of the study of infections and the related sciences of biology and molecular biology.

The first two chapters attempt to depict the transition of the virus concept in the light of our ever-expanding knowledge of life and the progress medicine has achieved in determining the causes and transmission of infectious diseases. Subsequent chapters are concerned with milestones in the history of experimental virology over the last century.

1 2000 Years Preceded the Emergence of Experimental Virology

The historical period which finally ushered in the field of experimental virology reflects how a religio-mystical concept of life and disease was gradually transformed into a rational one. The latter part of this pre-virology epoch, which extends into the late 19th century, ranges from a world at first observed and comprehended only through the doctrines of naturalist philosophy, and culminates in the development of experimental science employing highly sophisticated laboratory techniques. It marks the tenacious, century-long end of the religio-philosophical concept of the spontaneous origin of life, referred to as generatio spontanea or generatio aequivoca, abiogenesis, archigenesis, or heterogenesis in the literature of over two millennia.

The history of virology is linked to the beginning but also to the end of the miasma theory, when the unidentifiable, communicable-disease-causing substance contagium, expounded by Greek philosophers and European scientists and physicians, was firmly substituted by protozoa, fungi, and bacteria. The 2000 years in which three of the world's great religions emerged, in which the entire universe was permeated with a host of philosophical ideas, and in which the exact, natural sciences evolved, may also serve as a framework to broaden our outlook on the origin of infectious disease.

1.1 Antiquity and the Origin of the Plague

Let us begin with ancient Greece and its religious concepts of nature which date back to the Egyptian and Babylonian cultures, i.e. approximately to 4000 BC: Through the intervention of demons and gods, mankind was beset by maladies and pestilence.

Hippocrates, the Greek physician from Kos, can be traced back to the Doric nobility headed by Asclepios the son of Apollo, the god of healing. He was the first to attempt replacement of the religio-mystical concept of illness with rational observation and experience. Among the 74 books comprising the Corpus Hippocraticum are some about the prevalent diseases of the time, as well as treatises on the nature of man and the essence of gases. Hippocrates concluded in these writings that plagues occurred irrespective of man's way of life or everyday events. Instead, he found the origin of disease in alterations of the most common essentiality for all life, namely the air. During plagues the air carried vapours from swamps, stagnant waters, unburied, human bodies or animal carcasses. After a siege, for example, large crowds of people contaminated the air thereby creating miasmata which helped spread disease. The plague of Thucydides, the Attic plague (430–425 BC) is said to have occurred in this way.

According to Hippocrates, miasmata were inanimate hazardous gases. Strikingly enough, he maintained this at a time when life was thought to originate spontaneously, and great significance was ascribed to the effect of the sun on the ocean and the amount of moisture within the ground. The idea of invisible, spontaneously generated beings first appeared in the writings of the Roman cultural historian, Varro. He compiled the knowledge of that era in no less than 600 volumes. As for the source of disease, he supposed causative agents to be of animal origin and so small – animalia minutissima – as to defy human vision. In addition, Varro referred to infectious material as contagium, probably borrowing this term from the Roman poet Carus, who had described the origin of the Attic pest in 430 BC in the 6th book of his didactic poem *De Rerum Natura*. As a disciple of the Greek philosopher Epicurus, who derived his materialistic concept of nature from the atomistic materialism described by the Greek philosopher Democritus, Carus equated life-generating atoms – semina vitalia – with contagium.

The tiniest organisms mentioned by the Greeks are the maggots, grubs and small worms found in decaying meat. Aristotle also believed these to be the source of flies and diverse insects, as well as eels and other sea creatures. The causative agent of the endemic pest in Romagna, called mala aria, which Varro thought was animal in origin, must therefore have been much smaller. At approximately the same time, Celsus, a Roman writer of great versatility, summarized the teachings of Hippocrates and the Greek schools of medicine, which he had translated into Latin. In *De Medicina*, Celsus employed the word virus to mean the equivalent of the Greek poison ios.

Nevertheless, it may be erroneous to conclude that Celsus really introduced the term virus.

Before delving into the Middle Ages, it is essential to glance at Galenus, the most prominent physician in antiquity barring Hippocrates. He was familiar with the literature on the healing arts, including that of Hippocrates, and wrote 18 famous commentaries. In addition, he himself produced more than 200 original treatises on the healing and medicinal arts, which brought him the authoritative reputation he was to maintain throughout many centuries. In his position as personal physician to the Roman emperor Marcus Aurelius Antonius, Galenus personally experienced the severe plagues which followed in the wake of battle. These he generally labelled as the pest when mortality was high. One pest which broke out in the Roman army during the Parthian War in 165 A.D. was depicted in great detail. It has become known as the Antonius pest or the Galenus pest in the history of epidemic disease. Galenus claimed two causes for spontaneously generated epidemic or pest-like diseases: direct contact with the contagia or airborne miasmata. His two-fold explanation of the epidemic pathogens enjoyed unchallenged acceptance well into the Middle Ages for want of a novel or more exact description of contagia and miasmata.

Ios, i.e. virus, continued to remain a slimy poison of visible origin. This is illustrated by the dog foaming with rage depicted in Egyptian, Greek, and Roman literature, as well as in the constellation of dog-star Sirius, where it unites man and the gods.

1.2 The Cornerstones of Generatio Spontanea Begin to Quake

For centuries after Galenus, there was little reason to doubt the spontaneous origin of living organisms and diseases. Whether or not the pest, smallpox, black death or typhus were versions of the same disease remaind unknown, as did the morbilli. Various measures were recommended to prevent or to heal these diseases, and when those available here on earth proved useless, the gods and goddesses as well as the demons and the Saints were implored. They were promised not only the relinquihing of sin and wickedness, but also the construction of churches, and the conducting of pilgrimages and processions.

Infectious skin ulcers, better known as leprosy, which had been regarded as a contagious disease for more than 2000 years, became rampant as a result of war and the crusades. This led to countermeasures by governments, including isolation of the sick, and the construction of lazaretti on the Apennine peninsula, hospitals for incurables, and leprosaria in northern Europe. In many areas the afflicted were required to wear symbols or clothing designating their malady, stigmatized as leprous outcasts and deprived of any human contact, this effectively being buried alive. Desperation was evidenced by conspiracy and rebellion, as well as by persecution of the Jews, who were made responsible for the suffering. These consequences all entered the annals of European history, and seemed quite normal.

Greater than usual interest, however, was shown in an outbreak of a plague among French, occupation troops in Naples in 1494. This venereal disease spread rapidly through large parts of central Europe within 2 years, and soon thereafter throughout most of the world. Attributed to nationality and linked to politics, this disease was called Morbus Neapolitanus or Morbus Gallicus for hundreds of years. The more or less serious consequencies of this sexually transmitted disease sourred the authorities of many nations and cities to counteract its spreading. Even authors and poets were urged to combat this plague. Fracastoro, the best-known of them, wrote a piece of didactic poetry in 1521 entitled *Syphilis sive Morbus Gallicus,* in which Apollo punished a hero, a shepherd named Syphilus, by smiting him with the dreaded disease. Fracastoro, however, was not only motivated into setting down his thoughts about Syphilis – the name since retained for the disease – but he also began to ponder the causes of such infections. He did not look for these, as was common, in the position of the sun, moon and planets, and their passing through the constellations of the stars. In his publication which appeared in Venice in 1546 entitled *De contagionibus et contagiosis morbus et eorum curatione*, he advanced a unique theory on the origin of the types of pest known at that time. He concluded that there were living organisms, seeds of disease-producing agents, seminaria morbi animata, which when invested with certain attributes were able to evoke various types of disease.

Fracastoro reasoned that these seminaria were transferred in three ways: by direct contact, by casual carriers (fomites) such as textiles or water, and by the air over distances. When inhaled, these seminaria entered the chest, where they were carried to the heart by the humours of the body. Hence, Fracastoro surpassed the teachings of Hippocrates and Varro. His contagious substance seminaria contagionis developed spontaneously from sick organisms, whereas miasmata were thought to arise from dead material. That seminaria were transmissible and infectious was an accepted premise which the Parisian authorities acknowledged during the pest epidemic in 1533. They forbade the transfer of beds or clothing from house to house, and did not allow persons caring for the afflicted to visit people suffering from other diseases.

A different theory of the origin of living organisms was put forth by Harvey during the latter part of his life. Between 1598 and 1602 he had studied at the University of Padua, which was considered the most progressive centre of medicine and philosophy at that time. One of his teachers was Fabricius, born in Aquapendente in the Papal States in 1537. During the 82 years of his life, Fabricius greatly advanced the field of developmental physiology, Bursa fabricii reminding us still of his experiments with chicken embryos. After receiving a doctorate in Art and Medicine, Harvey returned to England with new knowledge about the anatomical-physiological relationship between the male and female sex organs. It was here that he worked as anatomist and physiologist, continuing his experiments with incubated eggs and motivating others to discover more about the origin of insects. In 1651 his book *Exercitationes de Generatione Animalum* was printed. Meanwhile, by coining his famous phrase "omne vivum ex ovo", he dealt the concept of spontaneous generation its first perilous blow.

Experimental support for Harvey's theory was provided by Redi, who was familiar with Harvey's works. In 1688 he prepared eight jars containing meat, four of which he left open, the other four tightly closed. The result was that maggots developed only in the uncovered bottles. To ascertain whether the presence or absence of air was responsible for this result, he repeated the experiment. This time he did not close the bottles but simply covered some of them with gauze. The gauze-covered bottles remained free of maggots and Redi concluded that maggots originated from the eggs laid by flies. This of course contradicted the spontaneous generation of maggots, but failed to shed more light on the spontaneous generation of disease.

Besides the Italian naturalists Cardan (1541–1576) and Fallopio (1523–1562) who were attracted by the concept of a contagium vivum, Kircher firmly placed the origin of disease-producing matter in life itself. In his book *Ars Magna Lucis et Umbrae* which appeared in 1646, he described innumerable creeping animalcula in decaying material, and reported worms in the blood of fevered patients. Similar observations had been made earlier by Borel, court physician to Louis

XIV. But in 1658 Kircher published his *Scrutinium Physico-Medicum Contagiosae Luis, quae Pestis dicitur* in Rome, which contained a particularly new idea. He described visible, living organisms taken from the body of a pest victim. These he observed with a microscope, the optical instrument invented in 1625 by Faber, physician at the court of Pope Urban VII. Kircher called the creatures contagium animatum, because he realized they were carriers of the plague. Although it is not possible to discover the power of his microscope or to ascertain which organisms he had seen, Kircher was, nevertheless, the first to describe a microscopic, disease-producing organism and to identify insects as possible carriers of disease. In the ensuing period, the greatly improved, compound microscope shifted the problem of generatio spontanea from the macro- to the microscopic world. Heading the list of inventors is van Leeuwenhoek, a non-professional from Delft who, perhaps to eke out his meagrely paid official and unofficial work, produced lenses and compound microscopes. He left no less than 247 microscopes with 419 lenses upon his death! Thanks to his friend the Dutch physician de Graaf and to Oldenburg, Secretary of the Royal Society of London, Leeuwenhoek's microscopic discoveries were confirmed in at least 120 letters sent to London and 26 to the Academy of Science in Paris after 1673. These include the levende Dierkens or animalcula in water (1675), spermatozoa in human seminal fluid (1677), and small animals in the scum on teeth. At the same time, an Italian botanist, Vallisnieri, began to pursue the development of insects and sheep. He is credited with the discovery of the correlation between the number of corpora lutea and the number of foetuses.

The first monograph on protozoa appeared in 1718 under the authorship of the microscopist Jablot, who taught mathematics and geometry at the Royal Academy of Painting and Sculpture in Paris. Not only was he interested in microscopically small living organisms, but he also constructed microscopes which were probably superior to those of Leeuwenhoek. He is well known for his experiments using hay infusions under various temperature and air conditions, for his observations of the development of small animals from eggs, and for his descriptions of the morphology of protozoa. Despite all these observations, the theory of the spontaneous origin of life remained unshaken. Two of its staunch supporters were Needham and Leclerc (known as Count de Buffon and Director of Jardin Royal des Plantes). Between 1745 and 1749 these men discussed and published works on the spontaneous origin of microorganisms based of experimental proof. Their claims stimulated Spallanzani, who was already deply involved in the problem of generatio spontanea, to carry out his own experiments. In 1768 Spallanzani refuted Needham and Buffon not only by experimenting with boiled infusions, described in *Saggio di osservazioni microscopiche concernenti il sistema della generazione dei Signori di Needham e Buffon*, but also by artifically inseminating lower and higher animals. His insemination in 1785 of a female spaniel with the sperms of a male resulted in three healthy welps and gradual comprehension of the significance of female and male sexual organs.

Of course, the conclusions about spontaneous generation which Spallanzani drew from his heating experiments were criticized. Nevertheless, the practical consequences of his findings paved the way for the introduction of food conservation by Appert in 1804, thus weakening and disqualifying the views of Spallanzani's critics. Appert won the Napoleon I prize for developing a practical, food-preservation method, and in 1810 he received 12000 francs to publish his book *L'art de conserver toutes les substances animales et* végétales. Appert found support from the Parisian physicist and chemist Gay-Lussac, who in 1804 had published *M*émoires sur l'analyse de l'air atmospherique together with the naturalist, von Humboldt. Gay-Lussac became a member of "Du comité consultativ des artes et des manufactures" in 1805 and had had to evaluate Appert's preserved foods. Carrying out experiments, he found, among other things, that conserved animal or vegetable matter began to ferment within a few days after being transferred to a different container. He concluded that oxygen is indispensable for fermentation.

Meanwhile, the discovery of further, minute forms of infusoria progressed rapidly, owing to great improvements in microscopy. By the end of the 18th century the Danish physician and naturalist O.Müller, had named approximately 150 species in the first systematic tabulation of bacteria, published in his monograph *Animalcula infusoria fluviatilia et marina* (1786). He divided the genus *monas* (Gr., for indivisible unit) into 10 species and the genus *vibrio* (L., for comma-shaped) into 31 species which included *bacilli* (L.,for rod) and spirochetes (L., *spiro* for coil and Gr., *chaite* for long hair).

By the beginning of the 19th century the number of known infusoria had increased to more than 500 species. This led Ehrenberg, a German physician and naturalist who had described over 300 of the infusoria, to publish a new monograph, *Die Infusionstierchen als vollkommene Organismen* in 1838. Three years later Dujardin, professor of zoology at the University of Toulouse, criticized Ehrenberg's work in an original document. In it he used the term bacteria for large, rod-shaped infusoria. The word thereafter being adopted for all microscopically visible organisms, which were in fact considered at that time to be minute animals.

Although spontaneous generation had been disproved by numerous experiments, belief in the phenomenon continued. The minuscule organisms which had been isolated from patients were considered to be the result and not the cause of the disease.

1.3 The Germ Theory is Born and Proven

Greek philosophers, as well as Greek and Roman physicians, described two ways in which pathogenic miasma and contagia were transmitted. It was firmly held that these agents were spontaneously generated, but their nature was the subject mostly of speculation.

In the preface to *Scrutinium...* Kircher suggested that the epidemic of childbed fever was due to worms causing putrefaction. Furthermore, it was assumed that diseases such as measles and smallpox were caused by other living contagia. This assumption also applied to syphilis as Andry wrote in *De la* génération des vers dans le corps printed in Amsterdam in 1701, and by the Swedish botanist von Linné in his *Systema naturae...* in 1735. As early as 1718, the Roman physician Lancisi advocated the draining of all swamps because he had concluded that the mosquito must be the carrier of malaria. Tragically, his advice went unheeded.

The Slovenian physician Plenĉiĉ presented a comprehensive concept of the cause of infectious diseases in his work *Opera medico-physica* in 1762. He maintained that airborne contagia, i.e animalcula minima or animalcula insensibilia, are able to reproduce inside as well as outside an infected body. He also presented the view that particular contagia enter a specific host. The conjecture about contagion throughout the 100 years from Kircher to Plenĉiĉ, was repeatedly reinforced in the first half of the 19th century. At the same time, however, other discoveries in the general field of infusoria demanded a reassessment of the nature of these minute organisms.

After 1833 a great deal of attention focussed on Bassi. He was a lawyer, educated in several of the natural sciences and in medicine, who heard Spallanzani's lectures contradicting the doctrine of spontaneous generation. Since 1807 he had been studying a silkworm disease, known as mal de segno or mal calcino in Italy and as la muscardine in France, which was causing extensive damage in both countries. The disease was thought to arise spontaneously, but Bassi looked for factors in the environment that might be responsible for the formation of contagia. The failures of his experiments led him to the idea that the disease was transmitted either by direct contact with dead silkworms, which bare a white efflorescence, or by the breeder himself via his hands or clothing. Spores of the fungi, which Bassi implanted in the silkworms, initially grew as hyphae inside the worms and continued their growth outside after the host had died. The fresh spores which were subsequently then formed explained the spreading of the disease, and Bassi determined them to be the sole carriers. He reported the results of his investigations at the University of Pavia in 1833, and to a nine-member commission of the Faculty of Philosophy and Medicine in the following year. His publications appeared in the period 1835–36.

Audouin confirmed and expounded Bassi's results on muscardine at a conference of the Paris Academy of Science in 1836. His microscopically controlled and anatomically documented contagion trials proved the transmission of the disease by spores and by hyphae. Thus, Bassi and Audouin first described the contagion mechanism of fungal spores and hyphae, and incidentally underlined Plenĉiĉ's cell theory, even though their contagious agent was not an animalcula. A Milanese botanist, Balsamo-Crivelli, also working with silkworm parasites, described them as Botrytis paradoxa in the family Muscedinaceae. To honour the discoverer of the cause of muscardine, he later renamed the species Botrytis bassiana.

In the years immediately following 1836, discoveries were made in the field of alcoholic fermentation. This evoked strong opposition by such noteworthy chemists as von Berzelius, von Liebig and Wöhler being convinced that fermentation was a purely chemical process, as had been proclaimed by Lavoisier and Gay-Lussac. The latter hypothesis was generally accepted in the fermentation of grape juice, which resulted in the production of carbonic acid and alcohol. Oxygen was considered necessary to initiate but not to carry on the fermentation process. The participating yeast was considered to be a plant product of unknown chemical nature which precipitates in the presence of fermenting sugar. Cagniard de la Tour began his research on fermentation in 1835. In November of that year, he concluded that brewer's yeast was a mass of tiny, globular bodies able to reproduce by budding. During the growing process these bodies released carbon dioxide and converted a sugary solution into an alcoholic liquor. He published the results of his research in 1838, but never again mentioned the subject of fermentation.

After the statement by Gay-Lussac in 1835 that oxygen was essential for both fermentation and the putrefaction process, Schwann initiated experiments which were to shed light on these reactions, as well as on spontaneous generation. He observed that heated air prevented fermentation, and by January 1836 he was convinced that fermentation was the work of living organisms. Publishing his final results in 1837, he recognized the close affinity between the growth of yeast and the process of fermentation. He concluded that yeast was a plant, naming it Zuckerpilz ("sugar fungus"), later termed Saccharomyces. Furthermore, he demonstrated that fermentation in a sugar solution requires the presence of a nitrogenous substance. A third researcher in this field was Kützing, who in the same year maintained that fermentation was a physiological process triggered by living organisms. This new knowledge about fermentation was confirmed by Quevenne and Turpin in 1838.

Of the three chemist friends mentioned above, Berzelius was the first to attack fermentation being a reaction directed by a plant cell. In 1836 he classified the process as a catalytic principle which combined and explained various chemical reactions. In keeping with his principle, yeast was dead matter and not a mass of organized cells. Three years later, von Liebig and Wöhler followed with an incognito satire which gave the impression that the mystery of wine fermentation had been solved. In the same year, von Liebig expounded his scientific ideas about fermentation in a lengthy publication which he firmly upheld in the years following. As a result of their publications, these chemists not only delayed the entrance of fermentation into the field of biology, but they also prevented the acceptance of the experimental findings of Cagniard de la Tour, Schwann and Kützing.

On the other hand, this trio motivated another group of scientists who were on friendly terms with J.Müller, an anatomist and physiologist active in Berlin. His scientific interests included the problem of infectious diseases, and developmental physiology. Schwann, Ehrenberg, J.Henle, Remak, von Kölliker and Schleiden were among this group. Stimulated by Bassi's and Audouin's findings on muscardine, Schönlein undertook to investigate a skin fungus he had observed in humans. In 1839 he wrote a letter to Müller mentioning a fungus which he thought might be the cause of dermatomycosis Porrigo lupinosa. In 1841 Gruby, working in Paris, described the morphology of this germ, naming it Microsporon audouini in honour of Audouin. Schönlein's assistant Remak produced evidence in 1845 of the infective potential of a fungus called favus which he then named Achorion schönleinii after his revered teacher.

Kölliker greatly added to the knowledge of fertilization in 1841. The theory had been set down in a letter and a book 14 years earlier by the Estonian-born developmental physiologist von Baer, who was a corresponding member of the Royal Academy of Science in St.Petersburg (Leningrad). He affirmed that every living creature stemmed from the union of a male and a female, and developed from an egg and not from a "moulding fluid." Having shown that the dog ovum developed in the Graafian follicle, he described how the male semen penetrated the egg to initiate development of the embryo. He established that male semen was an intrinsic part of the body, formed in the testicular tissue. Schleiden, botanist and anthropologist, and Schwann who had worked in Müller's laboratory between 1834 and 1838, developed a new cell theory. This was made possible by R.Brown's discovery of the significance of the cell nucleus. Schleiden and Schwann maintained that the formation of a new plant cell was possible only within a liquid cytoplastema containing sugar and mucous substances. Another participant in the cell theory was F.Unger. As early as 1833, in a study on plant exanthems, he had observed a relationship between rolling of the leaves and a parasitic fungi. He contradicted Schwann's and Schleiden's assumptions about cell development and asserted that cell reproduction resulted from cell division. These findings, which were submitted to the Academy in St.Petersburg in 1840, were received with acclaim. Remak supported Unger because he himself had come to the same conclusion, i.e. new tissue developed from cells which were already present.

A significant, theoretical contribution to the etiology of infectious diseases was made by J.Henle, who in 1833 became Müller's assistant. The following year saw the foundation of "Müller's Archiv", in which Henle was permitted to work and thereby familiarize himself quickly and comprehensively with the latest findings in biology and medicine. He became seriously interested in the causes of infectious diseases and in 1840 published his deductive studies *Von den Miasmen und Kontagien und von den miasmatisch-kontagiösen Krankheiten.* His extensive knowledge of contagious diseases in plants, animals and man led Henle to analyze visible, disease-transmitting organisms, the contagia, and inanimate poisons, the miasma, present in air mixtures.

He concluded that in those diseases transmitted by both contagia and miasma, the contagia and the miasmata were identical. Assuming this to be the case, he

"would have to admit that multiplication and reproduction of the infected matter, i.e. contagia might also be possible outside the infected body, ... and would have to be considered an individual entity ... hence, an independent animal or plant."

The second group, pure contagia diseases, wereas regarded by Henle as relatively independent and transplantable. The causal agent was unable to multiply outside of the diseased organism but was rather "inherited" by from one to another. He considered rabies, syphilis, tuberculosis, glanders and scabies to be diseases of this type. Henle did not know how to classify the contagia responsible for human as well as animal diseases, but he concluded that

"they belonged, at any rate, to the lowest and smallest but also to the most fruitful beings known."

The effect of Henle's publication on the scientific world of his time was negligible. He was unable to erase the theory of spontaneous generation or inanimate miasma, and his contagium was to be challenged as the sole causal agent for decades. The challenge, however, was not to be resolved by publications, but only experiments. Until approximately the mid-19th century, many important publications appeared in support of the germ theory in both medicine and agriculture, although often the experimental results could not be interpreted immediately. Gradually, the parts began to form a whole, allowing the germ theory to crystallize into the germ doctrine.

Vaccinations as well as aseptic and antiseptic measures were gradually developed to prevent certain diseases. Deriving his ideas from the ancient practice of variolation and the Middle-Age custom of transferring cowpox material, Jenner succeeded in developing a specific vaccination between 1796 and 1798. Even since 1758 the Scotch physician Home had been hoping for protection against measles by carrying out morbillization. This involved blood transfusion from measles patients to evoke morbilli, in the form of a mild skin disease. During transmission trials, Zinke, a physician working in Jena (1804), included rabies, and von Hebra, a Viennese dermatologist (1841), experimented with scabies, thus adding to the list of diseases caused by contagia. It was recalled that agents causing the pest had been experimentally transmitted from dead humans to dogs by Deidier during the 1720 to 1721 epidemic in Marseille.

From 1839 to 1847, the names of Gruby, Berg, Retzius, Malmsten, von Eichstedt und Sluyter appeared in connection with diseases caused by fungi. Goodsir meanwhile had described Sarcina ventriculi, a bacteria dividing into three directions. In 1853 the German botanist De Bary enlarged the knowledge of the contagia which cause burn and rust diseases in various plants.

The chemicals used in the field of medicine at that time to kill contagia included mercury, potassium hydroxide and potassium iodide. For agriculture purposes, calcium, copper sulphate, dried arsenic and urine, among other things, were used to disinfect the grain. Many people were convinced that contagia were actually the spores of fungi. In 1772 Priestley, the English theologian, philosopher, chemist and physicist, had determined the antiseptic property of carbonic acid. By 1825 the French apothecary Labarraque recommended using sodium chloride, sodium hypochlorite and calcium oxide to prevent putrefaction in the Paris morgue. A decade later, Bassi advised disinfecting the rooms and boiling the tools and instruments which had been used in muscardine cases. As disinfectants he employed calcium chloride and potassium nitrate.

O.Holmes, the Boston author and physician, who for 2 years had become acquainted with European medicine in Paris, set down some practical directives in his manuscript *The Contagiousness of Puerperal Fever* (1843) to ensure aseptic conditions. The object of his work was to avoid transmission of the disease. Among other recommendations was the prohibition of simultaneous delivery and autopsy of a mother who had died of childbed fever. In addition, a change of clothes after an autopsy was required, allowing at least a 24-hour interval before performance of the next delivery.

Semmelweiß introduced further protective measures against childbed fever in the Allgemeine Krankenhaus in Vienna in 1846 and 1847. These were not only aseptic but antiseptic, such as the disinfection of hands with a chlorine/water solution.

By the mid-19th century, there was sufficient experimental proof to make the theory of spontaneous generation, as it pertained to lower forms of life, untenable. Microscopic observation provided proof that lower forms of animals and plants transmitted and triggered disease in man, as well as in animals and plants. Subsequently, the ancient concept of contagia began to acquire a more tangible form as fungus spores, fungus hyphae and scabies mites. Fermentation and putrefaction were recognized to be biological processes significantly influenced by minute forms of life. The germ theory had already been proven in specific cases but still met with more opposition than acceptance.

1.4 The Germ Theory Crystallizes into the Germ Doctrine

Despite the evidence discounting spontaneous generation of lower forms of life, and indicating that transmissible diseases were caused by microorganisms, contradictive experimental observations continued to appear during the mid-19th century supporting generatio spontanea. Naturally, this was a great impediment to general acceptance of the germ theory. For example, it was not always possible to prevent the "spontaneous" development of infusoria by heating. Whenever prevention was effected, however, it was possible to argue that the chemical basis of the original material had changed, and this unfavourably affected spontaneous generation; in essence analogous to treat with acid. The cholera epidemic which had been spreading throughout Europe since 1830 lent credibility to the miasma theory. It also supported the notion of a specific poison, in addition to contagia, and changing factors in the environment which could influence the nature of the transmissible disease. If the concept of spontaneous generation was to be dropped and replaced by the germ theory, then conclusive proof would have to be provided.

The credit for providing this kind of proof goes to Pasteur. After researching crystallographic-optical problems until 1857, he subsequently devoted intensive efforts to fermentation and spontaneous generation. In 1865 he began delving into transmissible diseases, and 12 years later started combating these prophylactically. In all of his investigations, which sought solutions to practical questions, he unfalteringly searched for scientifically accurate proof. In this way, Pasteur not only discovered new methods highly significant for bacteriological and microbiological research, but he also motivated many other scientists to similar, successful activity.

In 1856 a Mr.Bigo requested Pasteur's help in the name of the dean of the newly founded, natural science department in Lille. He was having difficulty producing alcohol from beet. During subsequent investigations, Pasteur confirmed the discoveries made 20 years earlier by Cagniard de la Tour, Schwann and Kützing, and concluded that fermentation was a strictly anaerobic process. Furthermore, he was able to convince his scientist friends, who had been invited to his laboratory, of his conclusions. When in 1858–59 the Director of the Museum of Natural History of Rouen, Dr. Pouchet, published an article about generatio spontanea, Pasteur countered that spontaneous generation was a matter of faith that never been proven. Motivated by Pouchet's publications, the French Academy of Science in 1860 announced the Alhumbert Prize for experiments which shed new light on this problem. Pasteur participated in the contest and won the prize in 1862. His comprehensive work was published in 1861 and he was thoroughly convinced he had dealt generatio spontanea a fatal blow.

Air filtration through guncotton had been introduced by Schröder and von Dusch in 1854. Pasteur carried out filtration through asbestos and through tiny, glowing tubes, as well as experiments with newly developed culture flasks such as Balard's swan-neck bottle. He bred bacteria in heat-sterilized urine and milk, and carried out temperature trials with fungus spores and heat-resistant germs in laboratories in the city and at high altitudes in the mountains. Perishable liquids were heated to 50–60 °C for a short time (pasteurization). All this sufficed to secure Pasteur the prize but it did not dispel the idea of generatio spontanea. Nor was it to settle the quarrels with Pouchet and those which ensued with Pouchet's disciples. Furthermore, it provoked religio-philosophical and political discussions which severely encumbered progress. The subdued quarrelling, which had been festering for years, emerged into a heated issue in 1872, when the English bacteriologist H. Bastian published his 1100-page experimental findings *The Beginnings of Life* on the problem of the spontaneous creation of simple forms of life. As a result Pasteur was prompted to undertake new experiments and discussion, and the Irish physicist Tyndall, the Breslau phytophysiologist Cohn and others were all stirred into action.

In 1869 Tyndall had discovered that particles with the size of a wavelength of light could deflect a light beam. This effect, known as the Tyndall phenomenon, could be visibly demonstrated in a darkened room. Furthermore, by using this method, it was possible to determine whether the air contained only dead material or viable particles, i.e. particles growing in a culture medium, as well. Tyndall also participitated in the generatio spontanea discussion which Bennett and Bastian had been fanning in Great Britain since 1860. Convinced that methods other than microscopy were essential to clarify the problem, he obtained a "state of optically pure air" with the aid of a burner which removed or killed every sort of particle by heat. To study the development of living organisms in organic infusions he constructed a closed chamber, the insides of which were coated with glycerine to bind all refracting particles. This was in 1876. In addition, he improved the process of sterilization by treating heat-sensitive substrates at temperatures under 100°C on three consecutive days. This process, called Tyndallization, also lent itself to kill those forms of bacteria which he had termed heat-resistant. They were identical to those observed by Perty (1852) and Pasteur (1865), which Cohn (1872) had described as permanent cells, i.e. the spores of bacilli.

Meanwhile, further sterilization methods had been developed for various forms of bacteria under different conditions. It was, thus, possible to discuss the results of further experiments on spontaneous generation in the controversy between Pasteur and Bastian. Bastian defended the generatio spontanea with his own experimental results as well as with theoretical arguments. The conflict began in 1876 after Bastian had asserted in a note to the Académie des Sciences that he had detected the spontaneous formation of living organisms in urine. Within a week, Pasteur repeated Bastian's experiment and he too came to the same conclusion, but contradicted Bastian's interpretation of the result. Pasteur reminded Bastian of his memoir dated 1861 in which heat-resistant microorganisms had been detected in an alkaline medium and recommended that the experiments be repeated under increased sterilization temperatures, namely 50–100°C. Bastian declined the proposed experiments and refused to renounce his claim.

In turn, Pasteur experimented further with Joubert and Chamberland in the field of heat-resistant germs until Bastian consented to Pasteur's proposal in 1877 to air their dispute in the presence of a commission of the Académie des Sciences. The result was a repetition of the 1864 Pouchet drama, for Bastian was just as reluctant to repeat his experiment for the commission. Even as late as 1877, Pasteur's rivals were not able to prove the source of primeval genesis, and that year can be considered as a milestone when Pasteur, supported by Tyndall and Cohn, finally shattered the credibility of generatio spontanea. Furthermore, Pasteur played an active part in solving the related problem of the germ theory in contagious diseases. In 1865, Dumas had commissioned him to carry out investigations on the cause of a silkworm disease. Because of an easily detectable symptom, its similarity to pepper, it was named Pébrine disease. His two years of observation and experimentation convinced him that this disease as well as a similar illness, Flacherie, was caused by tiny organisms and was transmissible.

Pasteur had already proved that there were different types of fermentation, and following the results of his sterilization trials he was already probing into the field of medicine. The introduction of antiseptics and asepsis in surgery by the Englishman Lister actually can be considered a natural consequence of his earlier experimental results and recommendations. Membership of the Académie de Médicine in 1873 was a symbol of the public acknowledgement of Pasteur's endeavours. Thereafter, having researched the fermentation processes of wine and beer production, Pasteur felt encouraged – supported by physicians – to undertake closer investigation of transmissible diseases in humans.

He began with anthrax, a disease which at that time had been described in man as well as in animals. After Pasteur's publications on fermentation and generatio spontanea in 1863, the French pathologist Davaine studying the etiology of anthrax had demonstrated for the etiology of anthrax the causative organism to be rod-shaped. Subsequently, anthrax became the object of several discussions at the Académie des Sciences in which Pasteur had participated.

In addition to working on the origins of blood poisoning, after 1877 Pasteur increasingly developed an interest in the microbial causative agent of cowpox. His attempt to isolate the germ failed. At the same time, however, he began to expound and propagate the germ theory in discussions, lectures and publications. In all subsequent studies on the etiology or prophylaxis of communicable diseases, Pasteur and his co-workers proceeded from the assumption that germs were both the causative and the transmitting agents. He remained firm in this conviction, despite some doubts about the accuracy of his exhaustion theory in explaining immunity which arose while he was developing his attenuated rabies vaccine (with due respect to Jenner, he chose the term vaccine here). Pasteur's shift to a chemical theory of immunity to rabies, whereby an infectious, viable germ produces a lifeless substance which inhibits the germ's growth, is documented in the first edition of *Annales de l'Institut Pasteur* dated January, 1887. This publication, along with the founding of the Institute Pasteur in 1888 – financed by contributions from the entire world out of gratitude for the rabies vaccine – became a focal point of the germ doctrine, which had meanwhile found more widespread acceptance. Pasteur's successors as Directors of the Pasteur Institute, Duclaux and Roux, later founded more than 100 Pasteur Institutes throughout the world, thus enhancing the reputation of Pasteur and the glory of France.

Robert Koch in particular contributed greatly to the germ doctrine. While taking his final examinations in Hamburg in 1866, he witnessed a cholera epidemic and later upon the outbreak of the Franco-German War, he treated soldiers suffering from typhoid fever or dysentery in field hospitals. His interest, however, lay primarily in research and focused specifically on the epidemics caused by contagium animatum. Koch was familiar with the discussion begun by Davaine on the cause of anthrax and the problem of wound infections. Meanwhile, the Austrian surgeon Billroth had isolated Coccobacterium septicum from bullet wounds, and the German bacteriologist Klebs had isolated numerous germs from more than 100 autopsies at a battlefield hospital in 1870. Klebs thought these to be a single bacterial type which he named Microsporon septicum, because at that time all bacteria were considered to be derivatives of a single organism. Other researchers were isolating additional bacteria as presumed causes of wound infection.

Through experimental inoculations with disease-producing cultures or isolation of bacteria from infected material, it became clear that causative agents were not only to be found among fungi and protozoans, but also especially among the microorganisms which Perty and Cohn had classified as schizophyte plants. After 1858 these were named schizomycetes by the leading Swiss botanist von Nägeli who was working in Munich at that time.

In 1872 when Koch was appointed District Health Officer, he quickly procured the necessary instruments and equipment for carrying out his research on the etiology of anthrax. The results of investigations with Bacillus anthracis were discussed in Cohn's institute in Breslau and thereafter published in Cohn's 1876 *Contributions to Plant Biology*. Although he was virtually certain that results using rodents were not necessarily identical to those using ruminants, he considered his findings on the etiology of anthrax to be basically clarified though not unequivocally proven. During his experiments he had discovered the bacterium as well as its endospore, and had passaged the contagia in animals. In his final remarks, he foresaw obstacles in testing for similar results with typhoid fever and cholera because of the inadequate optical equipment and inability to carry out animal testing. He considered a further shortcoming to be the liquid media used for cultivation which did not permit the exclusion of unwanted germs. In an 1877 paper, he described a method for investigating, conserving and photographing bacteria. In 1878 a third new work appeared on the etiology of wound infections, which included a survey of the numerous important studies on the subject. None of these satisfied his requirement for valid proof. He demanded verification of a morphologically well-defined microorganism, the quantity and distribution of which could reliably explain the symptoms of various kinds of wound infections.

Koch augmented his own investigations of these infections by employing the dye aniline which had been introduced by the German pathologist and histologist Weigert. With the use of the Abbe illuminating apparatus he improved microscopy and microphotography, and introduced differential staining of bacteria with alkalis and acids. His animal studies proved that wound infections were caused by parasites and not by toxic substances, and was able to ascribe a germ of distinguishing morphology as the cause of each distinct disease.

After being appointed to the Kaiserliche Gesundheitsamt (Royal Department of Health) in Berlin, Koch's possibilities for further research on contagia animata improved immensely. In 1881 he published *Zur Untersuchung von pathogenen Organismen*, which appeared in a new series of reports from the Royal Department of Health. With this trend-setting paper he defined reliable pure culture as the prerequisite for determining a pathological microorganism, and the necessity for proof of contagiousness by repeated infection using either the same animal or a related species. In addition, he advised the following steps for improving experimental work:

1. Using Ehrlich's methods of producing heated blood preparations in various forms.
2. Employing the microtome for slicing fixed tissue.
3. Extending germ cultivation beyond using Cohn's, Mayer's, and Pasteur's culture solutions.
4. Transmitting germs by various forms of application.
5. Ensuring that instruments used had been sterilized by heat.
6. Using suitable, sterile culture media.
7. Using disinfected and germ-tight, closed containers.

He considered reliable pure cultures crucial for further experimental work. In early experiments Koch improved culture methods by substituting unreliable fluid cultures by slices of cooked potatoes kept in a damp room. Later he used stiffened gelatine, i.e. adding 2.5–3% gelatinous substance to the substrate, which made it possible to apply the culture by streaking to obtain single colonies. In Koch's opinion the best substrate for bacteria was nutritive gelatine produced from a mixture of blood serum and gelatine. Koch's lecture "The Etiology of Tuberculosis", presented to the Physiologische Gesellschaft at Berlin on March 24, 1882 demonstrated how the presence of a contagium could be proven without doubt by following basic, experimental, microbiological principles. Pure culture, meanwhile, had been improved in Koch's laboratory by Fanny Hesse (the wife of Koch's assistant Walther Hesse), who substituted stiffened agar for the gelatine.

In subsequent years, it was repeatedly proven at the Kaiserliche Gesundheitsamt, and after 1890 in the newly-founded Institut für Infektionskrankheiten under Koch's guidance, that a bacterium was the contagium which caused a specific infectious disease. Koch's institute became a shrine for his germ doctrine. The media for disseminating the ideas of his school were publications produced with his coworkers, as well as the *Zeitschrift für Hygiene* in 1886, and the *Centralblatt für Bacteriologie und Parasitenkunde*, which first appeared in 1887. Koch received great acclaim for his momentous work. Many of his coworkers, by merit of Koch's fame, became directors of institutes which, in turn, were centres of Koch's theory and the germ doctrine in the German-speaking countries. From distant corners of the globe, scientists journeyed to Paris and to Berlin, eager to find out about the new concepts on the causes of communicable diseases and to learn the new methods for microbiological experimentation.

As time passed, the divergent standpoints of Pasteur and Koch gradually faded into minor personal differences. Young scientists, such as the Swiss Yersin, helped iron out these differences by visiting both institutes and by disseminating their methods in other countries.

1.5 The Theoretical and Experimental Basis of the Germ Doctrine at the End of the 19th Century

More than a 100 years spanned the period between Spallanzani's first experiment and Pasteur's final refutation of the generatio spontanea. During that period, so many observations had been recorded that, by the end of the 19th century, there was hardly one serious opponent to the doctrine which had developed from the theory of bacteria as the cause of contagious disease. The number of localists, anti-contagionists, anti-Pasteurians, and anti-vaccinators had dwindled to the point where their argumentation became increasingly tenuous as new experimental results continued to appear. Research institutes were established in all parts of the world, where the problems of microbiology or bacteriology were studied, taught and researched in connection with hygiene and immunology. But, what exactly was known about pathogenic microorganisms and what sort of experimental equipment was available to solve the problems arising in this field at the end of the 19th century?

1.5.1 Theoretical Basis

Prior to 1876, when Koch began to define the requirements for unequivocal proof of the relationship of disease to the germs isolated from the pathological material, there had only been publications in support of the theory. By the mid-19th century, scientists were rather certain that fungi were pathogens in humans, animals and plants, and protozoa were pathogenic in humans. In 1846 the Philadelphia biologist Leidy isolated Trichinella hominis in the pig, a nematode which Zenker in Dresden in 1869 classified as a pathogen. Although the German physician Pollender had observed and described the bacterium which caused anthrax in 1855, he did not recognize it as the anthrax pathogen. Davaine, Pasteur and Koch pursued their studies in this field to clarify the etiology. The French surgeon Villemin described his successful transmission experiments with tuberculous material in 1878, and Koch proved a bacillus to be the cause of this disease in 1882.

The first indications of bacteriological diseases in plants came from the USA in 1878. Burrill had recognized these plant infections in corn and millet in the state of Illinois. E.Smith, a plant pathologist active in Washington, published numerous accounts of bacteria causing plant diseases as early as 1887. Thanks to his descriptions of bacteria penetrating the plant through natural openings and damaged parts, as well as transmission of contagia by insects, the germ doctrine gained in scope. At the close of the 19th century, it was known that bacteria were virtually the main cause of transmissible diseases. In general, the debate about the systematic classification of bacteria was carried out by botanists. The classification of plant bacteria undertaken by Perty and Cohn from 1852 to 1853 precluded extensive argumentation. Two further questions, however, were heatedly discussed: The significance of simple fungi as pathogens, and whether specific, morphological species could be defined.

In 1877 von Nägeli differentiated three groups in his book "*Die niederen Pilze...*": moulds, budding fungi (yeast) and fission fungi (bacteria, schizomycetes). However, he considered only the fission fungi to be infectious. A few of the erroneous conceptions which he as a leading botanist maintained have been extracted from his book:

1. Molds and yeast are closely related... whereas fission fungi are not genetically related to either of the other two groups...
2. Each species of fission fungi can appear in several morphological and physiological forms which undergo internal changes, quickly or slowly depending on external conditions...
3. The infectious matter of miasmatic diseases (miasmas) originates upon or inside the earth...
4. To cause miasmatic-contagious diseases (e.g. cholera, typhoid fever, yellow fever) two factors must be present: one from the patient and the other from the earth...
5. disinfecting fresh excrements... is superfluous...

Nägeli's views were disproved after only a few years because of the progress made in explaining the etiology of several diseases, e.g. anthrax in 1876, tuberculosis in 1882, cholera in 1883, typhoid fever and diphtheria in 1884, lobar pneumonia in 1885, and tetanus in 1890.

It was the "proofs" of bacterial plant infections from the USA, especially those put forth by E.Smith, which evoked more criticism than consensus in Europe. Nevertheless, there was soon enough proof of bacteria as causative agents also in the field of phytopathology. On the whole, the morphological and physiological specificity of microbial causative agents was no longer contested at the close of the 19th century. The implications were naturally widespread: microbiological and hygienic institutes were founded; methods for disinfection were developed which led to the establishment of disinfection centres; adoption of aseptic and antiseptic measures in hospitals; the publication of periodicals and textbooks for microbiology and hygiene in foreign languages.

An additional consequence was the development of bacterial vaccines for prophylactic purposes in man and animals. Paris in 1879 was the scene of this turning point. Having been recommended by Duclaux, Roux became Pasteur's assistant one month prior to Pasteur's receipt of blood samples from a cock which had died of fowl cholera. Roux was assigned to carry out the experiments, and in September 1879 after a 2-month holiday, he observed in a culture which accidentally had been left over clear signs of diminished virulence, i.e. destroyed pathogenicity and diminished reproducibility. Pasteur, who had been occupying himself with the problem of Jenner's vaccines, eagerly seized upon Roux's observation and developed an attenuated fowl cholera vaccine. He decreased the virulence by treating the cultures with oxygen or by changing the medium. After publishing his method of attenuation in October 1880, Pasteur assumed he had found a method for developing further vaccines. Thus, collaborating investigations into anthrax followed, carried out by Chamberland, Roux and Thullier. Attenuating the "vaccine" was achieved at 42° or 43°C., and the vaccination was repeated with various degrees of attenuated bacteria after two weeks, followed by a third injection of full-strength virulent bacteria 2 weeks later. Pasteur pursued further vaccine developments for pneumonia, swine erysipelas and rabies.

The approach used to create these vaccines was to begin with bacteria, i.e. infected substrates, which according to Koch were not pure cultures. Although Koch criticized some of the steps employed in Pasteur's method of producing vaccines, he was, nonetheless, convinced of their efficacy and acknowledged the principle of attenuation to achieve suitable strains for vaccines. Other methods of producing attenuated bacteria included changing the conditions of the cultures, using pressure or carbolic acid, or passaging in animals. As early as 1886, Salmon and T.Smith had demonstrated the possibility of using bacteria which had been killed by heat as an effective vaccine for hog cholera in the USA.

During experiments employed to change the physiological properties of bacteria, it became apparent that bacteria often underwent morphological changes. As a result, a substrate consisting of a morphologically pure culture could not be considered identical to a functionally or physiologically pure culture. Hence, a single colony was in effect nothing more than a physiological representative of a bacterial population.

In 1888 Roux and Yersin discovered diphtheria toxin at the Pasteur Institute in Paris, while Kitasato, a Japanese bacteriologist, focused his attention on the tetanus toxin during his work with Koch in 1888. The concept of the body's own defense mechanism against invading bacteria was extended to include terms like cellular and humoral immunity.

The Russian zoologist Metchnikoff began evolving his theory of phagocytes in 1883. This study showed that invading microorganisms could be ingested intracellularly. The leukocytes capable of doing this were called phagocytes and the process became known as phagocytosis. As for humoral immunity, it was assumed that there were substances in the blood and in the body fluids which could attack and destroy invasive microorganisms.

Salmon and T.Smith had already proved that humoral defensive substances could be induced also by dead germs in experiments with swine erysipelas vaccine. Approximately a decade later, it was already possible to characterize the various types of humoral protective substances inherent in the body, viz. the antibodies. Pfeiffer and Issaeff found a specific antibody against the cholera bacillus in Berlin in 1894. Since this antibody was in fact able to dissolve the bacillus, they called the process bacteriolysis and the antibody lysin. They too were able to prove the protective effect of immune serum against the active germ. In 1896 M.Gruber and H.Durham found agglutinating antibodies during their studies with cholera and typhoid fever. This type of antibody had been detected in 1889 by Charrin and Roger in their experiments with Pseudomonas. In the following year, the Austrian Kraus discovered the precipitating antibodies for cholera, typhoid fever and the pest.

Correspondingly, Behring and Kitasato found the anti-toxins for tetanus and diphtheria in 1890. As of 1892 both antitoxins were tested for passive immunization in humans, and Ehrlich undertook the task of developing methods to evaluate the protective effect of antitoxin immune sera. To carry out this task, he became director of the institute for serum research and serum testing in Berlin in 1896. Three years later, the institute was moved to Frankfurt on the Main River. The side-chain theory which Ehrlich proposed in 1897 was to have significant implications for further development in the field of immunity.

These numerous new findings caused a shift in approach away from a single etiological agent to reflections on the interaction of causative agents within the protective host organism. The concept of the pathological mode of action of the invading bacteria expanded. For a long time the most significant pathogenic factor was considered to be the increasing microorganisms which deprived the body of essential nutrients. It was learned that saprophytes, i.e. germs which live on dead or decaying matter, were able to excrete ferments and toxins just like parasitic germs, i.e. those which sponge on viable matter. Kleb's assumption, dated 1872, that bacteria caused disease by producing toxins had been proved for diphtheria, tetanus, cholera and staphylococci by the turn of the century. As a result, scientists generally accepted the principle that toxins excreted by bacteria were usually responsible for bacterial pathogenesis. Local as well as general symptoms were thought to result from bacteria production within or beyond the limits of an organ in the infected organism.

Thus, at the close of the 19th century, there was already a common frame of reference for microbiology, hygiene and immunology. Fungi, bacteria and protozoa occupied a defined position as infectious agents and as antigens, systematically classified by botanists, zoologists and physicians. T.Smith and Kilborne had published a monograph in the USA in 1893 about Texas cattle fever, proving that blood-sucking ticks could be carriers of infectious diseases.

1.5.2 Experimental Basis

In addition to the well-structured theoretical framework for microbiology established by the close of the 19th century, comparable progress had been achieved in the microbiological and especially in the highly interesting bacteriological fields of experimentation. The multiplicity of specific bacterial forms and functions was studied with reference to diverse forms of disease. The scope of differing sizes from the endospore to the spirochete was investigated. These factors, plus the obligation to validly prove one's theories, finally led to higher standards of laboratory methods, techniques and equipment. Textbooks on microbiology and hygiene, available in several languages, presented guidelines using liquid or solid substrates suitable for cultivating distinct forms of bacteria.

Development of suitable nutritive media began by cultivating yeast in a protein-free, sugary fluid containing yeast ash. It had been described by Pasteur in 1857 and designated Pasteur liquid by other researchers. An improved version is credited to the German agricultural chemist A.Mayer who published it in 1870. He determined the proportion of minerals in yeast ash and then produced a normal solution using the mineral salts. Cohn then modified Mayer's normal solution by adding potassium carbonate, magnesium oxide, calcium phospate, and ammonium tartrate, and then termed it normal bacteria nutritive solution. Nutritive bouillon, peptone solutions, nutritive solutions containing milk, and various normal solutions with or without biological additives represent examples of further developments up to the turn of the century. Other substrate additives such as fish glue, aqueous humour from cows' eyes, animal gelatine and agar etc. provided solid bases for stiffening cultures. Solid substrates ranged from Schroeter's boiled potato slices in 1872 to the agar plate used in Koch's institute in 1882, in which a transparent vegetable substance was used instead of gelatinous animal matter. Certainly it was possible to produce pure cultures using potato slices, although it was even better to use the single-cell dilution method found by Brefeld in 1872. He had developed this to grow mould, whereas Lister began to use the same method from 1878 to cultivate bacteria; the ideal was Koch's nutrient agar plate.

Meanwhile other methods emerged, two of which deserve mention: After 1873, Klebs worked with fractionated cultures in which a drawn-out capillary tube was placed at the bottom of a container filled with bacteria. The point of the tube was broken and the tube was pulled out into a new media. This process was repeated several times with the intention of ending with a single bacterium. Using a different method, Klebs and Brefeld placed the bacteria in a gelatinous medium. After cooling and stiffening of the medium, the bacteria grew in single colonies and could be isolated.

By the end of the 19th century, there were many different substrates available for aerobic as well as anaerobic bacterial cultures. To obtain sterile working conditions for bacteriological studies, reliable disinfecting and sterilizing methods had to be employed for rooms, instruments, media and devices. Von Berzelius had introduced Swedish filter paper to separate solid chemical particles from fluids. For bacteriological studies, other materials and methods were used. W. Schumacher used collodion membranes as small sacks in 1860; Tiegel filtered with clay cells in 1871 and, thereafter, various other filtration procedures were developed.

Staining methods for bacterial in vitro and in vivo preparations had achieved a high standard through Weigert's and Ehrlich's research activities. Besides carmine and hematoxyllin, additional materials such as acid and basic aniline dyes were available for many specialized staining techniques. To conduct microscopic observation as well as microphotography, optic immersion systems were developed using the apochromatic lens introduced by Abbe in 1886. A resolving power of about 200 nm and enlargements of 1:1600 were possible.

The results of testing humoral immunity led to serological in vitro tests to identify pathogenic germs. It all began with a method for proving cholera and typhoid fever by M.Gruber and H.Durham in

Austria in 1896 as well as by Widal in France. Kraus followed with serological precipitation in 1897. At the same time, the Belgian scientist Bordet began developing a method for proving lysines. It was described by him and his brother-in-law Gengou in 1901 as a complement-fixation reaction (CFR).

In summary, by the end of the 19th century, there were no apparent difficulties in explaining the etiology of bacterial diseases, in producing vaccines from bacteria and bacterial toxins, and in discovering more about the interaction of causal agents and their hosts by using immunological tests.

2 Microscopes and Culture Media Fail

By the close of the 19th century, it was no longer scientifically contested that infectious diseases in plants and in animals were caused by microorganisms. In order to isolate, characterize, and identify causative agents, laboratory techniques had been developed for classifying protozoa, fungi and bacteria. Each infectious disease had been shown to be linked to an agent specified morphologically and physiologically. If this happened to be a bacterium, identifying it was a simple matter, provided it had a minimum size of 200 nm and it could be cultivated on a synthetic medium. Transferring causative agents to suitable test animals was a standard measure of identification, as were immunological in vitro and in vivo testing, or determining disease-specific, histopathological changes. Two alternative or supplementary hypotheses had emerged to explain phenomena observed in diseases caused by bacteria: One reason for pathogenicity was the body's deprival of essential nutrients during germ multiplication. The other was the effect of bacterial toxins.

A list of 18 infectious diseases whose etiology had already been established in the 19th century, demonstrates the immensity of bacteriological research at the world's leading microbiological institutes at that time. However, more and more cases accumulated in which the infectious disease was characterized but the causative agent was not identified, despite imaginative and carefully executed experiments. This chapter will attempt to recount how the early failures to identify "bacteria" in effect paved the way to experimental virology.

2.1 The Virus and Koch's Postulates

To review the beginning phase critically, it is essential to clarify two concepts. One concerns the subject of our research, namely the virus, and the other relates to the so-called Koch Postulates as guidelines for experimental investigation.

2.1.1 The Use of the Term "Virus"

The early definition of the word "virus" as visible, poisonous matter was no longer in use. Pasteur spoke of virus vaccines in connection with vaccine production for the identified causative agent of fowl cholera, as well as for the non-identified germ in rabies. Koch, in contrast, never used the term "virus", but selected different terminology according to the state of knowledge. For his purposes, causative agents were bacteria, pathogenic bacteria, microorganisms or pathogenic microorganisms. In cases of hitherto unknown infectious matter, Koch employed "infectious material" or "microorganism not visible under the microscope."

Other bacteriologists, contemporaries of Koch and those in his circle, considered the toxic effect to be the most significant component of bacterial pathogenicity. To their knowledge, there were parasitic and saprophytic disease-producing germs. Parasitic germs multiplied and spread their poisons, while saprophytes solely produced poisonous excretions; both damaged the organism. The publications of Koch and his followers depict the virus as poisonous matter or fever poison in connection with known and unknown causative agents of the time. Hence, it can be concluded that the indiscriminate use of the word virus as substantive or adjective, i.e. without characterizing the contagium exactly, is without significance for understanding the history of experimental virology.

2.1.2 Koch's postulates

Whoever searches for traces of Koch's postulates in his early works – perhaps motivated by authors of books or publications – will likely come away with two impressions: Koch's language and his interpretation of his experimental results are clear and unequivocal; the Koch "postulates", interpreted by other authors as the sole claim for proving the causative agent of infectious diseases, do not exist.

2.1.2.1 "Koch's Postulates" Do Not Exist !

In 1882, by clarifying the etiology of tuberculosis, Koch was the first to demonstrate that adherence to strict experimental procedure could unequivocally prove the germ theory as well as the etiology of an infectious disease caused by a microorganism. However, he did not claim then, or later when he adapted his methods to new findings, that his "postulates" were *absolute* guidelines.

Koch's opinion after 14 years of etiological research, and his requirement that experiments should provide conclusive proof that an isolated bacterium is the cause of a given disease, were formally presented at the X. International Medical Congress in Berlin in 1890. This lecture clearly stated the difference between the two experimental stages: to prove the germ theory and, in turn, to demonstrate a bacteria as the causative agent for an infectious disease. As examples of the former, wound infections and tuberculosis were cited as the final stage of the first phase simultaneously ushering in the second phase. In response to the question of how to obtain irrefutable proof that an isolated bacterium had caused a specific disease, Koch explained the following:

1. "The parasite can be encountered in all cases under those conditions which correspond to the pathological changes and the clinical course of the disease.
2. The pathogen may not occur incidentally as a non-pathogenic parasite in any other disease.
3. The pathogen must be isolated and bred in adequate numbers in pure culture and must be able
4. to cause the disease anew."

Since Koch knew, after his 1884 experience with cholera, that it was often impossible to induce a disease experimentally in animals and yet not harbour the slightest doubt about the germ theory, he augmented these guidelines. His thoughts on the implications of the regular and exclusive occurrence of bacteria in infectious diseases without a possible animal experimental trial was reflected in his own words:

"Our contention is likely justified, even at this point, that if only the first two requirements for proof are fulfilled... the causal relationship of parasite to the disease is validly established."

Nevertheless, it is essential to explain how it was possible to arrive at Koch's postulates in a manner which contradicted Koch himself.

Koch's co-workers were well aware of his strict demands for each and every experiment, which as a rule was intended to prove something. These requirements for laboratory work might also be considered postulates in the sense of a challenge to observation. It was not Koch, but two of his co-workers, namely Hueppe and Löffler who wrote about the postulates from this point of view. In Hueppe's book *Die Methoden der Bakterienforschung*, we find the term "postulate" for various experimental steps which were considered by Koch as important in etiological investigations. This included, for example, the use of animals, the isolation of a single germ as a basis for obtaining a pure culture, and the problem of a solid and transparent culture medium. In 1883 Löffler examined the evidence for proof in a study on diphtheria which was published in 1884. He investigated the related contentions of individual scientists using as terms of reference three postulates which could be equated to Koch's proof of tuberculosis in 1882, as terms of reference. Löffler actually wrote about Koch's postulates some 10 years later in 1903, on the occasion of Koch's 60th birthday. There, however, he erroneously referred to Koch's study of wound infections completed in 1878. Löffler included the point concerning pure culture, although Koch had only mentioned this requirement in 1881 to 1882. Thereafter, it was possible to find no less than four different versions of Koch's postulates in the literature, none of which corresponded to Koch's modified, clear guideline of 1890.

2.1.2.2 Reliable Recognition of Pathogens using Koch's Proof

In his inaugural lecture held at the Academy of Science in 1909, Koch set forth his new methods as the key to progress in medical research on microorganisms. He conceded that these methods were inadequate for those putative causative agents for the transmission of disease which had remained totally invisible until that time. Koch further conceded that his methods were limited to microorganisms clearly belonging to the bacteria group; not included were the pathogens which he had already termed organized pathogenic agents in 1890. He assumed that these belonged to a different group of microorganisms which had defied his tried and proven methods of investigation. Included were the diseases measles, scarlet fever, smallpox, rabies, influenza, yellow fever and cattle plague. Thus, it would be an injustice to Koch's concepts if his key methods designated for bacteriological research were applied indiscriminately under the guise of "Koch's Postulates" to other types of pathogenic agents.

The aim of all etiological research was and is to characterize infectious disease accurately and to identify unequivocally the specific pathogen involved. In pursuit of that goal, not only bacteriological but also immunological, histopathological and molecular genetic methods have been successfully employed since the turn of the century. In the course of researching Koch's unknown "organized pathogenic agents", numerous valuable new methods emerged which would not have done so if the 19th century, so-called Koch postulates had served as the only guide. All of these methods deserve consideration when we pose the question: To what extent had the scientific claims of the time been proved? This applies most emphatically to the early stages of experimental virology. Here the literature lauds many a scientist as the "father" or the "founder", while many a claim remains unverified.

2.2 A New Type of Pathogenic Agent?

During the final two decades of the 19th century, many fungal and bacterial pathogens were identified as causing infectious diseases in plants and animals. After Laveran had described the malaria pathogen in Algeria in 1880 as a protozoon carried by an arthropod, and Manson had characterized the typical intracellular malaria tropica-halfmoon in 1893, research began to link diseases with visible intracellular changes or cell inclusions to protozoa or to still-unknown types of pathogens. This, for example, was the case with vaccinia and variola and their cytoplasmatic inclusions described by Guarnieri in 1892. It also applies to molluscum contagiosum seen by Henderson in 1841, and to fowlpox with its cell inclusions found by Rivolta in 1869.

2.2.1 The Discussion Begins

At the turn of the century there were publications about the etiology of infections in plants, animals and humans in which none of the authors was able to identify the causative agent. Several, however, had new theoretical or experimental ideas about contagia which will be of interest in later discussions about a "founder of virology."

2.2.1.1 Tobacco Mosaic Disease (TMD)

In 1876 the Heidelberg agricultural chemist A.Mayer was appointed to the Rykslandbouwschool Wageningen in the Netherlands, in order to set up an agricultural experimental station. At that time, a disease causing extensive damage to the tobacco crops in the provinces of Gelderland and Utrecht was becoming more and more widespread. Comparative chemical analyzes of healthy and diseased tobacco leaves and an examination of the soil were carried out at Mayer's experimental station in 1879. The conclusion was that the source of TMD was not to be found in the plant nutrients.

After numerous other possibly influential factors were excluded, Mayer began in 1882 to study destructive insects as the potential cause. It was in this connection that in 1883 he discovered the definite transmission of the disease to healthy plants through the sap of diseased ones, whilst it was not possible to infect plants with the sap of healthy cultures. New shoots were conspicuously favoured by the infection. Isolating the germs according to Koch's rules and subsequent inoculations proved unsuccessful, and led Mayer to exclude the isolated bacteria as causative agents.

Thereafter, he attempted in filtration trials to clarify whether the pathogenic matter belonged – as in alcoholic or lactic fermentation – to the yeast or bacterial enzyme group. He used single-layered filter paper, which allowed the pathogens to pass through, and then double-layered ones, which rendered them impermeant. His attempts to find the enzyme were in vain. However, while applying heat to kill the pathogens, he observed that the sap lost its capacity to infect as the temperature increased. Thus he concluded that the pathogen must be a particle with form, probably a bacterium which he had not been able to reproduce, and called the disease tobacco mosaic disease. His results were reported in 1885.

Two others also tried to detect the cause of this disease: the Russian botanist Ivanovski and the Dutch bacteriologist Beijerinck. In 1887 the Dean of the Natural Science Faculty at the University of St.Petersburg (Leningrad) sent Ivanovski and Polovtsev to the Ukraine and Bessarabia. There, they were to focus their attention on the tobacco disease ravaging the crops. Just as Mayer had observed earlier, they noticed that older stocks were particularly affected, and recommended the farmers to begin a crop rotation. Ivavovski felt the disease had been caused by two separate factors, and explained his conclusion to the university faculty in March of 1888.

In 1892 Ivanovski presented new experimental results on TMD to the faculty. His filtration results and the conclusions derived from them aroused great interest. He had crushed some infected plant leaves to produce a suspension which he then filtered through linen: the filtrate was infectious! The same results occured when the suspension passed through the Chamberland filter. Although he had not found either a specific bacterium or a specific toxin, he concluded that the agent must be one or the other. To explain the possible bacterial filtrate, he favoured the fine cracks in the filter rather than the possibility of minute bacteria. He disclosed further results later in 1899 in response to experiments which Beijerinck had presented to the Royal Academy in 1898 to 1899. These were published in two periodicals.

Beijerinck had already experimented together with Mayer and undertaken his own trials at the Delft Polytechnic School after 1897. In his lecture presented at the academy, he referred to the earlier trials in which it had not been possible to link the disease to a specific bacterium. His intention was to carry out filtration trials to determine whether TMD was caused by a contagium fixum, a concept which definitely departed from the views of contagia held at about that time. The filtrates obtained through fine porcelain filters were free of aerobic and anaerobic bacteria but fully virulent. Even after three months, the condition of the filtrate had not changed. In addition, he carried out diffusion trials with the crushed tobacco leaves on layers of agar. He wanted to test whether the infectious agent would remain as a solid particle on the surface of the agar or whether, as a water-soluble or liquid substrate, it would penetrate into the deeper layers of the agar. In preparation for these trials he observed the diffusion capacity of diastase and trypsin and then selected a diffusion period of 10 days. After this time had elapsed, the completely germ-free deeper layers of agar were just as infectious as the filtrate of the crushed, infected leaves. For this reason, he judged the contagium to be soluble or fluid and called it Contagium vivum fluidum. He concluded

"the contagium, in order to reproduce itself, must be incorporated into the living protoplasm of the cell, into whose reproduction it is, so to speak, passively drawn".

He was able to dry the contagium without loss of infectivity and to prepare an alcohol precipitate from the sap at 40°C which remained virulent. As a closing remark, Beijerinck referred to the diseases of the peach tree described as "peach yellows" and "peach rosette" by

E.Smith in 1894, mentioning that the causative agent of these diseases closely resembled that of TMD.

Beijerinck's research findings prompted Ivanovski to undertake further trials, the results of which were printed in several publications between 1899 and 1903. The 1903 publication contained a comprehensive summary of his results. In searching for an answer to the question of whether a new type of a causative agent had possibly been found, Ivanovski's following statements may prove meaningful:

- He confirmed the diffusibility of the contagium in agar.
- He contradicted the standpoint that diffusibility of a contagium was proof of its water-solubility.
- Filtration results were dependent upon the condition of the porcelain filter.
- The TMD contagium was not a soluble but a solid particle.
- The TMD contagium was a microbe.
- This microbe could only be a minute granule.
- TMD contagium was able to develop on an artificial substrate.
- Maintaining the infectivity of the filtered sap was linked to the development of spores.

Assessing the results of his experiments on the tobacco mosaic disease agent, it is not possible to find evidence on the nature of the causative agent. Ivanovski did not offer proof in his experiments or in his illustrations that the causal agent was a bacterium. Furthermore, there is no indication of his having suspected a new type of causative agent. Beijerinck merely substantiated his conviction that the causative agent was neither an aerobic nor an anaerobic bacterium.

2.2.1.2 Foot-and-Mouth Disease (FMD)

The symptoms of FMD had been known since Fracastoro described them as long ago as 1546. However, only since 1839, after an epidemic in England, had the clinical and pathological-anatomical symptoms been substantiated by Dun. Bollinger (1874) left a comprehensive description of this infectious disease which is primarily found in animals but sporadically also in man. Ideas on the nature of the causal agent ranged from fungi to protozoa as more and more new contagia were being discovered. Since FMD had been causing extensive economic damage to beef stocks, it was naturally of particular interest to develop an effective vaccine against the infection. This seemed justified after vaccines had been developed for other diseases at the

Pasteur Institute and at the institute of Robert Koch. Thus the Institut für Infektionskrankheiten in Berlin was commissioned to research the groundwork for a vaccine against FMD in 1897. The committee for this assignment was directed by Löffler from the University of Greifswald; his co-worker Frosch from the Institut für Infektionskrankheiten was later substituted by Uhlenhuth.

The results of their research were reported in four separate documents and a summary between April 17, 1897 and August 12,1898, and sent to the Minister of Culture. The reports included:

- The type of causative agent.
- Transmitting the disease to various species of animals.
- The mode of infection.
- The infectious material.
- The duration of activity of the agent.
- The destruction of the causative agent.
- The development of immunity.
- The possibility of vaccination.

As for the type of causative agent, the first report stated: *"Even the trial results show reliably that a bacterium which grows on a conventional substrate cannot be the etiological agent in FMD".*

The second report maintained that artificial immunization against FMD in animals was possible. In the third report, they considered the causal agent to be small enough *"to pass through the pores of a filter which is impermeable to the tiniest known bacteria, so small, that even the best modern immersion system renders the agent unidentifiable under our microscope".*

They reiterated the assumption Koch had already expressed in 1890, namely that pathogens of other infectious diseases also belong to this group of very minute contagia. In the fourth report, they decided for a corpuscular and against a soluble agent, whereby "soluble" was understood as a toxin and not a contagium vivum fluidum. By finding a practical, functional and protective inoculation treatment, having punctiliously followed Koch's principles, they considered their assignment to be completed. In order to develop a vaccine, it would have been desirable - although not absolutely necessary - to identify the contagium.

They characterized the FMD pathogen in the following manner:

- Corpuscular.
- Not to be cultivated on conventional substrates.

- Not identifiable under the microscope.
- Able to permeate the finest filters.
- Belonging to a new type of causal agent."

2.2.1.3 Bovine Pleuropneumonia

During the IX International Congress for Hygiene in Madrid in 1893, Nocard reported results on bovine pleuropneumonia, which he had investigated with Roux of the Institute Pasteur in Paris. At first, they defined the causative agent of being reproducible in vitro, microscopically visible, but not filterable. Citing the agent's filterability in a corresponding publication, they said this applied to Chamberland F- and to Berkefeld filters, but not to Kitasato or to Chamberland B filters. Although pleuropneumonia was later classified as a disease caused by a mycoplasm, Nocard's and Roux's presentation and published results exercised a great influence at that time on the experiments and the discussions about an assumed new type of causative agent. The culture methods employed were enhanced by the collodion sac culture in the peritoneum of young animals, described by Metchnikoff, Roux and Salimbenie in 1896, and the serum-containing peptone medium for the production of diphtheria toxin, developed by L.Martin in 1898. They reported it to be an in vivo method, while in reality it was an in vitro/in vivo culture, because the collodion membrane prevented the microbe contacting the host cells. Löffler, who had been present in Madrid and had had discussions with Nocard, felt that his theory of bacteria undergoing continous transitions in shape and size had been confirmed. This behaviour would necessarily affect their microscopic visibility, as well as their filterability and growth in culture, depending upon the type of filter and the culture medium. He was unable to discover any new aspect which might have changed his thinking about the assumed causal agent of FMD. For him the pathogen was a new type of very small bacteria. He had not managed to cultivate it, even when using the collodion sac in connection with Martin's culture solution, which had been so successful in pleuropneumonia.

2.2.1.4 Rabbit Myxomatosis

It was also at the IX Congress in Madrid that Sanarelli reported an infection which had been fatally attacking the rabbit stock in the Hygiene Institute in Montevideo since 1896. He called it rabbit myxomatosis. The main symptoms were neoplastic tumours which developed in certain organs at the transition of epidermis to mucous membrane. It was a transmissible disease but, although Sanarelli had clinically and histologically characterized the pathogen, he could neither cultivate nor microscopically identify it. Thus, he concluded:

"This etiologic agent does not belong to any of those organized causal organisms which we are presently inclined to consider the cause of specific diseases".

He had not reported filtration data and his lecture gave no indication of any concrete conception of the myxomatogenic virus as a new kind of causal agent.

2.2.1.5 African Horse Sickness

Since the close of the 19th century, there had been reports of a fatal disease found extensively in the horse and mule of South Africa. When the disease broke out in Natal in 1887, the English government commissioned Nunn with the first scientific investigation, and other researchers also became interested in the disease. Their reports appeared in various journals between 1892 and 1895 but received little public attention until the English veterinarian M'Fadyean published his research findings and reported them to the Royal Veterinary College, London in 1900. He put the pathogen into an albumin medium and passed it through the Berkefeld and the Chamberland F and B filters. He was, however, just as unsuccessful as A.Theiler in attempting culture with a synthetic medium; not even staining techniques rendered the pathogen visible. M'Fadyean then concluded that the contagium must be a microbe, but neither researcher engaged in further speculation on the nature of the pathogen.

2.2.1.6 Fowl Plague

An epidemic-like disease occurring in fowls had also spread through northern Italy in 1878 and, 20 years later, in Austria and southern Germany. Clinically this disease could be clearly differentiated from chicken cholera. In 1901 Centanni reported his experimental results on the etiology of fowl plague before the Academia medica di Ferrara and in further publications in 1902. Lode and J.Gruber as well as Maggiora and Valenti undertook investigation of the disease. The unanimous opinion of all the authors was that the causative organism would not grow on conventional media, would not reproduce using the collodion-sac method, and could not be identified under the microscope.

Lode and Gruber assumed the pathogen to be a small bacterium analogous to the FMD causative agent. Centanni mentioned the possibility that a filterable virus might be cultivated outside an organism by using "*tissue preferred by the microbe*". He speculated that this fowl plague pathogen developed within the living cell of the preferred tissue, but that these conditions could not be provided artificially.

2.2.1.7 Yellow Fever

As the American armed forces began to occupy Cuba in the course of the Spanish-American War of 1898, they suffered heavy loss of human lives from yellow fever. This epidemic disease was already well-known in many parts of the world at the time. It had raged for several centuries particularly in the countries bordering on the Gulf of Mexico. An epidemic which broke out in the USA in 1878 registered 74000 afflicted in 132 cities with a mortality rate of 27%.

The 1575 cases of yellow fever and 231 deaths in Cuba impelled the USA to send a Yellow Fever Commission to the island in 1900, under the direction of the bacteriologist W.Reed. The group included the bacteriologist Carroll, the entomologist and bacteriologist Lazear, as well as the Cuban pathologist Agramonte. There had been very few concrete clues to this disease up to this time. The assumption of the American research committee, namely that mosquitoes were the carriers of yellow fever, had not been proven. The conclusions of the dramatic investigation, which was overshadowed by Lazear's death from the disease, were presented and published by Reed, Carroll and Agramonte between 1900 and 1902. Recalling a directive which W.Welch, a bacteriologist working in Baltimore who had participated in the wound-infection discussions in Breslau in 1877, had given, the commission was familiar with the methods employed by Löffler and Frosch in connection with FMD. Their conclusions confirmed, among other things, that yellow fever was transmitted by Culex fasciatus (identical to Aedes äegyptii) as well as by the blood of the diseased. Berkefeld filtrates retained their infectivity, and disease-specific bacteria escaped detection both microscopically and in culture. Significantly enough, the causal agent which they had filtered was the first example of a new, human pathogen which, moreover, was transmitted by an insect.

2.2.2 The Pathogen Remains While the Definition of the Virus Changes

As a result of the experimental studies of the causal organisms described above, several pathogens began to stand out from amongst the hitherto unknown, non-identifiable group of fever poisons. These were characterized by the following three negative peculiarities:

- Microscopically not identifiable.
- Can not be cultivated on conventional media.
- Filterable - with restrictions

Finally, in 1902, after Centanni had determined the pathogen of fowl plague to be a filterable virus, a critical discussion started up over Roux's paper which was published in 1903. In it he had mentioned the "so-called invisible microbes" and heated the debate about what to call these contagia, and whether or not they actually represented a new type of causal organism. The basis for the responses to this question were the three above-mentioned experimental criteria and not the theoretical concepts.

2.2.2.1 Invisible Microbes

Roux discussed the invisible microbes with reference to the experiments and pathogens cited above. He augmented these with the progress M.Nicolle and Adil-Bey had made with cattle plague in 1902, with Borrel's findings on sheep pox in 1902, and with Marx' and Sticker's study on the Epithelioma contagium (molluscum des oiseaux). Although Roux had also included the unsuitable pathogen of bovine pleuropneumonia in his definition of "invisible", this was not the only argument against his choice. For the microscopic identification of germs at the limits of the resolving power of microscopes there were as yet no suitable staining-techniques. Apart from this, no one should have assumed that the 200-nm limit would not someday be surpassed. Roux countered all objections simply by using the term "the so-called invisible microbes".

2.2.2.2 Non-Culturable Microbes

By the turn of the century, an appreciable number of media had been developed, on which the familiar bacterial and fungal pathogens, but not the protozoa or the spirochetes could be cultured. Thus, each microbe failing to grow was simply considered as "non-culturable", because the appropriate culture medium had not yet been found. It seemed equally defensible to classify a non-culturable germ as a protozoa. Consequently, "nonculturable" served neither as an argument for or against a particular sort of causal organism.

2.2.2.3 Filterable Microbes

In 1906 Remlinger expanded the discussion of filterable microbes by including the pathogens of rabies and smallpox, among others. He did not fundamentally object to the term invisible or ultramicroscopic organisms; he simply regarded the use of filterability to be more reliable. In his opinion, this actually made the nature of the causative agent less definite; neither Remlinger nor Roux wished to be definitive as to whether the pathogens were microbes. That filterability was unsuitable for classifying microbes was confirmed in Remlinger's work, as well as in Roux's defense of his choice of the term "invisibility" three years earlier.

Adding to the problems of filteration arrising from the various kinds of filters and sterilization methods in use was the even greater problem of assessing the results. The various types of filters, for

Table 2.1. Filters favoured at the turn of the century

Material	Introduced	Introduced by .. and remarks
Clay	1871	Tiegel/Klebs: unsintered
	1891	Kitasato: sintered
	1893	Pukall: sintered mixture of kaolin and quartz
Plaster of Paris	1877	Joubert/Pasteur: unsintered
Asbestos	1884	Hesse: pressed layer
Porcelain	1891	Garros/Pasteur: powdered, biscuit-porcelain
	1884	Pasteur/Chamberland: unglazed
Diatomaceous earth	1891	Berkefeld/Nordtmeier: unsintered

example, had very different and difficult to ascertain sizes of pores. The Chamberland B, to name one at random, had a lower mean pore diameter than type F, and the Berkefeld filter *v* (i.e. "viel" or "many") was *more* permeable and type *w* (i.e."wenig" or "few") *less* permeable than type *n* (normal). The Kitasato filter was the least permeable. Beyond that, sterilization of the filters proved problematic. Heat produced tension in the material, which readily led to easily overlooked micro-cracks, and hence to false filtration results. Despite these shortcomings, most experts in the field agreed that filtration was the best approach to the characterization of a new causative organism, provided that various types of filters were employed for comparative results. Table 2.1 presents a survey of the filter types most frequently used in early bacteriology.

2.2.3 The Filterable Viruses are Recognized

Both leading microbiological institutes of the world, namely Pasteur's in Paris and Koch's in Berlin, rejected the above-mentioned pathogens as proof of a new type of germ, and no effort was spent on naming them. It was decided quiete simply to use the name of the toxin linked to the disease, i.e. virus, combined with the laboratory indication of its filterability, hence the term filterable virus.

This terminology became standard practice in publications and lectures early in the 20th century. Filterability appeared to take priority over culturability and microscopic features. Consequently, charts and tables published later often contained the names of pathogens for which there was merely evidence of communicability and filterability. Wolbach's compilation of 1912, for example, contained the pathogens for typhoid fever and scarlet fever, in addition to 30 other "filterable viruses", purely on the grounds that they passed through the Berkefeld filter. In contrast, two pathogens omitted from the chart were E.Smith's identified agent in peach disease (1894), and E.Baur's infectious chlorose in Malvaceae (1904); neither of them had evidence of filterability to recommend them. As for the typhoid- fever pathogen, which was later classified as belonging to the genus Rickettsiae, it has remained among the filterable viruses, even in present-day textbooks.

2.2.4 Who was the Founder of Virology?

As more and more disease-causing organisms became known as filterable viruses, emphasis shifted away from the more significant biological features of the organism to a technical laboratory criterion. When Borrel, in 1902, focused on the affinity of filterable viruses for particular host cells, this was not regarded to be significant for the characterization of the pathogens. Beijerinck's concepts were equally disregarded! As early as 1898, he had assumed in his reflections on colloid chemistry that "virus" might be a living, liquid contagium, and commented that it was probably taken up by a living cell, in which it then reproduced. Although protein-like crystals, i.e. organized biological material, had been known since Hartig's description of them in 1856, a liquid contagium hardly fitted into the concept of an infectious pathogen at that time. In short, Beijerinck's idea met with rejection and his model of multiplication was discussed no further.

The literature often cites someone as the "founder" of virology. Since it is customary – for non-scientific reasons – to associate the word "founding" with one particular person, in the case of virology such a contention can only be defended superficially. When Beijerinck is mentioned, the scales tip in his favour because his ideas at the close of the 19th century sound to us so modern. Speaking against him is the fact that he made no attempt, theoretical or experimental, to prove or even to defend his hypothesis of a Contagium vivum fluidum and its intracellular reproduction. To declare Ivanovski "father of virology" can only be credible if we disregard his publications. He did not observe the established rules for etiological experiments, and consequently arrived at false conclusions. In his opinion, the pathogen for TMD was a bacteria which could be photographed. On the other hand, no one can deny that he was the first to filter the causative agent of a plant disease, not discounting Mayer's use of filter paper in 1880.

As for Löffler, we must recall that he worked on FMD in collaboration with Frosch and with Uhlenhuth, and also wrote up the reports on their findings. Giving due tribute to Löffler's co-workers, together they concluded that the FMD pathogen belonged to a group of the smallest organisms, but without any clear demarcation to the bacteria. The criterion of "filterable virus" was considered significant by Löffler as a laboratory index but not for the characterization of a new causative organism. Nevertheless, it was the praiseworthy efforts of Löffler and his co-workers, who set high standards in their experiments on FMD, which were then applied to work on filterable viruses.

Even though these three researchers contributed greatly to the field of experimental virology, they personally made no claim to be "founders". There are really no grounds for equating them with other "founders" such as Pasteur or Koch, who experimented on, published, and heatedly debated their ideas for decades, and unequivocally proved that they were right.

Progress in virology did not emanate from the short-term efforts of particular scientists. It has come, seen more profoundly, from the consistent work of many astute researchers from all corners of the globe, interacting as teams over decades in different disciplines. These people and their achievements will be the subject matter of the following chapters.

3 Are Filterable Viruses Miniscule Bacteria?

If "filterable viruses" were simply small bacteria, it should have been possible quickly to provide uninpeachable proof of this using the then known, newly developed methods in bacteriology and immunology. Following the principles set down by Koch and later augmented by other scientists for producing bacteriological and etiological evidence, pathogens - providing they were stainable - were identifiable under microscopes with a resolving power of 200 nm. Moreover, laboratories had numerous culture media and staining techniques at their disposal, in addition to suitable laboratory devices and enhanced infection models for experimental animal trials. The methods employed by Löffler and Frosch while researching FMD were a model for experiments with filterable viruses. Despite all of this, the opinion prevailed for a long time that this kind of causal organism was merely a very small bacteria. On the one hand, it seems incredible for scientists to have assumed that the limits of technical aids such as microscopes and filters coincided with a dividing line amongst microbes. On the other hand, nothing was undertaken experimentally to investigate other hypotheses.

More than three decades were to pass before it was clearly recognized that filterable viruses were very different from small, or indeed any kind of bacteria. Experimental results in particular fields encouraged exchange of the term filterable viruses, synonymous with minute bacteria, for viruses as new causative agents. In describing the various paths leading to the use of the term virus, it should be mentioned, but then disregarded, that some pathogen types had been detected with properties close to those of the viruses, whose classification was equivocal. These included microbes like those discovered by Nocard and Roux in 1898 as the pathogen causing bovine pleuropneumonia. Later named Pleuropneumonia-like organisms (PPLO) and generally classified in 1956 as genus Mycoplasma, the microbe had a particle diameter of 100–150 nm and remained microscopically unidentified. Though filterable through layered filters as well as filter candles, its ability to reproduce on synthetic culture diverged from that of the filterable viruses.

A second group to be considered in this respect are the bacteria-like, gram-negative, pleomorph microorganisms called Rickettsiae. They were found to reproduce in the cells of mammals and arthropods, but with few exceptions did not grow on bacterial culture medium. They could be round in form, with a diameter of about 300 nm, or rod-shaped, reaching lengths of up to 2000 nm. Obligate, intracellular parasites, they seemed at first glance to belong to the filterable viruses. But their DNA *and* RNA composition, their reproduction by fission, together with their reaction to chemotherapeuticals, later placed them somewhere between bacteria and viruses.

The third group of pathogens to be separated from the filterable viruses was the chlamydozoa-strongyloplasma described by von Prowazek and Lipschütz. In 1907, von Prowazek found the first of stainable, grain-shaped trachoma inclusion, sheathed in a basic substance. Later, they were combined in a group called psittacosis-lymphogranuloma-trachoma as large viruses, in contrast to true or small viruses. In due course, they proved to be antibiotic sensitive and acquired a position outside of the virus group when it was discovered that they contained both types of nucleic acid.

Included amongst the filterable viruses, however, were microorganisms discovered by Twort and D'Herelle (1915–1917) and known as filterable viruses in bacteria, i.e. bacteriophages. Their virus character was rather more heatedly contested than acknowledged during the period covered by this chapter.

3.1 The Use of Physical Methods

Filterable viruses were distinguished by negative biological and physical features: It was not possible to culture them, and because of their small size they remained macro- and microscopically undetected. They were or were not filterable, depending on the pore size and quality of the filter used. The range of physical methods employed during the first three decades of the 20th century to characterize viruses included filtration, centrifugation, adsorption, electrophoresis, and optical methods.

3.1.1 Filtration

In the early stages, the same filter candles or filter layers used in experiments with bacteria were applied to filterable viruses. Whilst bacteria, as a result of their size, did not pass through intact filters, whose greatest drawbacks were clogged pores and germ-infected cracks, for the smaller, filterable viruses, new problems arose. Here, filtration results depended both upon the average and maximum pore size of the filter. In addition, there were annoying adsorption properties of the filters, and colloids which appeared, for example, as proteins in the solutions to be filtered.

The aims of filtering were twofold: to isolate the pathogen and to obtain the most accurate information possible about the organism. Both aims were achievable providing three prerequisites were fulfilled: the range of pore size about the mean had to be calculable; adsorption by the filter had to be excluded or measurable; and the organisms had to be as round as possible to enhance permeability. Since the filters in use at the time in bacteriology and in colloid chemistry had been developed in part for the purpose of dialysis, they did not necessarily fulfil these basic requirements. Their improvement or further development was, therefore, essential. Help came from the field of colloid chemistry which was simultaneously faced with the problem of separating colloids from crystalloids and specific colloids from colloid mixtures. High-pressure filtration had been practiced in this field since the English physiologists Starling and C.Martin had developed the technique in 1896, by which they were able to counteract the high filtration resistance arrising from the impregnation of Chamberland candles with stiffened gelatin.

New filtration possibilities emerged with the use of collodion. This was produced by dissolving nitrocellulose – better known as guncotton, the 1846 invention of the German chemist C.Schönbein in Basel – in a mixture of ether and alcohol. Evaporation of the solvent gave a thin, impermeable membrane through which the widely dispersed colloids and microbes could not pass. This membrane was used, for example, for sealing wounds, and in 1904, Malfitano first used collodion membrane sacks for filtration purposes. Since sub- or ultramicroscopic particles were unable to pass through the membrane, this process has since been known as ultrafiltration. Bechhold, a colloid chemist working in Frankfurt upon Main, improved filtration techniques by developing ultrafilters of varying densities. In addition, he devised a method for measuring not only the mean diameter but also the maximum pore size. He began in 1906 with hardened filter slices impregnated with gelatin. Diluted formaldehyde or glacial acetic acid-collodion in water was used for hardening. Just one year later, Bechhold-Koenig ultrafiltration devices appeared in which moulded, diatomaceous earth had been substituted for the filter paper. Such filters were able to withstand a pressure of 20 atmospheres. Although the ultrafilter represented appreciable progress, there was still one order of magnitude variation in pore size, and the sizes of filterable viruses quoted in publications after ultra- or another means of filtration were only relevant if they represented the means of repeated measurements.

Using a nitrocellulose base, Zsigmondy and Bachmann introduced membrane ultrafilters to the market in 1918. They obtained more consistant results by using dried colloid solution. The filters were produced with various pore sizes and, an even bigger advantage, they could be stored in a dry state. Elford, who in 1931 reported on the development of filters at the National Institute for Medical Research in London, also used nitrocellulose for his gradocol membranes. Aggregation of the substance into filter membranes of various pore sizes was achieved through the antagonistic solvent characteristics of acetone and amyl alcohol. His graduated, gradocol filters, with pores ranging from 3000 to 10 nm, were not used to separate different colloids but more specifically to determine the particle size of filterable viruses, regardless of their having a round or some other form. Elford contributed significantly to the more exact calibration of filters and to determination of particle size in relation to other relevant factors such as membrane thickness, the diameter and length of the pores, adsorption, etc., all of which were the subject of a correction formula.

The list of filterable viruses in 1915 – the year the bacteriophage was discovered – totalled 34, related to human, animal and plant diseases. By 1937 scientists had firm estimates of particle size for 23 "filterable" human and animal pathogens, 13 types of bacteriophage, and *TMV*. The particle sizes ranged from 8 to 230 nm. The smallest known germs, besides the bacteriophages, were the pathogens for poliomyelitis, yellow fever, encephalitis, *FMD* and *TMV*, ranging in size from 8–30 nm. Despite significant improvements in filtration techniques, the process continued to be problematic. The wide variation in pore diameter persisted, producing unreliable estimates of particle size. Isolating pathogens was connected with a high rate of germ loss because of adsorption by the filters, the adsorbing properties of filterable viruses known as biocolloids, and extraneous deposits on the filters themselves. In order to determine the mean or maximal pore diameter, three methods were used: the critical air-pressure method, the rate of water flow method, and the filtration standardized colloid suspension method.

3.1.2 Centrifugation

Centrifuges separating particles of different densities at low rotations per minute (rpm) had been employed since the mid 19th century, for example, in milk production. Colloid chemists began to show an interest in high-rotation centrifuges for determining particle size in finely dispersed suspensions which, in turn, led to various other developments. Electrically-powered versions first appeared on the market in 1889, and a water centrifuge boasting 200 rpm in 1924. G.Stokes, an Irish-born physicist and mathematician, had developed a formula within his law of friction which made it possible to measure the diameter of round particles according to their descending speed, provided they were not hindered in their sedimentation by any other factor. Measurement was possible if the density of the particle, the centrifugal acceleration at a given rpm, and the density and viscosity of the inert fluid were known.

Since laboratory centrifuges had been developed to meet the exacting requirements of colloid chemistry, experiments with bacteria proved unproblematic. The essential data were obtained and the necessary centrifugal force of approximately $500 \times g$ (equal to 500 times the force of gravity) was achieved for functional centrifugation up to about 4000 rpm. Filterable viruses, however, presented a much more complex situation. Their size, form, and density were unknown. Assuming the specific gravity of a virus to be in the range of bacteria, i.e. 1.1 or 1.2, it would take 60 minutes for a spherical virus particle with a diameter of 200 nm at $500 \times g$ (3000 rpm) to sink by 1 cm. In 1906, Russ centrifuged the causative agent of fowl plague at 6000 rpm and $2050 \times g$ for 110 minutes without success. Since it seemed justified to assume that filterable viruses were approximately equal in size to colloids, bacteriologists were just as interested as colloid chemists in improving centrifuges and centrifugation methods in order to extract purer substances and to determine particle size more exactly. High-speed centrifuges were produced individually, and thus, Friedenthal separated casein from cow's milk in 1911 in a centrifuge at 27500 rpm.

A first attempt to determine the particle size of colloids was undertaken by Dumanski in 1913. He examined relatively large colloids and compared his results with those achieved with the ultramicroscope. This comparison left him and others so disappointed that further efforts in this field were abandoned. In 1922, W.MacCallum and Ella Oppenheimer reported on the differential centrifugation of vaccinia material taken from a lymph in order to determine the specific gravity of the particles. Today their method would be called density gradient centrifugation. Looking for a sedimentation balance between the granules and a mixture composed of lymph, 2,3% glycerin and synthetic, Locke nutrient solution, they found values between 1.12 and 1.13 for vaccinia granules being probably inclusion bodies and not elementary bodies.

Significantly improved centrifugation methods for colloid chemistry and bacteriology-virology work may be attributed to Svedberg at the physico-chemical institute at Uppsala University. He was particularly interested in the smallest colloids and had co-developed an optically controlled centrifuge with J.Nichols. As a result of this invention, he was invited to an 8-month visit to the University of Wisconsin in Madison, U.S.A. in 1923. Returning to Uppsala with strong impressions of the technical possibilities he had observed in the U.S.A., he devoted his energies to the problem of sedimentation analysis of highly dispersed colloid solutions using centrifugal force. He supposed this could be best achieved by a significant increase in the rpm. and technical improvements to centrifuges in general, including vibration-free operation and the elimination of convection in the measuring equipment.

Thus, numerous rotors were developed for varying requirements: for sedimentation equilibrium determinations of high molecular weight substances as well as sedimentation velocity determinations of low molecular weight substances to obtain for the sedimentation constant S. The sedimentation process was photographically controlled and microphotometrically recorded, and molecular weights determined from the resulting data curves. The first ultracentrifuge which Svedberg developed together with Rinde in 1924 achieved 5000 $\times$ g. In 1926 Svedberg was a Nobel Prize winner for work on dispersion systems, and in the same year his high-speed oil turbine ultracentrifuge achieved 100000 $\times$ g at 19000 rpm; in 1934 he achieved 600000 $\times$ g at 120000 rpm using a smaller rotor.

In recognition of this accomplishment, Sweden granted 1.2 million Kroner for the construction of a new building, to which Rockefeller then donated 50000 dollars for equipment. The institute became a centre for international scientific cooperation, also in the field of virus research as Svedberg's interest soon began to focus on proteins. Since Svedberg's ultracentrifuges were not only very expensive, but demanded highly trained personnel to cope with the complicated technology, a parallel development began in 1925. The model for this was the smaller, fast rotation compression turbine of Henriot and Huguenard.

Bechhold and M.Schlesinger forged ahead with their own centrifugation methods for scientists who wished to analyse filterable viruses but who had no access to high-speed centrifuges. It was not their intention to build completely new equipment; they were far more interested in working out an experimental and mathematical method for determining colloid particle size using a maximum of 10000 rpm and later 15000 rpm. These Ecco-centrifuges, available on the market at that time, were supposed to lack the shortcomings of conventional centrifuges, e.g. heating of the apparatus, vibration and fluctuation in the rpm, the ocurrence and influence of convection, and the absence of exactly marked layers.

The basis for the formula published in 1931 to determine particle size was the decrease in infectious particles with centrifugation time and force of gravity. Without knowing the specific gravity of the cow pox pathogen, they found its diameter to be 210–230 nm, and that of fowl plague, 120–130 nm. A Coliphage 88 was centrifuged using a modified method for up to 4 1/2 hours at 7000–8000 rpm. Assuming a density of 1.2–1.5, the diameter of this phage was determined to be 79–90 nm. Thus, they concluded that this phage was significantly larger than those measured by Hetler and Bronfenbrenner in diffusion trials in 1929, and different from the phage used by von Angerer in turbidity determinations. Bechhold and Schlesinger considered their centrifugal method suitable for particles of up to 5 nm using 10000 rpm for 8 hours. When we compare the particle diameters of filterable viruses recorded by Schlesinger in 1933 with those measured by more sophisticated means much later, we cannot help but acknowledge his achievement.

Although the centrifuge experiments described by Bland (1928), by Ledigham (1931) and by Craigie (1932) were restricted to poxviruses, new centrifuges and centrifugation methods opened up the possibility of working with all the other viruses. As a result, scientists were in a osition to measure the density and size of viruses, as well as produce ure and concentrated preparations. The particle sizes determined by ltracentrifugation no longer had the broad range of variation esulting of filtration, and this mode of preparation of pure virus liminated the previously unavoidable losses.

.1.3 Adsorption

ven early researchers such as Roux and Yersin had treated diphtheria oxin with freshly precipitated calcium phosphate in order to bind the oxin to a salt. This enabled them not only to remove undesired ccompanying substances, but also to concentrate the toxin. The same ffect was achieved with aluminium hydroxide and animal charcoal. Villstätter used aluminium hydroxide of various types, and also aolin to purify and concentrate toxins. Other adsorbents soon in use vere metallic oxides, blood corpuscles, bacteria, globulin, carmine nd diatomaceous earth. These were also employed in adsorption xperiments with filterable viruses. To allow recovery of the adsorbed naterial, i.e. the elution of the adsorbent, solvents such as cane sugar r sodium phosphate solutions, or diluted ammonia were used.

In 1909 Kraus and coworkers delved into the question of whether rganisms might be able to eliminate filterable viruses by leucocyte dsorption. During their studies, they found that binding viruses to dsorbing substances presented a means for their purification. Studies n *FDMV* by Vallée and Carré, and work on the bacteriophage by Maisin followed in 1921. Levaditi and Nicolau published studies on erpes and vaccinia in 1923, and Margaret Lewis and Andervont vorked with the *Rous sarcoma virus (RSV)* in 1927. A study of nyxomatosis by Lewis followed in 1931, and in the same year, Lewis nd W. Mendelsohn purified *Chicken virus No 1* by adsorption onto aolin or charcoal. The following year, Doerr and E.Gold discovered hat fowl plague pathogen could be adsorbed by chicken and guinea ig erythrocytes. In 1932 Sabin reported experiments on the poliovi- us using an aluminium hydroxide gel.

New knowledge in the field of colloid chemistry, namely that urface active colloids could be adsorbed and, furthermore, could hemselves adsorb, helped to explain the considerable difficulties ncountered in filtering. As the factors impeding purification and oncentration of filterable viruses by adsorption and elution were ssessed, scientists began a search for more suitable adsorbents and solvents, and for optimal temperature and pH values. Since centrifugation was now being used more successfully than previously, adsorption was not considered particularly important for purifying and concentrating filterable viruses. Nevertheless, these experiments were to serve as the basis for later developments in the field of virus vaccines, encompassing the production of effective adsorbate and depot vaccines.

3.1.4 Electrophoresis

Bechhold recognized that colloids tended characteristically to migrate towards electrodes under the influence of an electric field. He also knew that this migration was influenced by the addition of acids or alkalines. In 1904 he observed that bacteria also had a positive charge and migrated within an electric field; in a neutral solution they were negatively charged and correspondingly moved toward the anode. In addition, he discovered that at a given pH value, the bacteria were inclined to aggregate and not just migrate. Since this phenomenon took place at an acid pH of 3–4,8, the clustering was termed acid agglutination. The point at which the minimum of H and OH ions were released and the least stable solvency could be expected was called the iso-electric point. The movement of colloids and bacteria in such electric fields was termed cataphoresis, which is synonymous with electrophoresis. In 1905 Billitzer described an experimental arrangement for electrophoresis in which he placed several glass vessels in a row and connected them with fork-shaped pipettes. He filled the vessels with a colloid solution which had been rendered salt-free by dialysis; the vessels at each end contained a platinum electrode. In work with filterable viruses in electric fields, it was hoped that improved filterability and easier separation of extraneous organic matter could be achieved by influencing the surface tension. Besides the determination of the iso-electric point, an enrichment process seemed to be possible. For electrophoresis experiments with filterable viruses beginning as late as 1924, U-shaped tubes and improved vessels were used. In 1924 von Angerer reported experiments with a bacteriophage; results from K.Koch (1926) and Todd (1927) followed in close succession. Olitzky and Boez experimented in 1927 with the agent of FMD, Douglas and W.Smith in 1928 with the pathogen of vaccinia, and Hindle and Findlay in 1930 with the pathogen of yellow fever. These and many subsequent experiments showed, despite numerous contradictory results, the migration of the smallest pathogens towards the anode and a range of iso-electric points for the various viruses reaching into the field of high (alkaline) pH values.

Olitzky and Long maintained in 1929, after immunization trials with electrophoretically purified *Vaccinia virus*, that immunity could only be achieved if the antigen was reproductive. Thus, scientists became motivated to similar immunological experiments with other filterable viruses with the aim of proving electrophoretically the presence of very minute amounts of a virus. Unfortunately, all their efforts were in vain. The hopes placed in electrophoresis for progress in experimentation with filterable viruses met with frustration due to the unsophisticated techniques available in the early forties. Nevertheless, one positive result was the new knowledge that the iso-electric points for filterable viruses lay much farther apart than those for bacteria. Scientists concluded that virus disinfection would be far more difficult than that for bacteria.

3.1.5 Optical Methods

The filterable viruses first discovered were ultramicroscopic, i.e. they were not visible nor recognizable microscopically because their size was beyond the 200-nm resolving power of the conventional light microscope. Since it had been established that one factor determining resolution was the wavelength of the light used, it was conceivable, that the morphology of particles down to 70 nm could be visualized by using UV light and an additional lens. Thus, quarz lenses and quarz prisms were inserted, and a microphotographic device attached to produce ultraviolet dark-field illumination. In 1934 Barnard and F. Welch described a method of ultraviolet photography. However, increasing the resolving power of conventional microscope, without the concomitant ability to stain filterable viruses was not likely to improve this branch of microbiology. Special methods for staining viruses were needed, and in this connection, it is important to remind readers that, until approximately 1929, the exact nature of the various inclusion bodies found in diseases caused by pathogenic filterable viruses had not been clearly established. Then in 1929, C. Woodruff and Goodpasture demonstrated that the inclusion bodies of chickenpox were composed of infectious particles, later termed elementary bodies by von Prowazek. Thus, for the purpose of this story, all publications on the microscopic detection of stained inclusion bodies are, in effect, useless.

Bacteriological methods of observation in hanging droplets or in a thin-layer between the cover glass and the lens of a normal microscope were not suitable for filterable viruses. The microscopy employing intravital staining, used by Lipschütz and von Prowazek proved to be just as ineffective. In 1925 Bechhold and Villa extended the use of the ultramicroscope, developed by Siedentopf and Zsigmondy between 1902–1903, to hydrophilic colloids based on the Tyndall effect, to hydrophilic, i.e., to biological colloids. They described their dark-field method in 1926 using a coli-phage and a poxvirus as examples. The procedure began with a pre-purification by filtering through a Chamberland candle, complete separation of the accompanying extraneous biological material by ultrafiltration, followed by a gold chloride coating, ashening, and a final gold coating. The ultramicroscope, however, was not really suitable for microbes smaller than 100 nm, and for filterable viruses its use was inconsequential. The same applied to dark-field microscopy, which detected only crude outlines. A preferred method was the Löffler corrosion as a part of the Carbol-Fuchsin-staining according to Paschen. Herzberg also worked with this method in 1933, the year he demonstrated intracellular virus multiplication modelled upon the *Canary bird virus*, which had been discovered in the same year by Kikuth and Gollub. An enhanced capability to recognize stained particles was developed in 1934 with a special form of victoria blue, which is taken up by lipoproteins.[1] Giemsa staining, which had been employed so successfully in bacteriology, rendered several inclusion body types visible, but no elementary bodies.

3.2 Biochemical Investigations with Filterable Viruses

After C. Woodruff and Goodpasture, while working with the pathogen for fowlpox in 1929, had found that inclusion bodies in animal viruses might be composed of numerous elementary bodies, scientists were prompted to examine these biochemically. The immediate aim of the early trials was to obtain pure and concentrated viral material. During an earlier period, Vinson had carried out biochemical studies

[1] This staining technique was originally introduced by Lustgarten in 1856.

on the virus particles of TMD for the same purpose. In 1927 he had reported the possibility of concentrating the infectious particles derived from purified pressed sap by using acetone precipitation. Beijerinck had achieved similar success with alcohol precipitation almost 30 years earlier. Collaborating with Petre, in 1929 and 1931, Vinson published an alternative method for purifying *TMV* by precipitation, which involved the addition of safranine, magnesium sulphate, ammonium sulphate, and iron acetate to the virus suspension. Amyl alcohol appeared to be the most suitable for elution.

It was well-known since Allard's work in 1916 that *TMV* could be inactivated quickly by alcohol or acetone; therefore, Vinson and Petre tested the precipitating action of these two agents in non-inactivating concentrations. They also found that considerable amounts of protein coagulated at temperatures between 85 °C and 90 °C without significantly impairing virus activity. In 1931 they were able to isolate active crystals by adding acetone and acetic acid to infectious sap which had been purified with iron acetate. The loss of activity which they observed after recrystallization led them to believe that the crystals were not pure virus. Vinson lectured on his results with infectious, crystalline *TMV* in 1928 and 1930 at meetings of the American Association for the Advancement of Science, even prior to co-publication of the work with Petre in 1931. Their specimens were not pure, only moderately infective, and consisted of 33% ash. They regarded *TMV* as capable of being precipitated like a chemical substance, and assumed from the nitrogen determinations and tests in electric fields, that the pathogen was proteinaceous in character, an enzyme by nature and reproduced autocatalytically.

The crystallization described by Vinson and Petre was considered to be such a significant achievement that the Journal of the American Medical Association devoted an editorial to it on August 20, 1932:

> "Possibly the reported successful crystallization of the etiologic factor of mosaic disease of tobacco may be regarded by future medical historians as one of the most important advances in infectious theory since the work of Lister and Pasteur. The announcement of the isolation of a crystallizable phytogenic enzyme necessarily throws doubt on the conception that poliomyelitis, smallpox and numerous other "ultramicroscopic infections" are of microbic causation."

This naturally aroused the interest of other scientists, prompting them to analyze *TMV* chemically. Thus, in 1931 a department of phytopathology was created at the Rockefeller Institute for Medical Research in Princeton. Its first director, Kunkel, came from the same department of the Boyce Thompson Institute in which Vinson was working.

His new colleagues at the Rockefeller Institute, Northtrop, Kunitz, Herriott and Anson, undertook to isolate and crystallize proteolytic enzymes. Other *TMV* purification and concentration methods emerging between 1933 and 1935 were based on preceding publications and did not aim at biochemical virus analysis.

But this was exactly the aim of M.Schlesinger. For him, purification and concentration experiments with bacteriophages were simply a preliminary stage of more profound biochemical research. In 1934 he investigated the metabolism and chemical composition of an E. coli bacteriophage which he had purified and concentrated according to his own filtration and centrifugation methods laid down in 1933. He demonstrated the bacteriophage was corpuscular, and that protein was its main component, and concluded on the basis of its 3,7% phosphorous content that this was a nucleoproteid. He deemed it possible that fats or lipids were also to be found in the phage, and definitely excluded polysaccharides. He supported the hypothesis that the nucleoproteid was the reproductive substance of the bacteriophage and interacted with the genetic material of the bacterial cell.

But sensational publication appearing in 1935 suddenly interrupted this chain of development from the filterable virus to the virus as a nucleoproteid which had been initiated by M.Schlesinger in 1934. This paper and lectures presented at the Second International Congress of Microbiology in London in 1936 left a profound impression, especially on European scientists. The author and lecturing participant was the biochemist Stanley from the Rockefeller Institute for Medical Research at Princeton, N.J. In 1932 he had left New York for Princeton, where in 1933 he concentrated his efforts on purifying *TMV* and isolated it as a crystalline protein. Neither his publication, "Isolation of a Crystalline Protein Possessing the Properties of Tobacco Mosaic Virus", nor his lecture gave an exact proof of the chemical structure of *TMV*. Far more significant was Stanley's conclusion that the development of crystals indicated the protein to be pure. Clearly defining the limits between his experimental success and that of Vinson and Petre, Stanley availed himself of Caldwell's 1934 interpretation that Vinson and Petre had worked mainly with inorganic material lacking any relationship to biological activity. As for the crystallization, his arguments were based on J.Sumner's urease crystallization in 1926. As Sumner had experienced, Stanley met with scepticism on the part of those who agreed with Willstätter that the enzyme protein was simply the carrier of protein-free, effective groups.

Although Stanley maintained in his publication that it was not just difficult but practically impossible to prove conclusively the purity of a protein, he defended his assertion from crystallization that *TMV* was a pure and autocatalytic protein with the following analytical data:

- 20% nitrogen
- 1% ash
- a positive biuret reaction for protein
- Fehling's and Molisch's test for carbohydrate was negative
- the loss of activity was directly proportional to the loss of protein.

The participants were taken by surprise when Stanley's findings were confirmed by Bawden at the Congress! Astonishment, not so much at the successful crystallization – this had been known since 1856 – but at the assertion that the crystals were pure protein capable of autocatalytic reproduction. The participants accepted this definition of a virus, supplemented by the finding that the presence of living cells was a requisite for reproduction. It was still accepted 10 or more years later, as textbooks substantiate, even after Stanley's conclusions had been corrected or disproved in 1937. Bawden and Pirie, supported by the British crystallographers Bernal and Fankuchen, and analytical data from three strains of *TMV*, had proved this pathogen to be not only protein but nucleoprotein in substance. When it came to describing their precipitate, they selected the word "fibrous" instead of "crystalline". In contrast to Stanley's crystal needles, 0,03 mm in length, they found the single particles of the precipitate too small to be exactly determined microscopically. The active material for them, as well as for Bernal and Fankuchen, was clearly "liquid crystalline", as they had already described it in a publication in *Nature* a year earlier. The following were their analytical values:

- carbon 49.3 – 50.0 %
- hydrogen 7.2 – 7.4 %
- nitrogen 14.4 – 16.6 %
- sulphur 0.24 – 0.59%
- phosphorus 0.45 – 0.55%
- ash 1.5 – 3.0 %
- hydrocarbon 2.5%

Even though Stanley's interpretation proved to be false, it was largely owing to his active experimentation and publication after 1935 that the virus was accorded such widespread scientific interest. Thereafter, viruses were also studied at institutes which approached microbiology not out of interest for pathogens but as a field which might unveil basic knowledge of biological principles. Thus, it seems justified to end the first stage of the history of experimental virology at that point in time when Stanley defined *TMV* as an autocatalytic protein.

3.3 Are Cell Inclusions Specific to Filterable Viruses?

Cellular changes resulting from infectious disease, such as inclusion bodies, had already been known before the discussion of filterable viruses ignited. As early as 1841, the first typical, epithelial inclusion bodies were described in Molluscum contagiosum by Henderson. In cases of fowlpox infection, Rivolta observed cytoplasmic inclusion bodies in 1873. The following year, Weigert recognized small inclusions in epithelial cells in smallpox, and Guarnieri described cytoplasmic inclusions in epithelial cells of the skin in pox and vaccinia infections. He interpreted these to be enclosed protozoa and called them Cytoryctes vacciniae.[1] The mesodermal and epithelial inclusions observed by Bosc in 1901 received further attention by Borrel in 1903. Simultaneously, Negri found abnormal cell structures in the brain cells of rabid dogs which he regarded to be specific to rabies and, therefore, a reliable diagnostic sign. In the same year nuclear inclusions were also described by Kopytowski in genital herpes.

Cornalia and Maestri had reported on inclusions found in the cell nucleus of the silkworm Bombyx mori L. as early as 1856. Verson defined their crystalline structure in 1872, and Bolle analysed them in 1893. They were insoluble in water, alcohol, ether and acetone but soluble in alkali and acids. They were difficult to stain without acid pretreatment. Ishimori observed cytoplasmic inclusions in the cylindric cells of silkworm intestines in 1934. Cell inclusions in plant diseases were first described by Ivanovski in 1903, when he found colourless, insoluble, crystal-shaped deposits in TMD. After treatment with an acid Flemming solution, they appeared horizontally striped and could be stained with eosin.

Cell inclusions found in human and animal diseases were considered to be the histopathological consequence of the infection or inclusions of the pathogen. Interpretations of the putative microbes

---

[1] Cytoryctes meaning gnawing at the cell.

ranged from bacteria via fungi to protozoa. It is remarkable that inclusions were clearly observed in diseases like rabies, smallpox and vaccinia whose pathogens remained undiscovered or were assumed to be caused by filterable viruses. When microbe-like particles were recognized in the inclusions or could be isolated from them, the particles were termed granulations élémentaires or elementary bodies, as Chauveau had so designated for vaccinia in 1883. At the beginning of the 20th century, only polyhedral-shaped inclusions were known amongst the insect viruses. Thus, E.Fischer chose the term polyhedroses in 1906, and Wahl used the expression polyhedra disease in 1909. Seventeen years later, Paillot detected a second inclusion form in the larvae of butterflies which he called granula. Having a diameter of 200–500 nm, they were only one third the size of polyhedras and thought to be the cause of granulosis.

As the number of newly-discovered filterable viruses increased, interest in inclusions began to mount. Borrel observed cytoplasmic pigeon pox inclusions in 1904. Paschen developed a growing interest in those occurring in smallpox and vaccinia; he termed them initial or elementary bodies and held them to be pathogenic in nature. Even today the literature refers to them as Paschen bodies. Disease caused by animal viruses with cytoplasmatic inclusions included Molluscum contagiosum (Lipschütz, 1907), measles (Ewing, 1909), myxomatosis (Splendor, 1909), African horse sickness (Kuhn, 1911), herpes simplex (Lipschütz, Luger and Lauda, 1921), ectromelia (Marchal, 1930), canary pox (Burnet, 1933), and hog cholera (Bendinger, 1934). The list of virus diseases exhibiting nuclear inclusions increased to more than 30 by 1935; in addition to insects, the victims included amphibians, fish, birds, mammals and humans. In insects, inclusions were found in blood, nerve, gut and germ cells.

In diseases caused by plant viruses Sheffield discovered crystalline as well as amorphous cytoplasmic inclusions in 1931. Goldstein designated them x-bodies in 1924; additional names include vacuolate bodies, ameboid bodies, plasmodium-like masses, and spherules. Subsequent to C.Woodruff's and Goodpasture's (1929–1930) proof that inclusion bodies in fowlpox consisted of myriad infectious particles, other scientists began to investigate with renewed interest the inclusions contained in other diseases caused by filterable viruses. By the mid-thirties, it was clear that there was more than just one explanation for the nature of inclusions. It had been established that there were:

- inclusions with elementary bodies
- inclusions with virus protein
- inclusions without elementary bodies and without virus protein
- inclusions without any perceivable relationship to viruses.

Thus, whilst inclusion bodies were a strong indication for a current or previous virus infection, the relationship could only be confirmed by evidence of elementary bodies or virus protein.

3.4 Culturing Filterable Viruses In Vivo and In Vitro

Since filterable viruses could not be grown on synthetic media, in vivo and in vitro experiments with pure cultures were not possible, and thus the isolation of a single pathogen was virtually impossible. As a consequence, the early stages of experimental virology mainly involved the classification of microbes as filterable viruses simply on the basis of filterability and the transmission of infection in vivo. This was inevitably accompanied by a bevy of potential errors. It is, therefore, small wonder that during the early decades far more pathogens were classified among the filterable viruses than could convincingly be proved as such. In vivo laboratory experiments were essential. This was not problematic in diseases of small animals, for these were readily available and easy to keep. As for plant diseases, the danger of the release of undesirable infection into the environment existed, - not only in outdoor trials. Experiments with large animals, like horses and cows, demanded strict isolation in separate rooms or areas, or a substitute experiment using a suitable laboratory animal. Laboratory animals were also required for all human pathogen experiments. A test animal was considered appropriate when the induced process of infection favourably compared with that of the natural host. Contingent upon the objective of the trial, it was desirable to use pure suspensions of the microbe whenever possible, for example, when producing immune serum. On the other hand, pure cultures were required for identification of the pathogen as a filterable virus. Pure suspensions of filterable viruses are not identical with pure cultures of bacteria which, as defined by Koch, are obtained by inoculating a single colony on a nutrient agar plate. By a pure suspension of filterable virus is meant a statistical population responding uniformly to filtration or to centrifugation.

3.4.1 The Search for Experimental Animals

Reviewing the pathogens considered to be filterable viruses at the beginning of the 20th century, either by filtration or from clinical experience, and for which trials in their natural host were problematic, the situation might be summed up as follows:

Foot-and-mouth disease:
Löffler and Frosch found cows and pigs to be appropriate experimental animals for infection and immunization trials in 1897. Significant progress was achieved in 1920 with the discovery of the guinea-pig as the most suitable test animal by Waldmann and Pape.

African horse sickness:
Not until 1932–1933 were Nieschultz and R.Alexander, working independently, able to establish virus susceptibility in the white mouse, the guinea-pig and the rat.

Yellow fever:
New impulses for yellow fever research came in 1927 after successful transmission of the pathogen to the monkey by A.Stokes and coworkers, and the successful passaging from the monkey to the mouse as described by M.Theiler.

Filterable viruses discovered later may be regarded in the following order:

Smallpox:
The cause of smallpox in the human was classified as a filterable virus by Negri as well as by Casagrandi in 1905, based on filtration results. At that time, the rabbit was already known to be a suitable experimental animal. Guarnieri had earlier described investigations on the cornea in 1892, which also had implications for diagnostic purposes.

Rabies:
In 1903 when Remlinger attributed the rabies pathogen to the filterable viruses, the rabbit had been known as an experimental animal for almost a century.

Herpes simplex:
Since 1912 this pathogen had been considered a filterable virus as a result of Grüter's successful transmission to the cornea of the rabbit.

Mumps:
C.Nicolle and Conseil had classified the pathogen a filterable virus in 1913. We have known the rabbit to be a suitable experimental animal as a consequence of investigations by Granata in 1916.

Poliomyelitis:
In 1908, Landsteiner and Popper experimented with two healthy Old World monkeys (Cynocephalus hamadryas and Macacus rhesus) which had remained healthy despite syphilis trials. In the following year, the filtration results obtained by Landsteiner and Levaditi in the Pasteur Institute, independent of Flexner and P.Lewis in the Rockefeller Institute, showed that the poliomyelitis pathogen was a filterable virus.

Influenza in the human:
The influenza filterable virus was discovered in 1933 by W.Smith and collaborators when they inoculated ferrets intranasally with infectious, human throat rinse.

On the whole, filterable viruses in animals behaved in principle like bacteria; for both, infectivity was shown to be host- and tissue-specific.

3.4.2 Is "Pure Culture" of Filterable Viruses Possible?

From the beginning, the most typical characteristic of the filterable virus was its inability to grow in pure culture on synthetic media. Since discovery of the various pathogenic, filterable viruses proceeded very slowly, hampered by their ready mixing with other germs in the host, it was clear that growing them in a pure state, even under the most favourable conditions, was grasping for the stars. Just what was needed to make in vitro culture possible was still unknown at the beginning of this century.

The impulse for new in vitro trials with filterable viruses was given by a publication of R.Harrison working at the anatomy department in the Johns Hopkins University in Baltimore. He had been concerned with the problem of growing tissue outside the organism, i.e. as a culture, since 1907. Then, in 1910 he described the growth of nerve tissue in vitro, and three years later, the first published tests of the in vitro cultur of filterable viruses according to Harrison's method began to appear. Edna Steinhardt and coworkers, for example, published a study on the culture of the rabies and vaccinia pathogens in vitro. They had been able to detect Negri's inclusion bodies but no increased virus in ganglion cells. They began to search for inclusion bodies and virus reproduction in cultures of small cornea pieces from vaccinia-infected guinea-pigs. They did not consider the granula which they induced to be proof of vaccinia inclusion bodies, but they did succeed in reproducing the virus by passaging and by eliciting vaccinia eruptions on rabbit skin.

Levaditi attempted to culture poliovirus in vitro in the spinal ganglia of infected monkeys. He embedded small pieces of ganglia in Grabitschewski trays containing monkey blood plasma, and passaged them every three or four days. In one trial, the fourth passage was still infectious, while in another, only the first two were. Flexner and Noguchi, in a series of trials with poliovirus, inoculated infected nerve tissue from humans or monkeys, and small pieces of rabbit kidney into ascites fluid. They sealed the cultures in test-tubes with a layer of paraffin oil, and were able to reproduce the poliovirus using infectious human as well as monkey material. This breakthrough ushered in a long series of attempts to grow filterable viruses in vitro, which were supported by constant improvements in in vitro culture using organ specimens, tissue, and cells. In 1911 Margaret and W.Lewis reported the culture of tissues from chick embryos in a solutions of NaCl, $NaCl_2$, KCl and NaOH. Rous and F.Jones showed 5 years later how tissue, which Carrel had stimulated with embryo extract and for which Burrows had devised plasma embedding, could be dispersed into individual cells by treatment with trypsin, and be maintained during passaging. The standard method was to embed growing tissue in embryo extract containing plasma on a depression slide.

Credit for the numerous innovations and improvements must go to Carrel. As a surgeon, he was familiar with aseptic laboratory practices and his special interest was organ transplantation, with a healthy curiosity as to whether tissue and organs could be grown outside the body. In search of the answer, he joined the Rockefeller Institute for Medical Research, New York in 1906. His name is intimately connected with the permanent culture of fast-growing cells, suitable media and forms of culture vessels. He also invented the roller-tube culture. This method produces a greater culture surface through rotation of the culture tubes, which in turn ensures a higher titre when growing viruses. The technique was perfected by Gey in 1933. Developments in organ culture ran parallel after their initiation by D.Thompson and colleagues in 1914.

Insect cell cultures can be traced to Goldschmidt's gametogenesis investigations in the testicular tissue of a moth in 1915. Since mitosis and meiosis experiments require the cells to survive for only a matter of hours, simple media such as lymph or lymph with a mixture of salts and sugars were the only prerequisites. These very media, however, proved to be the stumbling block for the subsequent long-term culture of insect tissue. Interest in them receded until 1935, when animal tissue extracts were added making long-term cultures possible.

Successful plant cell culture began in the year 1921–1922, when Molliard used roots and Kotte and W. Robbins worked with roots and pieces of stem. When P. White managed to grow pieces of tomato root in vitro in 1932, and Gautheret succeeded in getting a culture from the cambia of willow and poplar grown in a fluid medium, the in vitro culture of plant cells was given fresh impetus in botanical institutes everywhere.

The early phase of experimental virology was stimulated particular by two noteworthy technical strides. The first, described in 1928 by the Maitlands and known as the Maitland culture, is the culture of minced tissue in a liquid medium. For virological experiments, the tissue was infected either in vivo or in vitro. The authors of the second contribution were Alice Woodruff and Goodpasture. In 1931 they successfully inoculated chick embryos with fowlpox material and observed virus-specific inclusion bodies in the chorio-allantois membrane (CAM). Soon thereafter, they also discovered that, in addition to the induced inclusions, there were other typical changes in the membrane: lesions and haemorrhages varying in size and localization, as well as differing foci.

As early as 1911, Rous and J. Murphy grew *Avian sarcoma virus* in 7 to 10-day-old chick embryos. Tumour cells and cell-free filtrates served as material for direct inoculation into the allantois sack or onto the allantois membrane. Both methods were modified in numerous ways in the following decades, and this in turn, led to the rapid discovery of new viruses which could be grown in tissue explants or in chick embryos.

Before presenting a statistical summary of all "filterable viruses" discovered and cultured by 1935, it is important to mention two further, outstanding discoveries. Without them the science of virology would have been unable to make such meaningful contributions to the doctrine of infection prior to World War II.

Although cell culture studies represented a substantial advantage over animal experiments, they could not eliminate the need to confirm in vitro by an in vivo trial. Viruses which did not cause inclusions could only be detected and measured by the pathological changes induced by in vivo experimentation. The possibility of immediately assessing in vitro tests, i.e. when the virus induced either inclusion bodies or specific intracellular changes, bided its time – until 1928, when Rivers, Haagen and Muckenfuß were able to demonstrate this with herpes and vaccinia. Not only were the cytoplasmic inclusion bodies typical for vaccinia observed, but also numerous acidophilic nuclear inclusions were found in the herpes-infected cultures.

The spectrum of virus-dependent cell changes expanded with Hallauer's work on *Fowlpox virus* in 1932. He discovered early signs of morphological cell change, namely necroses, rounded cells with subsequent cell degeneration, and amorphous cell dendrites, in virus-sensitive tissue cultures as early as 3–4 days after inoculation. Since Hallauer consistently observed these changes, he seized the opportunity to substitute evaluation of the in vitro morphological cell alterations for the in vivo control tests. Thus, the cytopathogenicity and the cytopathic effect of viruses were established. Nonetheless, exactly how the cell changes came about still remained puzzling and mysterious.

From the very first successful, in vitro culture experiments with viruses, the opinion prevailed that living, but not necessarily growing cells were required for virus multiplication in vitro. In 1923 Goodpasture and Teague stated from the results of transmission experiments: "... that the virus of herpes febrilis is better adapted for growth in an intracellular rather than in an extracellular environment. There is no evidence that growth will take place primarily in body fluids, while there is the anatomical evidence that the intranuclear bodies so characteristic of an infection represent proliferation of this virus within nuclei". In 1929, also Haagen considered the herpes and vaccinia inclusion bodies to be indicative of the intracellular multiplication of the virus, but it was not until 1933 and 1936 that Herzberg proved, by means of tissue cultures of *Canary bird* and *Vaccinia viruses* that virus multiplication entailed more than the simple presence of living cells. He described the process of intracellular multiplication for both viruses, with concomitant drastic, possibly lethal alterations in the host cell.

Tables 3.1 and 3.2 give a comprehensive survey of early successful virus propagation in tissue explants and on the chorio-allantois membrane of the chick embryo until 1935.

Table 3.1. The primary culture of filterable viruses on tissue explants up to 1935

Year	Virus	Authors	Explant[a]
1913	Smallpox	Edna Steinhardt et al.	Gp-cornea
1913	Poliomyelitis	Flexner, Noguchi	Ra-kidney
1925	Herpes	F. Parker, Nye	Ra-testicle
1925	Avian sarcoma	Carrel	HE
1928	Fowlpox	Findlay	HE-skin
1928	Vesicular stomatitis	Carrel et al.	Gp-bone marrow
1930	Rabies	Stoel	HE-brain
1930	FMD	Hecke	Gp-embryonic skin
1931	Rabbit myxoma	B. Benjamin, Rivers	Ra-monocytes
1932	Swine fever	Hecke	piglet-bone marrow
1932	Fowl plague	Hallauer	HE
1932	Yellow fever	Haagen, M. Theiler	HE
1933	Rift Valley fever	Mackenzie	HE
1933	Louping ill	Rivers, S. Ward	HE
1933	Canarypox	Herzberg	HE
1933	Pseudorabies	Traub	Gp and Ra testicle HE
1934	Newcastle disease	Topacio	HE
1934	Influenza	Dochez et al.	HE
1935	Ektromelia	Downie, McGaughey	ME
1935	Rabbit fibroma	Faulkner, Andrewes	Ra-testicle
1935	St. Louis encephalitis	Syverton, Berry	ME

[a] Gp, Guinea-pig; Ra, rabbit; ME, mouse embryonic tissue; HE, chick embryonic tissue.

Table 3.2. The primary culture of filterable viruses on the CAM until 1935

Year	Virus	Authors
1931	Fowlpox	Goodpasture, Alice Woodruff
1932	Vaccinia	Goodpasture et al.
1932	Herpes	Saddington
1934	Rift Valley fever	Saddington
1934	Vesicular stomatitis	Burnet, Galloway
1934	Equine encephalomyelitis	Higbie, Beatrice Howitt
1934	Infectious laryngotracheitis	Burnet
1934	Fowlplague	Burnet, Ferry
1934	Newcastle disease	Burnet, Ferry
1935	Smallpox	Torres, Teixeira
1935	Influenza	W. Smith
1935	Measles	Torres, Teixeira

3.5 Filterable Viruses as Antigens

At the beginning of the 20th century agglutinines, precipitins, lysins and anti-toxins had been shown to be parts of the anti-bacterial defence or immunity mechanism of the body. Laboratory determination of these antibodies with the help of experimental animals and in vitro tests led directly to the development of active and passive bacterial vaccines, as well as to the establishment of serological laboratories in the fields of human and veterinary medicine. It was here that bacterial-serological research took root and infectious diseases were diagnosed. Naturally, the major prerequisites for diagnostic work were pure bacterial preparations and high quality immune serums.

Vaccines were needed to protect man and animals against infectious agents. Their production was possible with cultured suspensions or with material from animals. However, virological-serological diagnosis was an impossibility, as were reliable, epidemiological studies. Both required a suitable experimental animal, and an in vitro culture system was necessary for each virus. As for plant viruses, there was not the faintest clue that they might possess antigenic properties.

During the early phases of immunological testing with filterable viruses, the sole determinant was the neutralization test performed in the experimental animal. This entailed neutralizing the agent or the pathogenic properties of the agent concerned, by exposure to its specific combating substance, the neutralizing antibody, either in the test animal or during prior in vitro incubation. Back in 1889 and 1892, Tizzoni and Centanni in connection with rabies, and Sternberg working with vaccinia, had already focused attention on neutralization through the serum of immunized animals. Also in 1889, Babes and Lepp supplemented rabies vaccine treatment by using neutralizing serum. Lecturing in honour of Jenner at the assembly of the American Medical Association in Atlanta on May 5, 1896, Sternberg said: *"The blood serum of immune animals contains a soluble substance which is able to destroy the specific virulence of the Vaccinia virus, when it contacts it."*

At the same time, Tizzoni and Centanni were testing serum against the *Rabies virus* for mixed trials in vitro and in vivo, before and after infection. By January 1892, Tizzoni was in a position to summarize his findings:

1. ***The blood of immunized animals can provide protection from rabies before, as well as after the infection, provided it is administered within 24 hours.***
2. ***Blood serum behaves like an internal disinfectant which exercises a destructive effect on the adsorbed virus.***

Simultaneous vaccination was introduced by Babes and Lepp for rabies in 1889. They combined an active vaccination with an immune serum application. Nine years later, the same method was used for cattle plague by Kolle and Turner at Koch's experimental station in Kimberley. For neutralizing the pathogenic properties of the virus, a mixture of the agent and the serum containing the specific antibody had to be injected into different parts of the animal. In the absence of exact data either on the size of the virus or the antibody content, suitable amounts of virus and antibody had to be ascertained by trial and error. At the end of the vaccination process, a control was made of the efficacy of the vaccine, in which virus infections were performed at various time intervals. This method was considered exemplary for subsequent vaccinations.

A different method was chosen, for example, for rabies. In 1893 Tizzoni and Centanni stressed the importance of the virus dose. To substantiate the point, they mixed exactly weighed portions of medulla suspension containing viruses with various amounts of serum, and after a 24-hour in vitro incubation, they injected this mixture subdurally into rabbits. Babes searched for his own suitable rabies vaccine in a series of dilutions in which either the virus or the antibody content remained constant. Exact methods for effective and safe vaccine production depended in large part upon reliable aggregation tests. During these in vitro tests, the antigen reacted with its specific antibody. Individual tests, however, had to be devised for each antigen, notwithstanding the necessity of drawing parallels between the aggregation and neutralization test results.

3.5.1 Neutralization Tests

From the beginning, the difficulties of testing neutralizing or virucidal antibodies stemmed from the lack of suitable experimental animals. If an animal was available which was suitable for immunization and was sensitive to infection, then a neutralization method could be worked out in a relatively short time. Let us, for example, consider FMD. A neutralization test had been developed for cows which required that the animals were completely isolated to eliminate any danger to the rest of the stock in the area. Only as late as 1910 was Löffler finally able to meet this condition when he continued his work on FMD at the experimental station on the island of Riems. It wasn't until 1920 that scientists considered the guinea-pig to be a suitable experimental animal. In 1921, when FMD was introduced into the French department de l'*O*ise along with cows from Germany (*A*llemagne), Vallée and Carrée discovered that the newly manifest disease was not the result of varying degrees of virulence, but was due to the different immunological characteristics of French and German stocks. Thus, the O and A duality of the *FMD virus* became apparent, meeting at first with more scepticism than credibility.

A wave of investigation was unleashed in various countries, with particular attention focused on the neutralization test in guinea-pigs by Waldmann. With the aid of this test, Waldmann and Trautwein were not only able to confirm the French findings, but in 1926 they extended the immunological types of FMD to three, designating them A, B and C. Their typing was based on the neutralizing properties of serum from guinea-pigs which had been immunized with strains of *FMD virus* adapted to guinea-pigs on Riems. The simultaneous vaccination used by Löffler and Frosch for cattle was also carried out in guinea-pigs. The in vivo neutralization tests for animal viruses published during the first 30 years of viral history are summarized in Table 3.3. The value of neutralization tests, however, is best judged in the light of other findings.

Table 3.3. Neutralization tests with animal viruses in vivo up to 1937

Year	Virus	Animal	Authors
1910	Poliomyelitis	Monkey	Levaditi, Landsteiner
1912	Vaccinia	Rabbit	Henseval, Convent
1925	Herpes	Rabbit cornea	Gildemeister, Herzberg
1927	Bornea disease	Rabbit	Zwick et al.
1928	Yellow fever	Monkey	J. Stokes et al.
1931		Mouse	M. Theiler
1930	Ectromelia	Mouse	Marchal
1931	Influenza	Pig	Shope
1933		Ferret	W. Smith et al.
1937	Equine encephalomyelitis	Guinea-pig	Beatrice Howitt

Animal Viruses:

- The simultaneous-neutralization test in hogs was adapted for the control of hog cholera serum by many countries.
- Loewe and Strauß demonstrated neutralizing antibodies in the serum of herpes convalescents in 1921. Four years later, Flexner and Amoss found such antibodies in rabbits which had been immunized against herpes. Gildemeister and Herzberg (1925) tested herpesvirus-neutralization on the cornea of rabbits after prolonged in vitro incubation of the antigen-antibody mixture. Other authors in 1925 cited the optimal in vitro binding time for herpes to be 1–3 hours.
- It was 1926 when Olitzky and coworkers neutralized the *Vesicular stomatitis virus* in the horse with immune serum taken from the guinea-pig.
- Aycock and Kramer (1930) found that asymptomatic humans also had neutralizing poliomyelitis antibodies and, furthermore, that the number of antibody carriers increased with age. Burnet and Joan MacNamara (1931) supplemented their cross-immunity tests with serological cross-neutralization tests and thereby discovered that poliomyelitis was caused by strains of differing immunological features.
- Glusmann and Ssolowjewa (1931) found that vaccinated subjects developed neutralizing antibodies 10–20 days after final rabies inoculation.
- Andrewes observed neutralizing antibodies for *Rous sarcoma* as well as for *Fibrosarcoma virus* in the chicken in 1931.
- Two years later Ten Broek and Merill demonstrated that it was possible to differentiate *Eastern* from *Western equine encephalomyelitis virus* in the neutralization test when, instead of the intracerebral test, the guinea-pig was injected into the plantar and the white mouse intraperitoneally.

Plant Viruses:

In 1928 attention was focused on possible antigenic potency of plant viruses. Helen Beale showed that the injection of *TMV* into rabbits produced precipitating serum regardless of where the virus had been grown. One year later, Helen Purdy-Beale demonstrated that the infectivity of *TMV* was reduced when it was mixed with serum taken from rabbits immunized with *TMV*. The unequivocal proof for neutralizing *TMV*, *Tobacco ringspot* and *Cucumber ringspot virus* can be found in a study by Chester, 1934. This latter report also elucidates proof for species specifity of the three neutralizing sera, while rejecting strain specifity.

Bacteriophages:
In 1929 Bordet and A.Ciuca demonstrated the antigenicity of bacteriophages by injecting rabbits with phage lysate and thereby obtaining serum with neutralizing activity. Burnet was able to show the antigenic relationships of various phage strains by cross-neutralization tests in 1933–1934: the T-even strains of E.coli – T2, T4 and T6 – exhibited similar antigenic properties.

After three decades of experimental virology the development of neutralization tests presented no difficulties, and were executed with animal viruses in various quite different ways: the antiserum was injected 24 hours before the antigen was given; both were injected at the same time but in different parts of the body; an antigen-antiserum mixture with or without incubation was injected. These contrasting methods led inevitably to inconsistent and often to unsatisfactory results. Nevertheless, experts remained firm in their belief that viruses generally induced neutralizing antibodies.

3.5.2 Aggregation Tests

From publications, and particularly from books appearing at the beginning of the 1930's, the scientific community was clearly of the opinion that filterable viruses differed from bacteria and protozoas immunologically in that they induced neutralizing but not aggregating antibodies. This point of view completely disregarded the comprehensive literature on aggregations tests, as well as the wealth of bacteriological experience already established. It was known that both antigens and immune serum had to fulfil certain quality standards, and that the ratio of these components in a mixture had to be exact and consistent if immunological test results were to be positive and reproducible. These conditions were not always easy to fulfil for bacteria, but were incomparably more difficult for viruses. Thus, positive test results carried more weight than negative ones during the early stages of virology and serology. Conversely, experiments which proved negative did not necessarily preclude the reliability of an earlier positive result.

3.5.2.1 Agglutination and Precipitation

Although agglutination, i.e. an antigen clumping with its specific agglutinating antibody, and precipitation, i.e. the settling of a substance in solution or a colloidal antigen after association with its specific precipitating antibody, are quite different processes, the differences were not clearly delineated in publications; the term flocculation was often used for both types of reactions, and no distinction will be made in the summary that follows.

Animal Viruses:
In 1904 Freyer referred to the aggregation he observed as "precipitation". He had mixed vaccinia lymph with the serum of vaccinated human beings, and immunized calves and rabbits. In 1910 Paschen described the agglutination of vaccinia elementary bodies.

Burgess and colleagues successfully used the agglutination test to diagnose an outbreak of smallpox in Dundee in 1929. In the previous year, Bedson had discovered that the herpesvirus combined in vitro with its specific antibody.

Ledingham and coworkers augmented knowledge on agglutination with work on *Variola* and *Fowlpox virus* in 1931. In 1933 Hughes found yellow fever precipitin in monkey serum. In the same year, Amies described the agglutination of chickenpox elementary bodies and confirmed the serological relationship of varicella and zoster.

Insect Viruses:
In 1921 Aoki and Shigasaki published the results of their immunization studies with polyhedras of jaundice-infected silkworms. They immunized rabbits with inclusion bodies which had been washed 10 times and identified agglutinins and precipitins in their immune serum. These antibodies did not react with the epithelial cells of the organs from which the polyhedra had originated. Their conclusion was that polyhedra were not synthesized from cell material.

Bacteriophages:
In 1933 M.Schlesinger observed the agglutination of the WLL phage and Burnet the precipitation of phage C 16.

Plant Viruses:
In 1928 Helen Beale reported on precipitin induced in rabbits by *TMV*.

3.5.2.2 Complement-Fixation reaction (CFR)

In this test, which is essentially based on ideas and research of the Belgian scientist Bordet (1895), an invisible lysin reaction is made visible by a subsequent hemolytic reaction. This chain reaction occurs by virtue of a thermolabile component in the serum, which Ehrlich named complement. In the course of the test, the invisible lytic system (bacterium + specific lysin) and the visible one (hemolytic erythrocytes + specific hemolysin) vie with each other for the complement in the mixture. Hemolysis can only ensue if unused complement is still available. This method, which had been described by Bordet and Gengou in 1901 is known as Wassermann reaction for syphilis since 1906. Trials with filterable viruses soon followed, almost exclusively with animal viruses.

As early as 1906, Jobling obtained positive results with serum from vaccinated calves, closely followed by Sugai and Beintker with the serum of smallpox-infected patients. As for rabies, Nedrigailoff and Sawtschenko used the CFR as early as 1911, with salivary gland extract from rabies-infected dogs as the antigen.

Haring and Kofoid reported positive results with *Fowlpox virus* in 1912. Netter and Urbain, as well as M.Gordon, were able to demonstrate in 1924–1925 that tests with alastrim and variola antigen lead to the same results. In 1925 Kraus and Takaki diagnosed postvaccinal encephalitis caused by a herpesvirus. The following year D'Herelle used the CFR in his work with phages. In 1928, A.Ciuca demonstrated the type-specifity of the CFR in FMD. Nicolau and Stroian used it in 1929 for Borna disease in horses. Working with the results which Beaurepaire Aragão and A.Moses had obtained with yellow fever in 1928, Frobisher standardized this test in 1930. It was subsequently used for post-epidemic studies in 1932.

3.5.3 Immunity and Immunization

Immunity, i.e. an inherent or later-acquired resistance to infectious disease, had been known since antiquity. However, Pasteur was the first to begin searching scientifically for specific causes and effects. It was important to differentiate between natural and acquired immunity. To explain natural immunity, it was assumed that the infected organism simply denied the germ the appropriate conditions for growth and the development of pathogenicity. In contrast, acquired immunity, was thought to be the result of a confrontation between the host organism and the causative agent or the toxins it excreted. Moreover, the organ exhibiting obvious symptoms of disease was regarded as crucial in the development of the specific antibodies or antitoxins.

Institutes of hygiene and microbiology now commenced studies on the immunity status of both individuals and populations. This entailed further recognition of immunity to filterable viruses and the subsequent development of methods to enhance natural immunity. The variety of immunological responses observed after bacterial infections, and the recognized life-long immunity to filterable viruses such as smallpox, measles, cattle plague and hog cholera, led to the widely-held assumption that immunity to bacterial and viral infection were basically different.

3.5.3.1 Immunity

In 1929 Andrewes pointed out the significance of circulating and neutralizing antibodies of virus III, which had been isolated by Rivers and Tillet in 1923 after the third testicular passage in rabbits. Although it had proved possible to grow the virus in the tissue of immunized animals in the presence of normal serum, it did not grow in non-immune specimens with added immune serum. Since there was absolutely no trace of natural cellular immunity, Andrewes inferred that protection resulted from circulating antibodies. Henceforth, immunity was synonymous with the presence of specific, circulating, neutralizing antibodies. The antibodies were supposed to prevent symptoms or lesions due to viruses, but did not prevent virus penetrating an organism to multiply non-pathogenically in a suitable organ.

Progress in the field of induced, life-long immunity acquired after confirmed viral infection was hampered by the lack of new, convincing evidence. Clinical experience was often contradictory and, more often than not, lacked experimental detail. Diseases in this category included poliomyelitis, chickenpox and fowl plague. Immunity acquired by exposure to FMD and equine vesicular stomatitis was assumed to last only one or two years.

The multifarious forms of disease caused by *Herpes simplex virus (HSV)* reflected in the myriad opinions voiced about immunity to the virus. Only one thing was certain, namely that recovery from the disease did not provide long-lasting protection. Moreover, the virus was capable of persisting, i.e. surviving in an apathogenic form within the organism. The high percentage of neutralizing antibodies found in relapsed herpes cases, and the cell-mediated immunity substantiated in a variety of hosts, prompted innumerable questions which were to remain unanswered for some years.

3.5.3.2 Passive Immunization

Tizzoni and Centanni, pathologists in Bologna, had already provided proof in 1892 that immune serum could be used therapeutically for rabies. They had treated rabbits with rabies immune serum obtained from other rabbits previously vaccinated with virus taken from dogs 7 or 11 days post-infection. Intravenous, intraperitoneal, or subcutaneous administration of the serum prevented more severe rabies and the death of the animals. Very similar results had been achieved with an alcohol precipitate of immune serum which had been vacuum dried using sulphuric acid. These experiments were foreseen as a prelude to human trials, which obviously never materialized. Nevertheless, this directly addressed the main issue of the passive immunization of humans and animals. Often the first inklings of the protective effect of immune serum had been detected during laboratory experiments, but transmission to a natural host had always failed. Hence, though the literature on animal experiments was copious during these first three decades, promising studies on the practical application of immune or convalescent serum in humans were scarce.

Since 1899, it had been assumed that smallpox survivors and vaccinia-inoculated subjects alike had a similar substance in their serum which protected them from smallpox infection. Proceeding from this premise, Teissier and Marie decided to try immune serum doses of 25–100 ml in 13 cases of smallpox in 1912. Investigating cattle plague before the turn of the century, Kolle and Turner managed to obtain an effective immune serum which, with a 20 ml-dose, provided prophylactic protection for several months. Though it had been known since 1898 that the serum of cholera-diseased sows provided protection for their offspring, it took 20 years of trial and error in the epidemic-stricken countries to develop an effective immune serum for swine. Waldmann described its production and control in 1924.

Weech, as well as Mitchel and Ravenel, found that the injection of 45 ml of fowlpox serum within the first 5 days after infection prevented the outbreak of the disease in 1924–1925. In 1925 Waldmann also developed method using 2 to 3-year-old steers for obtaining effective hyper-immune serum targeted at all three types of FMD. In efforts directed against poliomyelitis, in 1910 Netter began to administer serum to patients during the convalescent period. Pettit reported positive results with immune serum from the horse in 1917. But many subsequent controversial or dubious results forced a reassessment of the polio question when, in 1931, Burnet and Joan McNamara pointed out the immunological differences between the various strains of the poliovirus. Only in 1953 the National Foundation for Infantile Paralysis committee on typing clearly defined the three distinct serological types.

3.5.3.3 Active Immunization

The term "active immunization" means the introduction of living (active) or dead (inactivated) pathogens or their toxins into an organism to induce the development of specific antibodies or antitoxins. In 1796, when Jenner first undertook the methodical trials to develop a vaccine which led to the protection of an 8-year-old boy against smallpox, he vaccinated the boy with infectious cowpox material, relying on the age-old practical experience of northern European farming population. Farm folk knew that they would not get smallpox so long as they had had the milkmaid's node, i.e. a bout of cowpox. To combat human pox with cowpox was Jenner's aim–and an ingenious idea! Furthermore, the knowledge of infectious germs which had accumulated so significantly in the 19th century, sowed ambitions to develop vaccines for other infectious diseases.

Pasteur was one of the first to appreciate Jenner's use of infectious biological material as a vaccine. The reasons for this in 1877 were simple: The germ theory had not yet been proved, and it was not yet possible to obtain a bacterial suspension– by means of so-called "pure culture"– which was completely free of extraneous biological material. 1879 marked the birth of the "vaccine age" at the Pasteur Institute. A culture of fowl cholera, which had been neglected during a 2-month holiday, was injected into chicks, and it was discovered that storage had destroyed the pathogenicity as well as most of the infectivity of the culture. Chicks showing no any reaction Pasteur proposed reinfecting them with a fresh culture. All re-infected chicks survived, and Pasteur reasoned that directed attenuation of the pathogen might lead to the development of vaccines. His methods for attenuation included increasing the passage intervals from two to three months, treatment with oxygen, producing cultures at higher temperatures, and various changes in the culture media.

In 1880 Pasteur began developing the rabies vaccine in collaboration with Chamberland and Roux. Because of Zinke's work 76 years earlier, it was known by then that the agent was transmissible to the dog and the rabbit. Galtier had reproduced Zinke's experiment in 1879, thereby observing that rabbits, quite apart from dogs, developed a form of disease called silent rage, which means rabies without the typical, severe symptoms. By continuous brain passaging in the rabbit, they shortened the incubation time from 14 to a fixed 6–7 days and named this pathogen virus fixe. Furthermore, after 30 passages, the pathogenicity of the virus had been so weakened that subcutaneously inoculated dogs remained healthy and were protected against infection from a street virus, i.e. they were immune. To inactivate the virus more gradually, it was dried over potassium hydrate (KHO) at 23–25°C during a period of 3–14 days. Vaccination followed as soon as possible after infection, and consisted of 20–28 subcutaneous injections at intervals of 1/2–1 day with bone marrow that had been treated with potassium hydrate for varying periods of time. Protective immunization began with the most inactivated material. Slightly injured subjects received 20 inoculations within 15 days. Severely injured were given 23 inoculations within 18 days and those with serious bite wounds and head injuries got 28 injections within 21 days. The bone marrow was dried for 3–14 days regardless of the immunization series, and a single dose ranged from 1–3 ml. France acknowledged Pasteur's laudable development of the rabies vaccine by founding the Institute Pasteur. United with daughter institutes throughout the world in the battle against rabies, the Institute Pasteur assumed a leading roll in vaccine problems. Pasteur passed some of the credit to Jenner by naming the material for immunization "vaccine" and the procedure "vaccination". The literature of the time used both the term "vaccination" and "Jennerization".

The original virus fixe was passaged about 1300 times between 1882 and 1927. After 270 passages, it began to show consistent characteristics which permitted the standardization of the process. After 1922 an increasing number of changes in the virus fixe were observed both in the original and in the branch strains; primarily they showed signs of adaptation to the rabbit brain, with a reduction in the incubation period of up to three days, as well as increased infectivity. The antigenic properties, however, remained constant.

Many critical investigations of the entire Pasteur immunization process, right from obtaining the virus fixe to the early administration of the vaccine, led to continous modifications in both the production and vaccination processes. What remained constant was the mode of subcutaneous abdominal vaccination modus, because all involved felt this to be the best safeguard against an infection of the nervous system. The changes were mainly in the degree of attenuation, the method of inactivation, and both the number and size of the vaccine doses. A few of these changes are worth mentioning in more detail. Babes had witnessed how Pasteur's method often failed with patients who had been bitten by mad dogs. He ascribed this to the drying method over potassium hydrate coupled with exaggerated attenuation. Therefore, in order to obtain more virulent material, he substituted heating for drying, shortened the attenuation time to 1–10 days and increased the number of inoculations to 56 within 31 days. Occasionally, immune blood was added to the vaccine. At the same time, Högyes tried to obtain an effective vaccine simply by diluting fresh infectious material. He considered Pasteur's method of attenuation to result only in reduction in the number of particles and not in the virulence of the particle itself. Högyes also decreased the number of both vaccination days and vaccinations. The name Remlinger conjures up the treatment of infected organs with ether in order to remove fats and lipids. Fermi, as well as Semple, developed inactivated vaccines by using phenol; just how avirulent, i.e. how safe they were, was questionable. A rapid ether method developed by Hempt resemled Remlinger's approach.

Later, statistical evaluation of the various vaccination procedures showed equal antigenicity, although live vaccines, despite greater production problems, were at the whole superior to inactivated ones. Vaccine-associated paralyses were naturally fewer with the use of inactivated vaccines. The successful vaccines against smallpox and rabies paved the way for further vaccines against pathogens which could not be grown on synthetic culture media. In contrast to the rabies vaccine, termed attenuated, which was partially or totally inactivated, pox vaccine was a pure live, i.e. infectious, but really an attenuated vaccine. Vaccines with completely inactivated infectious particles were referred to as inactivated or antigenic vaccines.

Immunization was attempted in various ways in the case of cattle plague. R.Koch adopted an idea practiced in South Africa to use as a vaccine the blood and bile from animals which had succumbed to the plague. A subcutaneous immunization with bile afforded protection for several months; its constituents prevented the development of the pathogenic characteristics of the agent. To develop live vaccine, attenuation was attempted by passaging in animals such as guinea-pigs, rabbits, goats and sheep – alas, without success. The production of inactivated vaccine was much more successful, using macerated organs and inactivation by addition of glycerin, a phenol-glycerin mixture, chloroform or formalin; prophylaxis was maximally one year. Not a single vaccine of this type served as the prototype for large-scale production. This was reserved for simultaneous immunization, which Kolle and Turner had developed shortly before the close of the 19th century. The treatment consisted of the simultaneous but separate inoculation of virus and immune serum to both sides of the body. As a rule, the source of the virus was the blood of recently infected cows. When inoculation was required in a town where infectious blood was not available, newly-infected sheep were sent to the site, because their blood was infectious for at least 8 days.

In 1918, Hess and L.Unger immunized 38 children intravenously against varicella using diluted pus containing virus from a pustule. Of these, 37 were given complete protection, while one child reacted with mild symptoms. However, despite this success the method was never a model for chickenpox prophylaxis.

The production of a vaccine against FMD, which seemed well within the bounds of possibility after Löffler and Frosch had published their "Summary Report" in 1897, actually proved far more difficult than was originally assumed. The wealth of studies with experimental animals contained not the slightest indication of a practical vaccine. Waldmann, who had adopted Kolle's and Turner's idea, proposed a simultaneous therapeutic and prophylactic vaccination in 1926 with a step-by-step method in infected cows. Cows exhibiting pathological changes or fever were to receive immune serum. Asymptomatic animals were to have an inoculation containing virus material from an infected cow of the same barn stock, as well as immune serum, whereas neighboring stock would only receive immune serum.

From 1928 to 1929, Hindle and Beaurepaire Aragão developed an inactivated vaccine against yellow fever obtained from tissue of infected monkeys; inactivation was with formaldehyde, phenol or glycerin. The effectiveness of the vaccine was tested in monkeys, and successful trials with humans first took place during the 1928–1929 epidemic in Rio de Janeiro. Of greater practical significance, however, were the live vaccines which appeared somewhat later. Sawyer and coworkers described such a vaccine in 1932; the immune serum obtained from infected mouse brain and human immune serum was each injected subcutaneously. Prophylaxis against severe infection was said to exceed four years. The large serum dose required of 50 ml per person was reduced by Pettit and colleagues using horse immune serum, of which only 10 ml was required.

In summary, before World War II there were numerous examples of effective vaccines produced as live or inactivated vaccines against pathogens which had been classified as filterable viruses. The realization of this goal depended greatly on the development of appropriate attenuation methods and reliable inactivation. Simultaneous vaccination was meaningfully employed whenever safe inactivation or attenuation could not be guaranteed because of deficient control methods. The efficacy of vaccines was assessed from results with experimental animals, or laboratory tests which determined induced humoral antibodies. Humoral immunity took priority over tissue immunity. The improvement of production and control methods, while ensuring the efficacy and reliability of immunization, was the indisputable goal of the then experts in the field.

3.6 Filterable Viruses as Pathogens

If filterable viruses were nothing more than very small bacteria, their virulence in the host organism would have caused effects identical to those of bacteria. Virulence means all properties of a contagium which enable it to thwart the resistance of the infected organism. Transmission of the ability to multiply - called infectivity - as well as the pathogenic properties - called pathogenicity - were considered the defining elements of virulence. Variation and mutation were known mechanisms of bacteria by which they changed their virulence. Immunity reinforced the resistance of the host to the virulence of the infectious agent.

It was generally known that infective bacteria in an organism could be confined as a localized focus, or disseminated throughout the entire body. It was also known that the port of entry might undergo change due to the invasive bacteria. Local reactions with characteristic signs, inflammations, various pathological changes in the nearest lymph glands etc. were known morbid reactions of the infected host. The typical pathological changes of organs were interpreted as a systemic reaction, whereby bacterial toxins had probably entered the organs and interacted with their cells to cause a pathogenic effect. Fever was considered the result of pathogenic bacterial action and accompanied acute infectious phases with a characteristic pattern in time. Exactly why fever erupted was not known in any specific case, but discussion was rife on the subject of causal factors. These included toxins secreted and set free by degrading bacteria, or those which were by-products of bacterial metabolism and the decay of body protein. The differing courses of disease were ascribed to various pathogenic causes.

Bacteriological laboratory techniques enabled scientists to ascertain quickly the reasons for pathological changes, provided they were caused by known pathogens. In contrast, diagnostic difficulties threatened when a disease was the result of a filterable virus. Tumour development was even more enigmatic and infectious agents were no longer excluded as causal; this was especially the situation after comprehensive investigations, such as the collective study undertaken throughout the entire German Reich in 1900, disclosed endemic foci.

3.6.1 Infectious Diseases Caused by Viruses

Until the 1940's, transmission of infective virus to a host organism was thought to follow the same pattern as for bacteria:

- direct contact
- dust and droplet infection
- insects as carriers or intermediate hosts
- wounds from bites of infected animals
- transmission to a foetus

Direct contact with a virus was considered the exception, whereas dust and droplet infection were most frequently encountered. These infections supposedly were caused by sick as well as healthy people who continuously excreted the pathogens. Examples cited in this connection were poliomyelitis, herpes, measles, rabies in dogs, FMD in cows, cattle plague and hog cholera. Arthropods were the only known vectors or intermediate hosts of both human and animal virus diseases.

Reception of the disease was considered to occur via the mucous membranes of the upper respiratory and digestive tracts, and further progression was earmarked by characteristics of incubation time, viremia, systemic reaction and organ manifestation, with specific tissue and organ affinities. Frequently an affinity was observed for the organs stemming from the ectoderm, i.e. skin, mucous membranes, and nervous system, as well as the salivary and mammary glands. Rapid virus multiplication took place in those organs for which there was an especially strong affinity.

Distribution throughout the body followed via the bloodstream or the lymphatic system. Migration into or along nerve pathways to the central nervous system was discussed and described for *HSV* in rabbits by Goodpasture and Teague in 1923, and for *Rabies virus* by Goodpasture two years later. The viremic phase was found to occur at the onset of the disease and end when the virus was taken up by the disease-specific organ. This was the most contagious phase, although transmission was also possible during other phases. The clinical course of acute viral disease in humans showed a characteristic double-peaked fever curve occurring in various phases. It was found that, by passaging in other hosts, the pathogenicity and organ-affinity of the virus could often be altered, but the immunogenic properties remained constant. *HSV*, the *Salivary gland virus* in the guinea-pig, and virus III in the rabbit belonged to the latent viruses, which could remain in an organ until activated by some, probably unspecific, stimulus activated. Goodpasture (1929) predicted *HSV* latency in sensory ganglia. As a result of intracellular multiplication, virus- and organ-dependent infiltrations around small blood vessels were often observed, as well as proliferation of the epithelium with subsequent degenerative changes.

Polyhedral and granular cytoplasmatic inclusions were typical pathological consequences of infections caused by insect viruses. The course of an infection in the caterpillar stage of the silkmoth (Bombyx mori L.) was described in 1921 by Aoki and Y.Shigasaki. The nuclei of all epithelium cells were found to contain inclusion bodies which later entered the bloodstream. This resulted in a thickening of the blood and the loss of elasticity of the epithelium tissue, which became so brittle that caterpillars died of the slightest injuries. Escherich and Miyajima in 1911, and Wahl in the following year, confirmed the assumption made in 1892 by von Tubeuf that there was a latent as well as an acute form of polyhedra disease in the nun moth.

Pathological changes caused by plant viruses were often adopted as names for the diseases, which accounts for the large variety in nomenclature: light and dark areas in the leaves were termed mosaics, and then there was chlorosis, mottling, stripes, streaks, curling, bushy tops, thorns and spikes. To these can be added: rosette-shaped and evil-smelling plants, stunted-growth, large sprouts, necrosis, hard and fibrous growth and sterility. Most changes resulting from plant viral infection were macroscopically observable. Examination of the cellular inclusions, whose real nature was unknown in 1935, required the aid of the microscope. It was believed that viruses entered the plant through injuries to the leaf, and that the infection did not remain local but spread systemically, often causing malformation on the upper or lower part of the leaf. Each virus was capable of causing several, though never all of the symptoms, and so similar symptoms on various kinds of plants did not necessarily mean the same type of virus was at work.

3.6.2 Tumour Development

The term cancer to describe the developmemt of a tumour can be traced to Galenus, who was originally referring to the translucent blood vessels observable on the hard tumours of the mammary gland which, in their branching nature, appeared similar to the legs of crabs. Over the centuries, the term became more widely accepted to mean rank-growing tissue and ulcers in humans, animals and plants. Until well into the 19th century, the outwardly visible signs of a disease sufficed to classify it as a cancerous growth or an ulcer. Thus, it is not surprising that syphilis was grouped with cancer (Ger. "Schanker" and Fr. "chancre"). Progress in microscope development, coupled with improvements in histological preparation in the course of the 19th century, made it possible to differentiate and define tumours with greater accuracy. Thus, if at first a tumour was simply considered a malignant change in the epithelial tissue, it was subsequently found that other types of tissue could be involved in the growth of a cancer: malignancies as well as benign growths, epithelial as well as mesenchymal tissues with their numerous histologically differing structures. Why this was so, was of course unknown.

In 1898 Sanarelli had reported on the transmission of the myxomatogenous virus to the rabbit. Since he did not manage to supply a microbiological characterization of the agent, it was assumed that it belonged to an as yet unknown group. In the advanced stage of the disease, he noted small, subcutaneous tumours and hyperplasias, especially on the ears and extremities at the point where skin tissue changes to mucous membrane, and particularly in the lymph glands. Although Sanarelli was not able to clear up the etiology of this rabbit disease, myxomatosis was in any case disregarded as a forerunner of the virus-caused tumours. It was not until 1911 that A.Moses provided the results of the missing filtration experiments with this agent. Just as little credit was given to the Danish pathologists Ellermann and O.Bang who had provided the first evidence of the virus etiology of leukemia in chickens in 1908. They had transmitted cell-free filtrates of erythromyeloblastic leukemia and, apart from organic changes typical of leukemia, they had observed non-leukemic blood counts and coined the terms aleukemic leukemia and pseudoleukemia. Transmission of pseudoleukemia in turn caused classical leukemia. Later, they introduced the term leukosis to include the leukemic and aleukemic forms.

By contrast, Rous's findings completed at the Rockefeller Institute New York were acknowledged. Since 1910 he had been working with a group of chickens in isolation because of spindel-celled sarcoma. In 1911 he described the transplantation and cell-free transmission of fibrosarcoma. This was labelled chicken tumour 1 and was followed in 1912 by chicken tumour 7, an osteochondrosarcoma. As a result of 45 various transmissions of chicken tumors, Rous was convinced that different types of tumour could be caused by different viruses, or by a single one.

Rous and J.Murphy further probed the question of tumourigenesis in 1911–1912 by injecting minced tumour-pulp into the uterus of pregnant rats. Since almost all of the embryos were killed by this procedure, they modified their experiments by using 7 to 10-day-old chick embryos of the Plymouth Rock breed, which were known to be particularly susceptible to sarcoma. The eggs were opened after 7–10 days incubation. These researchers worked successfully with fresh and dried material, and with Berkefeld filtrates of the tumour extracts. The newly formed tissue showed preference for growing on the chorio-allantois membrane, and upon examination turned out to be spindle cell sarcoma similar to that found in chickens! In 1910 Fujinami and Inamoto, working at the pathological institute of the Imperial University of Tokyo, reported the transplantation of Rous sarcoma tumours in chickens. It required four additional years to transmit the sarcoma through filtrates. In 1932 Shope, who was working at the Rockefeller Institute in Princeton, discovered small fibrous tumours under the skin of a cottontail rabbit he had just shot. At the institute, he succeeded in transmitting these tumourous growths via filtrates to other wild cottontails and to domestic rabbits. It was not a fatal condition and most of the tumours regressed spontaneously. After observing a striking similarity between the lesions in the initial phase of his fibromas and those in Sanarelli's myxomatosis, Shope hastened in the same year to investigate the immunological relationship of the two rabbit diseases. In 1936 he reported cross-immunity whereby a fibromatose condition rendered the rabbit immune to myxomatosis. He was also able to demonstrate that the few rabbits surviving myxomatosis were immune to *Fibroma virus*. In 1933, the cell-free transmission of a tumour again crowned Shope's endeavours, namely a papilloma in the cottontail rabbit. Furthermore, he was able to infect domestic rabbits but re-isolation of the tumour agent only proved possible from the cottontail rabbit.

Working at the pathological institute of the University of Pennsylvania, Lucké reported in 1934 on a transmissible virus in frogs which induced carcinoma in the kidneys, particularly in the leopard frog (Rana pipiens). In the same year Furth, and E.Stubbs managed to induce sarcoma and erythroleukosis with cell-free material from aviary carcinoma cells which had been cultivated for 67 days in vitro.

By the mid-thirties, opinion was devided on virus participation in the development of tumours, with scepticism prevailing. Some saw justification in the findings with chicken tumours for the assumption that viruses were involved in causing animal tumours. They were even more prone to assume tumour growth in humans was analogous, when in 1933 the cell-free transmission of rabbit papillomas was successful. Others contested an exogenous factor in tumour development and reasoned that tumours were induced by transmissible mutagens. Eventually, the virus followers saw their views confirmed when in 1935 Elford and Andrewes, using recently accepted filtration and centrifugation methods for the *Rous sarcoma virus*, determined the particle diameter to be 75–100 nm. It should be mentioned that Carrel had assumed intracellular multiplication an entire decade earlier.

3.7 Viruses are Not Just Very Small Bacteria

In various parts of this chapter, I have attempted to summarize the significant results of almost 40 years of experimental research with filterable viruses. All investigations had had the goal of clarifying the nature of the virus. The characterization of viruses either as ultramicroscopic and invisible, or as filterable microbes – and, therefore tiny bacteria – was simply makeshift and clearly not final.

Although it was not possible to answer the question of the nature of the virus any better in 1935 than 40 years earlier, it was nevertheless clearly recognized that, despite certain similarities, viruses were by no means minute bacteria. It might have been possible to revert back to the definition of the virus as a reproducible contagium, providing one conceded that it was no longer limited only to the negative context used at the beginning of the century. Barring a few divergent opinions, the following is the generally accepted definition of the virus in the mid-thirties:

- Not identifiable through the available microscopes
- Not reliably retained by filters developed for bacteria
- Diameter as a rule less than that determined for bacteria
- Behaves immunologically like other microorganisms
- An autonomous microorganism
- Reproduces only in connection with living cells
- Pathogenicity not developed through toxins.

Thus, after 40 years of experimental work little had really changed in the physical information on the virus. In contrast, the biological aspects added in that time were of great significance.

3.7.1 Linking Virus Reproduction to Living Cells

It was clear that viruses could not be cultured in vitro on synthetic media. At the same time, however, it seemed increasingly feasible that viruses were to be expected everywhere living cells existed, i.e. in the entire realm of nature. This reasoning was substantiated by the ever-increasing number of viruses being discovered. Their number was no longer limited to the few types which had been observed in man, in his domesticated animals and in the plants he used. Their organ and tissue affinity, which caused striking, characteristic signs during a virus infection in humans and in animals, prompted initial attempts to classify viruses according to their tropisms (Table 3.4). Mosquitoes, ticks, mites and plant lice were discovered to be carriers of animal viruses.

Plant viruses had been found to occur in tobacco, tomato, cucumber, cauliflower, bean, sugar cane, corn, banana, aster, dahlia, mallow, strawberry and raspberry. The numerous, newly discovered

Table 3.4. An attempt to classify animal viruses known in 1936

- *Epidermal and epithelial viruses:*
  Herpes simplex
- *Dermotrope viruses:*
  Pox, measles, varicella, equine stomatitis, FMD, rabbit myxomatosis.
- *Dermo-neurotrope viruses:*
  Herpes zoster
- *Neurotrope viruses:*
  Rabies, poliomyelitis, encephalitis, Aujeszky disease, Borna disease, hog distemper, Louping ill.
- *Organotrope viruses:*
  Influenza
- *Viruses for acute common infections:*
  Fowlplague, Dengue fever, hog plague, yellow fever.

plant viruses led to extensive experiments to distinguish between new types and varieties of those already known. Their carriers were known to be plant lice and biting insects.

Insect viruses infected silkworm (Bombyx mori L.), the honey bee (Apis mellifera), the nun moth (Liparia monacha), caterpillars of the gypsy moth (Porthetria dispar), the crane fly (Tipula palidosa), the large cabbage white (Pieris brassicae L.), and the turnip dart (Euxea segetum).

The viruses which parasitize bacteria, i.e. the bacteriophages (bacterium intestinale or protobios bacteriophagus by D'Herelle) were found not only in intestinal bacteria of the coli-dysentery-salmonella group but in almost all aerobians. They could not be isolated in anaerobians and acid-proof bacilli.

3.7.2 Viruses are Autonomous, Organized Particles

Interpretations of filterable viruses ranged from Beijerinck's contagium vivum fluidum, and Löffler's and Frosch's small bacterium to Stanley's autocatalytic protein. The nature of viruses, therefore, was assumed to embrace lifeless toxins and enzymes as well as living microorganisms. Those investigations undertaken on the island of Riems with enzymatic methods for *Rous sarcoma* and *FMDV* gave not the slightest indication of the toxic or enzymatic nature of viruses.

Bacteriophage activity was linked to corpuscular particles. The assumption that *TMV* was an autocatalytic protein was quickly disproved and its particle structure was demonstrated by Takahashi and Rawlins as early as 1932. They identified *TMV* as a rod-shaped particle by the colloid chemical Tyndall method, as proposed by Freundlich in 1922, and by the polarization microscopic determination of double refraction, described by Ambronn and Frey in 1926.

Despite this evidence refuting the concept of a lifeless virus, opinion on the virus as a living microorganism was definitely divided. When the diameter of a virus closely approximated that of a bacterium, i.e. the *Vaccinia* or herpesviruses which measure about 200 nm, it seemed safe to assume an organized, living structure. This, by the way, in spite of Bronfenbrenner's failure to demonstrate respiration in the *Rabies virus* and a herpesvirus in 1924 and 1926 with the aid of a highly sensitive microrespirator. Opposition to the view of the virus as a microbe grew stronger in direct proportion to the number of viruses found with diameters of less than 100 nm. This applied to poliomyelitis and *FMD virus* as well as to the bacteriophages, which had also eluded Bronfenbrenner.

But it was precisely from the research on bacteriophages that we acquired evidence confirming viruses as being autonomous microbes! They could be separated from their host, and a single particle could disintegrate a bacterium by multiplying. Furthermore, immunological studies on bacteriophages and their host bacteria showed no confirmable immunological relationship between the virus and the host organism. Similar investigations were undertaken by Aoki and Shigasaki with silkworm jaundice. After 40 years of research, there was absolutely no indication that viral disease resulted from intracellular processes in which unorganized structures, such as toxins, were produced and triggered cells. On the contrary, the virus appeared to be submicroscopic, organized, reproducing parasite which needed to penetrate cells in order to multiply. Viral disease was thus taken to be the result of this cellular invasion, but exactly what took place after the invasion was unknown. Scientists who were loathe to accept virus etiology for tumour induction opted for the mutagen hypothesis by J.Murphy, particularly after the transformation of pneumococci had been established by Griffith in 1928.

Besides the intracellular process noted in viral infection, there was additional proof of a difference between a bacterium and a virus. It had already been furnished by D'Herelle but substantiation was not due until it was supplied by Burnet in 1929. His dilution experiments unequivocally showed the stepwise multiplication of phages. He termed the initiation of multiplication the lag phase, in which particles could not be found. During the subsequent step-by-step multiplication within the bacterium, each step did not lead to duplication of the germ, as is the case with the bacterium, but to a 6 to 60-fold increase of phage. The climax of intracellular phage reproduction was reached when the bacterium burst under the excessive pressure of the newly synthesized particles.

As for the chemotherapeutic aspects of viral disease, parallels between bacteria and viruses were absent. Motivated by the success of arsenic preparations (notably atoxyl, the sodium salt of p-amido-phenyl-arsenic acid) used for treating Trypanosoma since 1905, Friedberger and F. Sachs in 1909 attempted to treat rabies in rabbits in the same way. Subsequent anti-viral experiments in animals using, for example, quinine, hexamethylenamine and hexamethylenetetramine, were equally unsuccessful, although according to Ehrlich, a substance with inactivating, anti-bacterial properties should also be effective in the chemotherapy of viruses. It should be mentioned that this initial experimental phase with its false theoretical basis included investigatory work with poliomyelitis and fowlpox. Only after the discovery of the antibiotic properties of sulfonamides and penicillin did any new approaches enhance the field of anti-viral chemotherapy; the further development of this topic will be described in Chapter 5.

After almost 40 years of research on the virus, microbe hunters had learned a great deal about the behaviour of this type of agent, but its true nature continued to be an enigma.

4 On the Way to a Definition of the Virus

In 1935 when Stanley presented his crystalline autocatalytic protein to the scientific community, the reaction was surprisingly not entirely antipathetic. He met with great interest, even approval, in those branches of science were potentially able to incorporate protein crystals into their own theoretical reasoning. Proteins were at that time the primary substance of living cells.[1] A crystallized protein was for biochemical purposes a pure protein; thus scientists considered the crystallized *TMV* protein to be a pathogenetic molecule, perhaps even a "ferment". This broadened concept of chemical molecules as pathogens also met with support from medical scientists, as was evidenced on February 6, 1941, when Stanley was awarded the Gold Medal of the American Institute in New York. It was Rivers who conferred the honour and the tenor of his text recalled the distinction bestowed by the Journal of the American Medical Association upon Vinson and Petre for their achievement in the same field no less than 9 years earlier.

Rivers stated:

"As much as many bacteriologists hate to admit it, Stanley's proof that *TMV* is a chemical agent instead of a microorganism is certainly very impressive. Moreover, everyone admits that the agent of tobacco mosaic is transmissible indefinitely in series from plant to plant, a fact beyond dispute, indicating abundant multiplication or reproduction of the virus. Inasmuch as reproduction is usually considered an attribute of life, great confusion and consternation have been caused. In fact, the results of Stanley's work had the effect of demolishing bombshells on the fortress which Koch and his followers so carefully built to protect the idea that all infectious maladies are caused by living microorganisms or their toxins. In addition, his findings exasperate biologists who hold that multiplication or reproduction is an attribute only of life".

---

[1] Gr.protos, first

European physicists headed by Bohr began in 1932 to search for parallels to the new knowledge of quantum mechanics and nuclear physics which would explain the biological phenomena that had either eluded explanation up to that time, or were simply not generally known. This applied, for example, to biological reproduction as well as to the transference of genetic characteristics to succeeing generations. They hunted for hereditary molecular structures, genes, about which geneticists had rather concrete ideas after the work of Father Gregor Mendel had been rediscovered by deVries, Correns and Tschermak in 1900. Thus, Sutton, working at Columbia University in 1902, linked cytology to genetics when he accepted the inherited elements of Mendel's law to be parts of chromosomes.

Cytogenetic studies with corn and Drosophila endowed the following three decades with many important findings on the genetic elements, i.e. the genes of the cell nucleus. It was also known that chromosomes multiplied by reduction division and that they carried in their genes, rather like a string of pearls, the traits which could be inherited as such or be modified by mutation either induced or spontaneous. In 1922, T.Morgan determined the size of the gene, which has a definite position in a chromosome, comparing it to size of the hemoglobin molecule, and Muller termed the gene the basis of life in 1926. The geneticist Demerec, a native of Yugoslavia working at the Carnegie Institute in Washington D.C., summarized what was known about genetics in a lecture in 1933. His approach to the gene question was:

1. Genes are located in chromosomes.
2. Each gene must divide at every cell division.
3. The gene reproduces by formation of a new gene next to the old one rather than by the division of the old gene.
4. A gene is ultramicroscopic.
5. Mutation is a chemical process.
6. Every change within the gene is the result of a direct hit of a photoelectron.
7. A complete set of genes is a prerequisite for the life of an organism.
8. The assumed structure of the gene is thymonucleic acid. (Fig.1).

This assumption was confirmed in 1944 by a transformation trial with pneumococcus DNA by Avery and colleagues.

A German, Delbrück was one of those European physicists who had turned to the field of molecular biology. Motivated by Stanley's *TMV* findings, he devoted a monograph to this virus, which

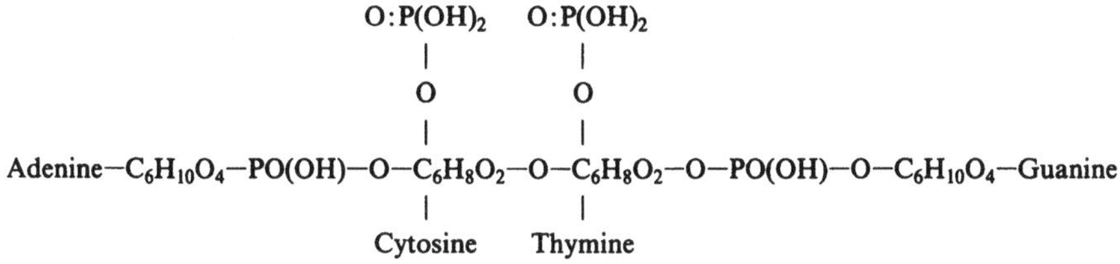

Fig. 1. Demerec's 1933 model of the gene structure

he had assumed to be a living molecule, entitling his book "Riddle of Life". Delbrück was not the only scientist who thought it possible to experiment quantitatively with this crystallized protein to learn more about reproduction and identical reduplication. While on a short tour of study in the USA, he realized the difficulty of working with plant viruses and shifted to bacteriophages in 1938. Based on his own calculations of their size, he was rather certain that bacteriophages were genes, which seemed better suited for quantitative analysis of heredity than *TMV*. His experiments on bacteriophages served as the groundwork for a physically oriented, molecular discipline severed from cytogenetics, which later became known as molecular biology. This term was not Delbrück's choice. In a letter to *Science* in 1938, Weaver, while working at the Rockefeller Foundation, had coined the new term for the emerging branch of knowledge which drew simultaneously upon physics, chemistry and biology. The goals of molecular biology, however, were first defined by the English crystallographer Astbury in his Harvey Lectures of 1950–1951:

> It is concerned particularly with the forms of biological molecules, and with the evolution, exploitation and ramification of those forms in the ascent to higher and higher levels of organisation. ... It must ... inquire into genesis and function.

The new branch of science received an additional impetus when the Austrian physicist Schrödinger who had emigrated to Dublin, began in 1943 to lecture on "The Physical Aspects of the Living Cell", based on Delbrück's "Atomphysikalisches Modell der Genmutation", written in 1935. What Schrödinger called "Delbrück's Model" was incorporated into his own book entitled, *What Is Life?* With this text he not only aroused interest and enthusiasm, but he clearly motivated young scientists to pursue hypotheses and experiments in molecular biology, many of whom carried out pioneering work in the new field.

In order to attempt a sound description of experimental virology between 1935 and 1957, it seems essential to link the research results of various scientific disciplines. There appear to be three separate yet clearly linked approaches: viruses as causative agents of diseases, viruses as biochemical structures, and viruses as genetic entities. Owing to the efforts of Doerr, the Archiv für die gesamte Virusforschung was established in 1939, and exercised a unifying effect by publishing findings from many disciplines which might otherwise have gone their separate ways. However, this aim to unite the disciplines of science, was jeopardized by the outbreak of World War II, and both the scope and depth of virus research was soon to gain proportions exceeding the limits of even the most ambitious journal of research.

But let us return to the chronology of experimental virology. As early as 1934, M.Schlesinger, while working with bacteriophages, was the first to conclude that viruses were nucleoproteids in nature, and contained a generative substance which in some way communicated with its counterpart in the host cell. Nevertheless, despite numerous publications providing relevant contributions to the discussion, it was more than a decade later before this knowledge was applied to our understanding of "filterable viruses". D'Herelle assumed that the bacteriophages discovered in different pathogenic bacteria were simply variations, i.e. different genetic entities of a single species. This view of the phage could not be confirmed, but in 1929 Sertic recorded the first traces of phage variants and mutants. Then in 1936, Burnet and Dora Lush proved that the phage was generative; they discovered a stable morphologic mutation on a *Staphylococcus-phage C*, which was linked to inability to induce phage resistance in the phage-sensitive Staphylococcus.

The transformation of *Rabbit fibroma virus* to *Myxoma virus* described by Berry and Helen Dedrick in 1936 – analogous to Griffith's work of 1928 – went unnoticed. Thus, only a handful of scientists, coming almost exclusively from the ranks of bacteriophage research, who supported the definition of a virus set down by R.Green in 1935. He regarded the virus as the smallest genetic entity, equipped with typical reproductive factors necessary for generating life. In 1938, Laidlaw expanded this definition, referring to the virus as borrowed life, which for him signified the most advanced form of parasitism. Hence, two diametrically opposed concepts had crystallized. For some researchers, viruses, including bacteriophages, were living entities because they practiced the most significant, basic function of life, namely the ability to reproduce. For others, by excluding the bacteriophages, viruses were the lifeless link between the inorganic and biological spheres. The ideas for the origin of the virus led into the realms of metaphysics, and the question of the essential part played by the living cell in virus multiplication prompted two hypotheses:

1. The conditions indispensable for virus reproduction can be found only in the living cell, providing an intracellular nutrient substrate.
2. The host cell provides the enzymes necessary for virus metabolism and multiplication that the virus lacks.

Two further decades of experimental science were to pass before a generally acknowledged definition of the virus crystallized. Only then did the new findings make it feasible to pursue both basic research as well as applied research in the institutes of virology.

The development of laboratory techniques which began in the 1940's proved to be a boon for virus research, including developments in electrophoresis, separation methods, as well as X-ray diffraction, and both fluorescence and electron microscopy. Electrophoresis and other separation methods are coupled with the name of Tiselius, a Swedish physical chemist, who, since 1925 had been working with Svedberg. Together they published an electrophoresis method in 1926 which they felt provided the possibility to separate minimal amounts of colloid particles of various sizes, e.g. different proteins, with greater precision than hitherto. Moreover, Tiselius also applied himself to chromatography, adsorption, ultracentrifugation and crystallography. Comparative experiments undertaken in Svedberg's institute soon convinced Tiselius of the advantages of electrophoretic methods, especially for separating serum proteins. After a period of study at Princeton University from 1934–1935, he began developing his own method of gradient electrophoresis in a U-tube, which considerably facilitated the separation of proteins in a mixture. Following up the separated protein bands using Toepler's Schlieren method, he discovered the fast-migrating albumin band and the three slower-migrating globulin fractions of blood serum, which he termed α, β, and γ. Tiselius also discovered that antibodies were globulins located in the γ or between the β and γ bands, an invaluable finding for immunological investigations. Additional separating procedures developed or influenced by Tiselius include paper- and zone electrophoresis, column adsorption, column methods for chromatographic ion exchange, as well as the introduction of the adsorbent hydroxylapatite.

The fluorescence microscope, an instrument well-suited for virus research, was a further development of the luminescence microscope invented by Lehmann at the Zeiß Werke in 1911. Lehmann's microscope exploited only the primary or non-specific fluorescence of microbes by using filtered, ultraviolet light. Bommer adapted the device for histological studies by first treating the tissue with fluorescent material called fluorochromes, which reflected the invisible, ultraviolet wavelength as visible, secondary radiation. Staining techniques were improved by Haitinger and Linsbauer, and in 1937 Hagemann reported his method of staining viruses: after treatment with primulin, a thioenyl dye, the virus appeared as a self-illuminating image against a dark background. This permitted a 3600-fold magnification, and using only slightly toxic fluorochromes, such as acridine orange, virus-infected tissue could be vitally stained. The fluorescence-antibody technique, to which Coons devoted much effort, was further improved for virological and for virological-serological investigations. Working with M.Kaplan in 1950, Coons described a direct staining method in which labelled, specific antibodies were bound to their antigens. In 1954, he published together with Weller the indirect method, whereby an antigen is first linked to its unmarked specific antibody. This antigen-γ globulin-complex is then used as an antigen for obtaining anti-γ globulin antibodies; these are labelled and linked to specific antigens for investigative purposes.

X-ray diffraction can be traced to Friedrich and Knipping at von Laue's institute. In 1912 they discovered that X-rays could be diffracted by a crystal. Diffraction depended upon the structure of the crystal and its orientation to the incoming X-ray beam. Thus, patterns typical for a given crystal can result on a photographic plate, producing a Laue diagram. Since this discovery, X-ray diffraction has been a part of crystallography and other methods for recognizing the structure of matter. After it was shown that *TMV* could be depicted in crystal form, it seemed appropriate to apply this method also to virus research. Wyckoff and Corey were the first to report X-ray diffraction patterns of *TMV* protein crystals in 1936.

A long line of theoretical and experimental physicists working at university institutes and in the electro-technical industry contributed to the development of the electron microscope (EM). This instrument works on the principle that beams of electrons can be bundled in certain electric or magnetic fields, and can be used to form an enlarged image of an object on a fluorescent screen or photographic plate. Magnetic lenses produced high quality images and thus made such microscopes into desirable instruments. The EM developed by E.Ruska and von Borries was introduced by the Siemens Company in

Berlin in 1938. It permitted resolution of structures down to 10 nm and a magnification which soon exceeded 100000. In the same year, the first electron micrograph of a "submicroscopic virus" was produced. Kausche, Pfankuch and H.Ruska displayed these images of the rod-shaped *TMV* at the 95th Congress of German Scientists and Physicians.

This chapter, which began in 1935, cannot avoid consideration of the retarding effect on virus research of the unsettled political conditions in Central Europe culminating in World War II. Scientific bonds between individuals and between institutes were seriously impaired in Europe as well as in other parts of the world, if not severed completely. The fruitful exchange of ideas and experimental findings was annihilated, and worse still was the hardship of restoring those ties 10 year later.

4.1 Viruses as Pathogens

In researching on animal viruses as etiological agents, many significant findings were obtained with just a handful of infectious diseases, which proved, however, to be of special interest for public health. Meaningful results were also obtained with a few plant and insect viruses.

The first causative agent that should be mentioned among the animal viruses is the poliomyelitis pathogen. A conspicuous increase in the frequency of this disease since the turn of the century, accompanied by a rise in the number of fatalities in Europe and in the USA, spurred of the research efforts of many countries. This applied particularly to the chemotherapeutic and immunological approaches. Despite concerted efforts, however, there was no success.

The results of the vaccination programme in the USA in 1935 were disappointing. Brodie from the Public Health Service of New York and Kolmer in the Dermatological Research Laboratory in Philadelphia were adversaries of a common foe. Both had developed vaccines from the spinal cord of infected monkeys; Brodie's was inactivated with formalin, while Kolmer's was attenuated by monkey passaging and addition of 1% sodium ricinoleate, which was assumed to have a mildly inactivating effect. The failure of both vaccines called for a sober but critical reassessment of the knowledge or lack of it about the poliomyelitis pathogen. Interest again shifted back to chemotherapy and to basic research of fundamental questions. Even the National Foundation for Infantile Paralysis (NFIP) founded in the name of President F.D.Roosevelt in 1938, began supporting research in this direction. Initiating the Penny Parade Program, the NFIP after the end of World War II became a motivating force for polio research, and in turn for clinical virology on a global scale.

The intensive concern aroused by 20 million deaths during the pandemic outbreak of influenza in 1918 stimulated researchers to intensify their efforts in this field. By 1933 W.Smith and colleagues had succeeded in isolating a human influenza strain from throat washings transferred to ferrets. In addition it was found that the pathogen could be neutralized in serum taken from convalescents. The same authors demonstrated the pathogenicity in mice in 1935, while Burnet adapted the pathogen to the chick embryo, successfully inducing the typical cellular changes. These were amongst the findings which paved the way for influenza research on a world-wide basis. Such studies became even more widespread after 1941. Working independently, Hirst as well as McClelland and Hare had discovered hemagglutination (HA), i.e. the capacity of the influenzavirus hemagglutinin to bind to erythrocytes. Hirst's findings proved to be a motivating factor in global influenza research, and not only from the perspective of the disease itself. From the start, hemagglutinin appeared to be identical to the infectious particle. However, as early as 1947, P. von Magnus discovered hemagglutinating but non-infectious forms which he called incomplete viruses. It was possible to inhibit HA with specific antiserum and, to a lesser extent, with unspecific inhibitors present in normal serum. Thus, the virus could by elution, which amounts to an enzymatic reaction, free itself from the erythrocytes.

HA proved to be a model for testing the interaction of a virus and its host cell. It was learned that several viruses require an enzyme for this interaction as does influenzavirus. However, some do not, as Burnet first discovered in 1936 with an arthropod-borne encephalitis virus. In 1942, F.Nagler also found a case of HA with hemagglutinin no longer bound to *Vaccinia virus*. Many of these results served as the basis for the development of influenza vaccines, which were of vital interest just then at the beginning of World War II, in view of the hardships caused by influenza during the previously world war.

The yellow fever pathogen was the subject of renewed attention in 1932, when data from Espirito Santo, Brazil, indicated a second cycle of jungle- or forest yellow fever threatening vast areas of the Americas and Africa. The success in curbing the urban form of the disease by combating the mosquito vector Aedes aegypti had been undermined by frequent changes in the cycle mosquito- monkey- man, and a reassessment of the disease was required. This also applied to those countries which had previously been spared outbreaks of yellow fever but which had large jungle or forest areas with both mosquitos and monkeys, such as India and the tropical countries of the Orient.

The unresolved complications also in the development of an effective vaccine for FMD, and the enormous economic losses in many countries due to repeated epidemics and endemics, forced international cooperation in research on this scourge. As for plant viruses, it was obvious that further studies on *TMV* were practical because of the groundwork during the past 50 years, and with insects, significant experiments were carried out with nuclear polyhedral viruses.

In addition to the question of whether a suspected virus was the etiological agent in a given disease, it was also essential to plot the entire course of infection, beginning with the uptake of the pathogen by the host, its dissemination within the organism, and the reaction and damage induced in cells, tissues and organs.

4.1.1 New Knowledge about Animal Viruses

Animal models for experiments and methods of quantifying viral content were essential for experimental investigation. Assays of this type had only been used so far by Calmette and C.Guérin from 1901–1905 for smallpox, by d'Herelle with his plaque method for bacteriophages, and in local lesion test for certain plant pathogens. This latter test was described by F.Holmes in 1928 for *TMV* as an improvement upon the quantitative method worked out by McKinney in 1927. In local lesion testing, the infection at first was carried out with needles but one year later Holmes modified this standard inoculation to rubbing the leaves with virus dilutions. If no visible lesions resulted, the leaves were treated with warm 95% alcohol and stained in a potassium iodide solution. After removal of the excess iodine with 30% alcohol, necrotic foci appeared as blue rings.

The smallpox method described by Herzberg in 1935 was similar to the Holme's leaf method and involved rubbing squares of shaved and scarified rabbit skin with O,O25 ml of diluted lymph. Quantification of pox foci began 24 hours later by dripping a 1% fluorescein solution onto the specimen to facilitate counting. Further new developments in the in vitro determination of animal viral content were presented by the Walter and Eliza Hall Institute in Melbourne after 1935. Burnet and colleagues devised a quantitative pock counting method in which viruses grown in chick embryos induced pox-like foci on the CAM. Burnet and his team worked with pox and influenzaviruses, the pathogens of infectious ectromelia (mouse pox), infectious laryngotracheitis in fowls, myxomatosis in rabbits, and louping ill in sheep and goats.

In 1936 Kidd and coworkers published a method for quantifying *Shope papilloma virus*. The procedure involved shaving the abdominal skin of a rabbit roughening it with sandpaper. The virus suspension was then rubbed into the dried skin and within 3 weeks it was possible to count the papillomas which had developed. A method described in 1949 by Bernkopf for growing viruses in the CAM of the de-embryonated chick embryo resulted in low virus yields, but the use of synthetic media meant that the experimental conditions were more defined than in the complete chick embryo. A very significant improvement that proved crucial for quantitative virology was introduced by Enders and colleagues in 1949. F.Robbins had been able to show in tissue-culture experiments in 1948 that the Lansing strain of poliovirus, which was adapted to the mouse, would multiply in a culture of human, embryonal, intestinal tissue. In an analogous procedure, virus were also successfully cultured in human embryonal nerve, skin, connective and muscle tissue. Proof of virus multiplication was obtained in subsequent infection experiments with mice and monkeys. Far more important, however, were the changes in cell morphology they could trace, and the discovery of the unstainability of infected cell nuclei. In a 1950 study, these two scientists, together with Weller, presented the "cytopathogenic effects" which were later referred to as the cytopathic effect (CPE) of cytopathogenic viruses. They added to the list of virus-induced cellular changes:

1. Piknosis of the nucleus
2. Cell fragmentation
3. The death of the cell
4. Inhibition of intracellular migration
5. Reduced acid production.

Since CPE's can be inhibited by specific immune serum, the following could now be realized:

1. The identification of polioviruses indirectly in vitro
2. The quantitative determination of poliovirus content
3. The typing of poliovirus strains by neutralization
4. The testing of anti-polio substances

Inhibition of acid production and secretion by infected cells later turned out to be the basis of a widely-used colour test: The absence of a change in the indicator phenol red in the virus-infected cells pointed to a CPE without the use of a microscope.

Proof that the poliovirus, which until that time had been strictly classified as a neurotropic virus, could multiply in vitro in non-neural tissue triggered analogous tissue-culture trials with other types of virus. It should be mentioned that the aim of the Enders group was to demonstrate the multiplication of the poliovirus in the intestinal tract. Clinical indications noted since the beginning of the 20th century justified this assumption, helped by after histopathological and experimental data after 1909. However, the authoritative conclusion of Flexner reigned as late as 1949, namely that the poliovirus entered through the nose and penetrated the CNS via the bulbus olfactorius. Back in 1935, Armstrong and W.Harrison, conforming to Flexner's premise, had undertaken chemotherapeutic trials by administering viruses and chemical substances intranasally. A sequel of these trials was the unsuccessful use of a zinc sulphate spray, known as the Schultz-Peet prophylaxis, during the 1937 polio epidemic in Toronto.

In 1938 Trask and coworkers were able to show the existence of the virus in the intestinal tract, though without impairment of the intestinal mucosa. Evidence of the extraneural multiplication of neurotropic viruses was previously given in 1947 by C.Evans and R.Green in the case of a poliovirus. F.Robbins repeated this in 1948 with the Lansing poliovirus strain, and C.Evans and Velma Chambers with the neurotropic *MM virus*, the *Encephalomyocarditis virus* in the paws of hamsters. Thus, contrary even to recently published textbooks, it was not Enders and colleagues who first demonstrated that a neurotropic virus was able to multiply in extraneural tissue. They did success, however, by virtue of their results, in persuading other scientists to reassess the pathogenesis of poliomyelitis and the culture of its agent. The discovery of polio viremia was one of the consequences.

In 1952, improved methods for growing viruses were reported in a publication by Moscona. He abandoned the former practice of mechanically mincing tissue and adopted the Rous and F.Jones (1916) enzymatic digestion of tissues into individual cells using trypsin. This was the origin of monolayer cell cultures, which facilitated staining and recognition of virus-induced CPEs. Moscona's method gained further in importance as synthetic culture media were developed, for example Medium 199 by T.Morgan and colleagues in 1950.

The technique of monolayer culture was used in 1951 by G.Gey and coworkers to establish the first permanent HeLa cell strain from a cervix carcinoma. Dulbecco meanwhile, having worked in phage research and having searched intensively for a quantitative test for animal viruses, improved upon their technique to develop the plaque test, reporting on the test in 1952 in connection with *Western equine encephalitis virus (WEEV)* and *Newcastle disease virus (NDV)*. Then in 1954 a publication co-authored with Marguerite Vogt appeared on the plaque test for poliovirus. In 1956, Vogt and Strickland published the basic principles of a plaque test neutralization method using as examples *WEEV* and poliovirus. Plaques occur in cell colonies which are damaged or destroyed by the multiplication of virus (so-called plaque-forming units, PFU); the damaged cells are unable to take up a neutral red stain being an indicator for living cells. An agar overlay of the infected culture prevents secondary infection. Within a few days, countable plaques are visible in direct proportion to the number of infective viruses in the virus suspension.

Larger cell populations in suspension culture became possible with the work of Earle and colleagues (1954) using L cells, a permanent fibroblast cell strain taken from the mouse. Monolayer cultures could be easely grown at will from the floating, constantly multiplying cells. In 1956 Puck and H.Fisher refined the technique with a feeder layer system: an additional layer of non-dividing cells stimulated the growth of the test cells by providing essential growth factors. The feeder layer was a HeLa monolayer culture irradiated at 4000 r to block growth; the test cells were added later.

The hem-adsorption test described by the group of Shelokov in 1958 led to the discovery of certain viruses which did not cause CPE in cell culture. Crucial for this test is the ability of viruses to accumulate on the cell surface during the multiplication stage; the cell is then changed in such a way that virus-induced foci bind themselves to erythrocytes. In serological tests it was found possible to inhibit this process using specific antiserum. The method proved to be of particular interest in paramyxovirus studies. Table 4.1 presents a survey of the extensive possibilities for experimentation

Table 4.1. The improvement of methods to isolate and to quantify animal viruses in vitro and in vivo up to 1957

Virus	Year	Test system[a]	Authors
Coxackie	1947	Suckling mouse	Dalldorf, Sickles
	1954	CPE	Stuhlberg et al.
Yellow fever	1927	Monkey	A. Stokes et al.
	1930	Mouse	M. Theiler
	1959	CPE	Hallauer, Porterfield
HSV	1912	Rabbit	Grüter
	1932	Mouse	Saddington
	1933	CAM	J. Dawson
	1953	CPE	Stuhlberg, Schapira, Enders; Scherer
Influenza	1933	Ferret	W. Smith et al.
	1934	Mouse	Andrews et al.
	1936	CAM	Burnet
	1955	CPE	Haas, Wulff; Nada Ledinko
FMD	1920	Guinea-pig	Waldmann, Pape
	1951	Suckling mouse	Skinner
	1956	CPE	Wesslén, Dinter
Measles	1911	Monkey	J. Anderson, Goldberger
	1935	CAM	Torres, Texeira
	1954	CPE	Enders, Peebles
	1958	Suckling mouse	Imagawa, J. Adams
Mumps	1916	Rabbit	Granata
	1952	Suckling mouse	Kilham, H. Murphy
	1954	CPE	Gertrude Henle et al.
Poliomyelitis	1908	Monkey	Landsteiner, Popper
	1939	Mouse, cottontail rat	Armstrong
	1950	CPE	F. Robbins et al.
	1952	Suckling hamster	Moyer et al.
Rinderpest	1938	CAM	Kunert
	1942	Rabbit	Curasson
	1957	CPE	Plowright, Ferris
Rabies	1804	Rabbit	Zinke
	1930	Mouse	Hoyt, Jungeblut
	1955	CPE	Sellers
Vaccinia/Variola	1892	Rabbit	Guarnieri
Vaccinia	1931	CAM	Goodpasture et al.
Variola	1935	CAM	Torres, Texeira
	1953	CPE	Noyes

[a] CAM, on the chorioallantois membrane; CPE, cytopathic effect in cell culture.

with virus types causing disease in the humans either directly or by transmission from animals (zoonosis). Laboratory aids range from an appropriate animal model, to the pox test on the chorio allantois (CA), to CPE in the plaque test in primary and permanent monolayer cultures without, or with an agar overlay in the plaque test.

To obtain models for animal experiments in the laboratory patient and imaginative experimentation was often required, as is shown in the following two examples: In 1935 Brodie attempted to adapt the poliovirus to rabbits, rats, guinea-pigs and mice. When the trials failed, these workers did not give up but concentrated on trials with mice. Their method included repetitive irradiation with low-dose X-rays and virus inoculations intracerebrally and intraperitoneally. It was not until the 17th passage in the mouse that typical polio lesions were found by testing in the spinal cord of monkeys.

During the post-war period, when monkeys were not easily obtainable, H.and P.von Magnus, working at the State Serum Institute in Copenhagen, decided to switch to mice. Trials were carried out with the TO (Theiler-Olitzki) strain of the *Encephalomyelitis virus*, thought at that time to belong to the polio group. Right at the beginning of the work with mice, they learned that the majority of the animals were already infected with this virus and consequently large-scale tests were impossible. However, they managed eventually to obtain a colony of virus-free mice by immediately isolating the newborns and raising them with surrogate rat mothers.

The concept that viruses and bacteria enter the body in the same way for a time remained in place. As better methods ensued for providing the presence of a virus or an antibody in an organ or organism, it at least became possible to observe the pathway and the various phases of virus infections.

4.1.1.1 First Contact With a Virus

Infected organisms readily release various types of virus through the skin, in secretions, excretions and blood. It is thus possible to transmit the virus by direct or indirect contact, by inhalation, through contaminated food and by active or passive insect vectors. Flies and ticks, for example, may be passive virus carriers, while mosquitoes are active, in certain cases operating in rotation with birds as biological vectors. Tables 4.2 and 4.3 present a compilation of virus contacts.

Table 4.2. First contacts in animal virus diseases

Direct contact:
Herpes genitalis
Genital warts
Diseases of skin (e.g. smallpox)
Indirect contact:
by virus-soiled hands (e.g. poliomyelitis-diseased children)
by nursing persons
by objects of daily life
Dropinfection by virus-containing secretions of the mouth:
Influenza, common cold, atypical pneumonia, adenovirus infections, measles, rubella, chickenpox, mumps, smallpox, poliomyelitis
Transmission by food and drinks:
Poliomyelitis, infectious hepatitis
Passive transmission by flies:
Poliomyelitis, smallpox
Active transmission:
by mosquitoes: urban-type of yellow fever, Dengue fever, St. Louis encephalitis, Rift Valley fever, West Nile fever
by ticks: Russian spring-summer encephalitis, Louping ill (e.g. sheep in Scotland and northern England)
by monkeys: Russian spring-summer encephalitis, Louping ill (e.g. sheep in Scotland and northern England)
by monkeys: sylvatic type of yellow fever
by foxes, dogs, cats and small rodents: rabies

4.1.1.2 Tissues and Organs In Which Cell Infection is Initiated

Depending on its site of entry into an organism, the virus may encounter specific chemical groups called receptors on the surface of cells to which it has an affinity. This type of cell is found in the mucosa of the respiratory and the digestive tracts, lymph tissue, blood cells, endothelium and in the venous space of the spleen and bone marrow. Although at one time erythrocytes were considered to be only the means of virus transport, leukocytes possess a greater affinity for tight binding of virus. As early as 1914, Todd and R.White were able to demonstrate *Cattle plague virus* in leukocyte fractions. This was confirmed for *Fowl plague virus* by Todd in 1928, for poxvirus by W.Smith in 1929, for *HSV* by Jamuni and Holden in 1934, and for influenzavirus by Smorodintsev and Shishkina in 1941. Whilst scientists acknowledged phagocytosis of virus by leukocytes,

Table 4.3. Virus reservoirs in domestic and in laboratory animals

General illnesses:
Anaemia of horses
Horse influenza
Equine abortion
African horse sickness
Rinderpest
Rift Valley fever
Swine fever
Aujeszky's disease of dog, cat, cattle
Dog and cat distemper
Fowl plague
Newcastle disease
Pox-like illnesses:
Horse pox
Vaccinia
Sheep pox
Goat pox
Camel pox
Swine pox
Rabbit pox
Monkey pox
Avian pox
Vesicular illnesses of the skin and mucous membranes:
Vesicular stomatitis
Foot-and-mouth disease
Blue tongue
Ectromelia
Illnesses of the respiratory organs:
Equine and bovine infectious bronchitis
Malignant bovine catarrhalic fever
Swine influenza
Virus pneumonia of ferret
Infectious bronchitis of fowls
Laryngo-tracheitis of fowls
Illnesses of the alimentary tract:
Diarrhoea of cattle
Nairobi disease of sheep and goats
Canine hepatitis
Gastroenteritis of swine
Salivary gland disease of rodent
Illnesses of the nervous system:
Borna disease
Western equine encephalomyelitis,
Rabies in various animals
Louping ill of sheep and goat
Teschen disease of pig
Lymphocytic choriomeningitis
Marek's disease of fowls

they were aware that this was very different from bacterial phagocytosis. As for the cells in the tissues mentioned above, it was certain that they took up the virus and thereby presented the opportunity for intracellular multiplication.

Viruses producing respiratory disease bind to receptors directly by way of the nose and throat mucosa. Some viruses which manage to reach the bloodstream through a skin lesion find suitable cell during subsequent circulation. This latter type includes the mosquito-transmitted Yellow fever virus and especially those viruses transmitted by insects which can damage the CNS. The transmission of rabies through the bite of a rabid dog falls into separate category: the virus is admitted to host cells either by phagocytosis or by droplet adsorption (pinocytosis). Incipient viral infection in the nose and throat mucosa was shown in the case of influenza; mumps and German measles (rubella) were thought to follow a similar course. Influenza can begin anywhere between the epithelial cells of the mucosa and the bronchioli; the infection spreads to cells of the same type in the surrounding tissue and the symptoms of the disease are a consequence of cell destruction. Whilst for the most part only slight, these symptoms can increase in severity due to specific virus properties, secondary bacterial infection, an extra-large virus dose, or a weakened immune response.

Invasion of the *Measles virus* not only takes place in the epithelial cells but also in the lymphatic tissue of the respiratory tract. Poliomyelitis infection commences in the lymph nodes of the nose and throat and is passed on to the intestinal tract. The predilection of *HSV* is for the mucodermal areas, i.e. at the lips, nose, eye and genitals. With infections resulting from stings or bites, viruses enter the epithelial or connective tissue where they multiply and are further spread directly or via the bloodstream. Multiplication of the virus at the initial sites may be inapparent, although it can also be accompanied by visible initial lesions or by coughing, elevated temperature, organic disturbances and pain, as well as by other mild symptoms of disease. The initial lesions can be proliferative, inflamed, necrotic or vesicular.

In many cases, virus infection terminates at this point. This termination of the virus to multiply depends on the severity of the primary infection and the virulence of the viral strain, but rather more on the natural or aquired defences of the organism. Extensive experience with passive immunization against various viral diseases indicates for humoral antibodies a very significant role. A direct relationship between the antibody level and protection against severe infection was found by Isabel Morgan with poliomyelitis in 1949. For months she immunized monkeys with intramuscular injection and determined the serum-neutralizing titre in an intracerebral test in mice. The immunized monkeys were given a paralytic loading dose of 10 PD_{50} intracerebrally. Protection of the monkeys against paralytic poliomyelitis rose in direct proportion to the serum antibody levels. Complete protection was found in animals which had a minimum neutralization titre of 1:1000.

4.1.1.3 Secondary or Generalized Symptomatic Infections

If the body's defences are not able to curb the infection in its initial phase, the virus not only infects cell after cell but also avails itself of three systems for dispersion in the organism: the blood stream, the lymph system, and the nervous system. These modes of distribution allow the virus to reach organs in which it may multiply, or remain latent or persistent while the organism remains free of symptoms. The presence of poliovirus in the blood (viremia), shown by D.Bodian in the monkey in 1951, and by Dorothy Horstmann in humans in 1952, scientists were convinced that, in general, the bloodstream is the most significant means of disseminating virus in the body. Despite untiring effort, however, they were able to prove this for only four diseases: poliomyelitis, yellow fever, herpes simplex and smallpox. For other diseases such as influenza, German measles and mumps, there were merely strong arguments for the existence of viremic phases. This phase formally belongs to the incubation period, i.e. the time during which there are no organ-specific symptoms but simply slight premonitory signs of disease. Although the initial phase of a viral disease may be considered a part of the incubation period, when the symptoms are minor, this term is commonly used to denote the time from the entrance of the pathogen to the first symptoms in the organs typical for the disease.

The subsequent prodromal period with its premonitory symptoms as a rule lasts only a few days. It is the precursor of the full-blown disease with its typical clinical and specific histopathological changes. Table 4.4 shows examples of the incubation and prodromal periods of several viral diseases known before 1960. Table 4.5 presents the common target organs involved during symptomatic infection and the proportion of cases with secondary infection during the progress of nine viral diseases.

Table 4.4. Examples of incubation and prodromal periods

Illness	Incubation time (days)	Prodromal time (days)
Influenza	Hours to a few days	
Yellow fever	3– 10	
Measles	9– 11	3
Poliomyelitis	8– 36	
Smallpox	5– 21	2– 3
Chickenpox	11– 21	1– 2
Rubella	16– 18	1– 2
Mumps	18– 21	1– 3
Infectious hepatitis	10– 45	4–11
Rabies	30– 90	2– 4
Serum hepatitis	56–182	

Table 4.5. Symptomatic infetions with animal viruses

Illness	Target organs	Part of cases with secondary infection (%)
Smallpox	Skin, mucous membranes	Near 100
Mumps	Parotid gland, gonads	About 80
Measles, Rubella, and Chicken pox	Skin, mucous membranes	Near 100
Poliomyelitis	CNS	About 1
Infectious hepatitis	Liver	Low
Yellow fever	Liver	Varying
Rabies	CNS, salivary glands	100

In most cases, viral disease affects specific target organs and exhibits typical symptoms. The direct fluorescence antibody technique facilitated detection of virus in the involved organs. Using this method, in 1950 Coons and colleagues found the *Mumps virus* in the salivary glands of infected monkeys, and in 1955 W.Noyes located the *West Nile virus* in neurons of the brain as well as in the spinal cord of monkeys.

4.1.1.4 Uncommon Interactions Between the Virus and the Host Organism

Peculiar interactions between virus or viral components and the host organism have also been discovered, one of which was persistence. It was previously known, for example, that *HSV* could persist in an organ from which it might potentially penetrate the CNS. In 1941 Gregg observed and described chronic infection of the foetus to be the cause of teratogenic effects of the *Rubella virus*. Prior to this, deformities caused by viruses had been reported for variola by Madsen and Greenfield (1934), for varicella by Shuman and Campbell (1939), and for measles by Dryer (1940). Subsequently, a relationship with teratogenicity was shown for influenza by Landtman (1947), infectious mononucleosis by H.Miller and colleagues (1949), poliomyelitis by the group of Baskin (1950), hepatitis B by J.Stokes and coworkers (1951), coxsackie B by Gear and Measroch (1952), and mumps by Ylinen and Yärvinen (1953).

Gertrude and W.Henle published their findings on the virus toxin of influenza in 1944. In the pox group, toxins led to variolosa and to secondary, hemorrhaging variola. Damage caused by released virus allergens was also observed in cases of measles. Pyrogens were first reported for the influenzavirus in the rabbit by Wagner and colleagues in 1949, and by Fastier for the *WEE virus* in rabbits and mice in 1952.

The matter of latent virus infections, i.e. those which run their course without obvious symptoms but which become perceptible upon experimental provocation, has been previously adressed, and good examples are the *Encephalomyelitis virus* and *Lymphocytic choriomeningitis virus*. The former virus was discovered during the development of yellow fever vaccine in mice at the M.Theiler institute in 1934, and the latter was isolated from monkeys and mice during a St.Louis encephalitis epidemic in 1934 by Armstrong and Lillie in 1934, and in mice by Traub at the Rockefeller Institute in 1935. Additional viruses belonging to this latent group are the *Ectromelia virus* in the mouse, isolated by Gönnert and Kikuth in 1940, as well as latent hepatitis in the blood discovered by the J.Stokes group in 1954. There were indications that latent forms existed for other viral diseases, but definite proof awaited suitable demonstration methods. Monkey organs were used to generate cell cultures for the production of vaccines against poliomyelitis and these were tested for the presence of latent viruses. They were detected but only after induced CPE or the inoculation of susceptible animals. In 1954 Wood und Shimada found

Herpes B virus in poliovirus suspensions from such cell cultures. Sabin and Wright had isolated this virus from a human patient who had succumbed after a monkey bite in 1934.

In 1955 Rustigian and colleagues isolated latent viruses from monkey cells which they termed *Foamy virus* simply because the induced CPE of this virus in cell cultures looked like foam. Finally Hull and his collegues from E.Lilly Laboratories discovered several so-called simian viruses (SV) after 1956. The subject of latent viruses brings up the field of tumor viruses and interference problems.

4.1.1.5 Tumour Virus Infections

The term "tumour virus" seemed justified after the successful cell-free transmission of a rabbit papilloma in 1933, the determination of the size of *RSV* in 1935, and the multiplication of this virus in cell cultures. Furthermore, additional tumour viruses were being continuously reported. In 1936 Bittner solved a problem to do with the influence of female sex hormones that had been discussed since 1916. At that time, Lanthrop and L.Loeb had already found that mammary tumours in hybrid strains of mice tended to be linked to the femal parent rather than to the male, but had not been able to show any hormonal influence by removing the ovaries. However, in 1928 W.Murray managed to implant ovaries into castrated mice, and in 1932 Lacassagne administered estrogen to normal male animals for 5–6 months, using the RIII strain preferred for such trials. Jackson Memorial Laboratory in Bar Harbor, Maine, experimenting with the same tumour type, set up three hypotheses about the origin of these mammary tumors: Transmission through the cytoplasm of the ovum, transmission during development in the uterus, or transmission via breast milk. Bittner and coworkers investigated the last hypothesis and were able to verify it in 1936. The successful transmission of Bittner's milk factor to female animals by Andervont and colleagues in 1942, and also via sperms of male animals by Mühlbock in 1950, revealed that this factor was a filterable virus. In the electron microscopy, it was a virus having a diameter of 50–70 nm, as determined in 1954.

The method by Duran-Reynals (1942) to transmit a chicken tumour to ducks no older than 24 hours was adapted by Gross for leukemia in mice. In 1951 he reported that he had been able to transmit cell-free material in newborn mice. Using the *Mouse leukemia virus*, he induced a wide spectrum of leukemia and lymphoma tumours, and with various types of filters determined the diameter of these viruses to be 70–100 nm. In 1953 Gross augmented the tumour spectrum of mouse leukemia viruses. While transmitting filtrates of the AK inbred strain, he not only found leukemia tumours but also parotid tumors, and termed the causative agent *Parotid tumour virus*. At the NIH in Bethesda, Maryland, Sarah Stewart and Bernice Eddy transferred leukemia extracts from AK mouse experiments to kidney cell cultures of rhesus monkeys which were routinely used for growing poliomyelitis and other virus strains. Whereas their attempts with *Leukemia virus* failed, they did manage to grow the *Parotid tumour virus*. With this cell culture material, they were able to induce a broad spectrum of tumours in mice, rats, hamsters, guinea-pigs, rabbits and ferrets. The highest parotid tumour frequency was found in the salivary gland, while other tumour types occurred in most other organs. The name of the agent was, therefore, altered to *Polyoma virus*. Its diameter, determined after filtration, was approximately 40 nm, and the EM values for a spherical particle were 30–60 nm.

From 1954 onwards publications appeared by Graffi and Bielka, who had undertaken the cell-free transmission of various types of mouse sarcomas. In 1956 Charlotte Friend isolated a virus which was shown to cause erythroblastotic-like syndromes in mice. Thus, even before 1957, there was much evidence and sundry indications that viruses were actively involved in induction of tumours, the development of which followed a course similar to that of latent virus infection.

4.1.1.6 Interference

Growth-dependent, protective effects among virus strains have been known since 1935. At that time, Hoskins discovered that monkeys inoculated with a neurotropic yellow fever strain were protected from subsequent infection by a lethal viscerotropic strain. When Andrewes (1942) observed a similar protective action between WS and NWS strains of influenza in chick-embryo cell suspensions, he borrowed the term interference from the field of physics for this interaction. By 1957 there had been many examples of experimental interference among homologous and heterologous virus strains. During interference, the influence of the virus which had been administered first ranged from complete prevention or inhibition of the challenge virus, to the absence of macro- or microscopically observable pathological changes in cells or tissues of the host organism in vitro or in vivo. The various forms of interference that were discovered are not to be confused with auto-interference or with double infection. The development of incomplete viruses is the result of auto-interference upon infection with high doses of virus. Double infections can be explained by binding different viruses to two different cell receptors; Katherine Anderson had reported on this in 1942 with reference to fowlpox/herpes, fowlpox/laryngotracheitis and vaccinia/herpes in chick embryos.

It appeared that the phenomenon of interference precluded immunological mechanisms, but the possibility of verification by highly sensitive immunological testing was not excluded. Thus, the relatively crude methods available early on for determining viral content, acquired immunity, and pathological changes detected only the effects of strong interference. This was found to take place between active viruses, and between inactivated and active viruses in experimental animals, in embryonic eggs, and in cell cultures. Under given experimental conditions, the initial multiplication of a virus could not be influenced, only the further synthesis of virus was prevented in cells which had become resistant. The interference mechanism, unclear as it was, opened up an additional possibility for protective vaccination, and interference experiments with inactivated and living viruses in the chick embryo proved all the more cogent.

4.1.1.7 Immunological Problems

Essentially, there were now three outstanding problems:

1. To attain high-titre, uncontaminated virus suspensions to be used as suitable antigens for serological tests and the production of more effective immune serums.
2. To assess the development and importance of neutralization tests using cell culture techniques.
3. To ascertain how meaningful immunity, achieved with humoral antibodies, was with reference to virus diseases.

It was especially important to improve virus suspensions for CFR. Before cell culture techniques had been introduced for growing viruses, only animal material was available which often offered only a low or unsuitable amount of virus. In order to enhance antigenicity, the organs with the greatest virus content were singled out and treated with heat, phenol, freezing and thawing, or ultrasonically. It was also possible to enrich the virus additionally by ultracentrifugation. In order to purify virus suspensions, Casals (1949) introduced acetone-ether extraction, and Oker-Blom and Nikkilä (1955) used bentonite. Cell culture techniques offered a way of preserving high titres and pure antigen preparations. In this way, CFR could be used for a wide range of serological tasks as well as for classifying newly discovered viruses. Hilleman and Jacqueline Werner (1954), for example, found the soluble, group-specific, adenovirus antigen from material secured from HeLa cell cultures.

Aggregation tests were improved by new culture, purification, and concentration methods. Influenza serology was refined in 1938 when Magill and Francis used material from mice lung for agglutination tests. W.Henle and Leslie Chambers carried out their agglutination tests in 1941 with virus-containing fluids from chick embryos, and in 1953 K.Jensen and Francis demonstrated influenzavirus precipitation using the gel-diffusion technique. This method had first been described by Oudin in 1946 and introduced by Ouchterlony in 1948 to analyze bacterial toxins and precipitins.

Neutralization tests with cell cultures supplanted those involving animals. Since Enders and colleagues for the cytopathic effect, like Dulbecco and Marguerite Vogt for the plaque test, had found evidence that virus-induced effects could be compensated by neutralizing serum, the use of neutralization tests, therefore, required a suitable cell culture for each virus. The literature of the 1950's not only abounded with suitable cell culture systems, but also pointed out that each kind of virus represented specific problems to be solved. These included, for example: Finding the optimal incubation period for the antigen-antibody mixture; determining whether the virus concentration or the serum dilution should be constant in the test; discovering how the toxic components of the host cell could be eliminated.

The metabolic inhibition or colour test on plastic plates allowed large-scale testing; this test had first been described back in 1943 by C.Huang for *WEEV* using chick-embryo tissue cultures, and in 1950 by F.Robbins and coworkers for poliovirus on human embryonic tissue. In 1954 it was employed for neutralizing poliovirus using trypsinized monkey kidney cells by J.Salk and colleagues. The inhibition of HA by specific serums in the hemagglutination inhibition (HI) test was used extensively for the myxoviruses, and the 1957 influenza pandemic evidenced the value of the test for typing and diagnosing, provided certain laboratory conditions were maintained. These included, for example, the ability to exclude inhibitors which behaved like antibodies during the test, thereby giving false results. It was particularly relevant for diagnostic purposes to retain reliable titre differences and to be able to determine which indicated a positive result. Biometrics helped calculate the necessary scope and confidence limits of the various tests and compute the antibody and virus contents. L.Reed, Münch, and also Spearman and Kärber deserve recognition in this regard.

Viruses are particulate antigens, but in most virus infections a specific soluble antigen is also produced, which can be detected in serological tests. This was first demonstrated for yellow fever by Hughes in 1933 and, 2 years later for smallpox by J.Craigie and Wishart in the fluid of vesicles and in extracts of the dried crusts. In the following two decades, soluble antigens of several viruses were demonstrated using CFR.

Meanwhile the concepts had been changing of the significance of humoral antibodies for the protection of an organism from later infection by a similar virus. It was clear that the presence of antibodies in the blood alone was not adequate. It was also known that reliable protection required both cell-mediated and humoral immunity. In 1956, Good and Zak discovered that children with innate agammaglobulinemia recovered normally after measles and proved to be just as resistant to re-infection as those with intact humoral immune systems. Scientists were still convinced that a person was immune against a specific disease provided he had an adequate amount of a specific antibody to neutralize the virus before it attacked the organ associated with the disease; at least this was assumed in cases of disease which spread via the blood. Herpes-immune and vaccinia-immune humans had been shown in 1926 by Teissier and colleagues to become infected through skin scarification, despite having antibodies. Latent *HSV* carriers who had humoral antibodies were infected through the skin by 4–7 inoculations before their organisms became resistant. In 1940, Francis was able to demonstrate the protective effect of nasal secretion in influenza, proving that virus-induced inflammation was linked to the exudation of antibodies which entered the tissues and thus inhibited further spreading of the virus.

Hence, there was no doubt about the value of humoral antibodies; the levels required, however, could only be determined by performing protection tests in experimental animals, which naturally might include additional defensive factors.

4.1.1.8 Immunization

More appropriate methods for culturing viruses and determining antibodies led to a better quality of heterologous serum for passive immunization. Moreover, new control measures and purification processes which were developed facilitated trials with human γ-globulin fractions. These were initiated by J.Stokes and Neefe as well as by Havens and Paul (1945) for infectious hepatitis and by S.Anderson and McLorinan (1953) for rubella. Trials with γ-globulin for measles and poliomyelitis (1952) indicated the possibility of its effective use after an infection. Administration immediately upon infection with measles prevented the disease, whereas a later injection merely weakened the symptoms.

Culture methods gained in importance as scientists strived to develop vaccines for active immunization. The search for strong immunogenic virus strains to yield inactivated vaccines was encouraged and much later, new possibilities for attenuation appeared which aided the production of live vaccines. The ideal procedure was to vaccinate with suitable harmless, unchanged strains by way of the natural portals of infection, e.g. by inhaling the *Rubella virus*. However, this was rarely possible and work with live vaccines had, unfortunately, been burdened by several incidents. It began in 1907 with the outbreak of FMD in the environs of Greifswald which prompted Löffler to interrupt his experiments with FMD vaccine. It was not until 3 years later, in October 1910, that he was able to resume his work at the Staatlichen Forschungsanstalt on Riems Island, in the world's first virus institute. The other setback for immunization occurred during World War II (1941–42), when U.S. soldiers received yellow fever vaccine containing normal human serum. Tens of thousands contracted hepatitis as a result of serum contamination.

Simultaneous vaccination, which was used primarily for disease prevention in animals, e.g. hog cholera, cattle plague, African horse sickness and canine distemper, required proper proportions of vaccine virus and antibodies. When it was impossible to determine these component ratios in preliminary trials, the efficacy of the treatment always proved unsatisfactory. The growing tendency away from this sort of vaccination encouraged the development of inactivated vaccines; ample justification for this trend could be found in positive experiences in veterinary and human medicine. In the following description of this development, emphasis is placed on specific diseases. Not only was the interest of the state in large-scale vaccination for certain diseases crucial, but also the impulse for virus research itself which resulted from the development of vaccines.

Foot-and-Mouth Disease Vaccine

The list of disappointing approaches and results during the development of an active FMD vaccine is long: pig lymph, physically and chemically attenuated virus, FMD lymph attenuated with cow pox lymph, simultaneous vaccination, and the inability to extrapolate vaccination results from the guinea-pig to the cow. Since the live-virus vaccine reported in 1936 by Schmidt-Jensen and coworkers was adsorbed by aluminium hydroxide, it was soon excluded as a large-scale vaccine on safety grounds. The following year, the development of a formol-adsorbate vaccine was undertaken, based upon successful experiments with formol vaccine reported by Vallée and colleagues (1925) and Bedson and colleagues (1927). In addition, it was possible to carry on where earlier scientists had left off with the development of a diphtheria toxoid vaccine: in 1925 Ramon and Glenny had independently succeeded in completely detoxifying diphtheria toxin using formalin. Ramon had also discovered that horses which had shown an inflammatory reaction at the site of injection often developed more antibodies than those which had not. He also found in 1926 immune-stimulating substances which he called adjuvants. These adjuvants, which were administered in combination with the antigen, included agar, tapioca:[1] starch oil, lecithin and saponin. Glenny and M.Barr (1931) added alum and Schmidt added aluminium hydroxide to diphtheria toxin. A single dose of Schmidt's adsorbate vaccine led to reliable immunity.

This served as the basis for Waldmann's adsorbate vaccine for FMD. In 1931 his coworker, Pyl, used aluminium hydroxide to test whether adsorption and elution, which Willstätter and von Euler had used in fermentation, were suitable for purifying and concentrating *FMD virus*. He managed a 1000-fold enrichment and achieved elution by shifting the pH to 7.6. With the observation that formalin alone did not reliably inactivate *FMD virus*, it was decided to enhance the effect by heating. The resulting vaccine proved to be both safe and effective and stood the test during the pandemic of 1938–1940. It allowed for the plurality of the pathogen, and was produced from material of

---

[1] granulated starch from the bulb of the Euphorbia plant found in tropical forests

primary aphtae injected subcutaneously in a dose of 60 ml. Its efficacy and safety were virtually guaranteed by a special testing procedure. The vaccine soon replaced the Löffler hyperimmune serum, which had been used in annual amounts of 150,000 litres although it afforded only 8–10 day protection.

Influenza Vaccine

In 1933, on the heels of successful transmission of *Influenza virus A* to ferrets, W.Smith and coworkers initiated their immunization trials with ferrets and mice. Francis demonstrated the presence of a neutralizing antibody in the serum of convalescents in 1934 and, working with Magill, he investigated the antigenicity of *Influenza virus A*. They found the content of neutralizing antibodies to be age-dependent and, moreover, were able to determine the antigen differences of the various epidemic strains. After intra- and subcutaneous transmission by inoculation of living virus, they found that their test volunteers exhibited neutralizing antibody levels similar to those of patients who had run the course of the disease. In 1938 they transmitted the causal agent directly from humans to mice and in 1940, working independently, they discovered *Influenza virus B*.

In 1937–1938 Fairbrother and Leslie Hoyle concluded that virus inactivation by heat was suitable, as a result of their immunization trials in mice and ferrets with mouse lung tissue containing virus. Shortly thereafter, in a 1940 publication, Fairbrother and A.Martin reported equally reliable immunization results in mice with vaccine which had been inactivated with phenol, or formalin or by heating to 50° or 57°C and then injected subcutaneously and intraperitoneally twice at 7-day intervals. F.Horsfall and colleagues tested various combinations of vaccines in a larger series of trials in 1940. These vaccines had already been tried experimentally in ferrets and, subsequently, in human volunteers, and resulted in antibody levels that would provide at least a 3-month protection period. Above all, they were eager to ascertain the optimal efficacy of the constituents of their formaldehyde-inactivated complex vaccine, which also contained *Canine distemper virus*. This complex chick-embryo influenza-canine distemper vaccine proved to be optimal and, according to the antibody levels, afforded a 5-month protection period.

The successful culture of the virus in the chick embryo in 1936, and the discovery of HA, accompanied by the development of the HI-test in 1941, represented further milestones in influenza research which were to prove highly significant for developing a vaccine. At the beginning of World War II, it was recalled that the influenza pandemic of 1918 had brought great military-political danger, although it was not clear at that time whether the 1918 causal agent had been a virus, and whether the 20 million casualties were to be ascribed to a virus or a bacterial, secondary infection. This provided a strong impulse for influenza research in general and active immunization in particular. Investigational work on inactivated vaccines took priority in the USA and the UK, while from 1940–1944 the Australian group headed by Burnet focused on live vaccines. Francis directed the vaccine trials undertaken in the USA in 1941 by the *Commission on Influenza of the Armed Forces Epidemiological Board* (AFEB). Five commercial companies made available various vaccines for initial trials in 1941.

From 1942–1943, the trials were extended to larger groups totalling 8000, who would receive subcutaneous injections of concentrated influenza A+B vaccine. To enrich for antigen, chick embryo tissue containing virus was frozen and then thawed or chicken erythrocytes were exposed to adsorption and elution. Since the 1942 influenza wave did not materialize, and thus no conclusions could be drawn about the possible protective action of the vaccine, it was decided to infect a proportion of those already vaccinated with active virus. These and subsequent natural infections demonstrated the efficacy of the vaccine, but immunological investigations showed a marked decrease in antibody levels 4 months after vaccination, when the serum levels dropped to one-third of the maximum. The success of inactivated vaccines in the USA and the failures with live vaccines brought the trials in Australia to a halt. Nevertheless, Smorodintsev and Zhdanov pursued these studies further in the Soviet Union.

To achieve even greater protection, various groups added adjuvants to their vaccines. For example, Friedewald (1944) achieved a 4-fold increase in antibody production over controls by adding complete Freund's adjuvant. This mixture containing paraffin oil and tuberculosis bacilli, which enhanced antibody development, had been concocted by Casals and Freund in 1939 while they were testing paraffin oil as an adjuvant for these bacilli. Beginning in 1942, increased and prolonged antigenicity was achieved for various antigens, as evidenced by levels found in rabbits and guinea-pigs. In 1945 W. and Gertrude Henle developed the influenza A+B–mineral oil–Falba[1] vaccine for humans. In the same year, Salk ran tests on a

---

[1] bees's wax +paraffin + Oxysterine

calcium phosphate–adsorbate vaccine in mice and, 2 years later, he tested a more concentrated mixture for humans, to which merthiolate had been added as inhibitor of secondary germs. His objective was to endow protection using smaller amounts of virus than in the vaccines employed in the large-scale trials of 1943–1944. Salk's vaccine provided at least a 7-month-long elevated antibody level without undesirable side effects caused by adsorbent. The increased protection, furthermore, resulted from only a single dose. This improvement was accounted for by elution of the antigen from the virus adsorbate over an extended period of time, a phenomenon known as permanent antigen stimulation. During the following decade many other substances were tested as adjuvants, with aluminium phosphate, aluminium hydroxide and calcium phosphate being preferred for mass production.

The basic material for these vaccines was the virus-containing fluid from chick embryos, inactivated primarily with formalin and phenol. The efficacy of the vaccines was determined either in mouse-protection tests, or by neutralizing or HI– antibody counts in mice or humans. The known side effects were caused by viral lipid toxic reactions or by allergic reaction to chick protein. The protection rate for the prevalent vaccines used in the 1950's was quoted at 60–80% after 1 or 2 injections, providing the freshly produced vaccine contained a suitable virus strain, i.e. a strain homologous to or at least immunologically related to the endemic virus. Since vaccines were only stable for a maximum of 1 year, it was not possible to produce supplies much in advance of use, and in order to expedite production of suitable vaccines as required, the WHO established influenza centres throughout the world. They assumed epidemiological control of the population and undertook the typing new strains using the HI test.

Poliomyelitis Vaccine

Although World War II of necessity had efforts on influenza control, it conversely prevented progress in the battle against poliomyelitis. The situation in the newly founded National Foundation for Infantile Paralysis (NFPI) was depressing: epidemics were becoming more widespread whilst the intervening epidemic-free intervals were getting shorter, and effective defence measures were yet to be discovered. The death of President F.D.Roosevelt in 1945 aroused new interest in the Penny Parade, which made 2.5 million dollars available to the NFIP for research purposes. These were not limited to studies in USA, and it may be recalled that international polio conferences financed by NFIP were motivating forces for virus research in many countries. Since there were no real approaches to effective prevention and therapy at that time, workers began searching for polio links in nutrition. Alas, in vain!

As for development of a vaccine, two basic questions were clear to the NFIP: it was crucial to know if there were different types of polio and, if so, how many. The green light for fruitful cooperation among the scientists sponsored by the NFIP was given at a conference in Baltimore in 1946. Two years later, a committee had worked out a standardized programme with the neutralization test to start polio typing, which required about 30000 monkeys within a 3-year period. Initially, 100 strains from about 250 which had come from every imaginable anatomical source were selected and then tested. A further 96 strains were tested later; 86% of all strains came from the USA. Results published by the committee after 1951, showed three different immunological types: prototype 1 was the Brunhilde strain, prototype 2 the Lansing strain, and prototype 3 the Leon strain. Of the strains examined 82% belonged to type 1, 10% to type 2 and 7% to type 3.

With the introduction of the in vitro culture of poliovirus on extraneural tissue in 1949, and an established cell-culture method in 1952, two basic laboratory methods became available to support further polio vaccine development. Additional encouragement took the form of the antibiotics which had meanwhile emerged, and the demonstration of polio-viremia in the monkey and in humans in 1951–1952.

The NFIP had commissioned Salk to work out a vaccination programme in 1950. Including 10 years of collaboration with Francis, he had been active in the influenza programme during the war and thereafter in polio typing. He was considered to be *the* expert in the development and production of inactivated vaccines. By 1952, in addition to the human trials which Howe had carried out with an inactivated vaccine, there were Koprowski's results with attenuated vaccine. The experience which was collected tended to favour further development of the inactivated form.Taking a lead from Brodie's work in 1935, Salk inactivated with formalin and made use of the knowledge accumulated by Syverton and coworkers in growing all three types of polio in cell culture. R.Parker, from the Connaught Laboratories in Toronto, aided Salk in producing large quantities of poliovirus suspensions and Rhodes assisted with his clinical experience.

In March of 1953, Salk's team published their first results with just 100 volunteers. Whilst the group then continued with small vaccination programmes in order to evaluate several possible vaccination factors, the "Vaccine Advisory Committee" of the NFIP, founded in April 1953, addressed itself to further planes for the "Salk-type vaccine". Although the results of early trials seemed encouraging enough, the opinion that further experiments should be planned in Salk's habitually cautious manner prevailed. Amongst others he considered the possible use of an adjuvant. NFIP, however, decided to carry through a field trial with a Salk-type vaccine in October 1953 which meant vaccinating several thousand children. Sabin and Melnick, who had presented results with their attenuated strains, attempted in vain to have their vaccine included in future planning.

In December of 1953, the plans for the first large-scale field trial had been finalized, with the aim to produce reliable findings on the efficacy and safety of the vaccine. Francis untertook to direct the trial on specific conditions: 600000–700000 children were to be vaccinated; in addition, there was to be a placebo as well as an observation group; grouping would be coded; serological controls were to be documented; and vast sectors of the USA were to be included. The virus suspensions were produced by Connaught Laboratories using monkey kidney cells, with various US firms undertaking inactivation and production of the vaccines. Safety tests were employed to ensure non-infectious virus and the absence of bacteria; batches with suspected contamination were excluded from the trials. On April 12, 1955, the day on which President F.D.Roosevelt had died 10 years earlier, Francis presented a brief report on the success of the field trial. It was based on 1,829,916 children participating in 44 states, 200,745 of whom had received the vaccine. Barely 2 weeks had elapsed, when reports began to appear that several of the vaccinated children had contracted poliomyelitis – a total of 204 vaccinated or contact persons, with 11 fatalities. Recovering from the shock, scientists began to search intensively for the cause. It was found in the inactivation process, which obviously had permitted live virus to remain in virus aggregates and thereby escape detection by the safety tests. It was hoped that filtering after exposure to formol the infectious aggregates would in future be eliminated.

After these field trials were concluded, a vaccination plan was proposed of an initial vaccination of 2 injections with a 14-day interval, a booster shot 4–6 weeks later, followed by refreshing doses annually. Implementation of this vaccination plan, it was reasoned, would confer 90% protection from paralytic poliomyelitis. The effectiveness and safety of the vaccine would be subject to government regulations for production and testing. Through these measures, a repeat of the tragic error of the past was to be eliminated and the risk rendered calculable and minimal. Subsequent large-scale polio vaccination programmes in many countries have confirmed the conclusions from the above field trial in the USA, as far as effectiveness and safety are concerned. The experience gained in the production, testing and use of Salk-type vaccines, which varied only slightly from country to country, served as a basis for the development of inactivated vaccines against other virus diseases.

Yellow Fever Vaccine

Newly developed virus culture methods paved the way for progress in vaccine prophylaxis for yellow fever which was then reaching epidemic proportions. Beginning in 1939, a dry vaccine developed at the Pasteur Institute in Dakar, termed "Dakar vaccine", had been used. The vaccine consisted of desiccated, minced monkey brain which was injected intracerebrally with a neurotropic standard virus. It was found essential to use fresh vaccine from an ampule either for injection or scarification. The latter method required dissolution of the vaccine in gum arabic, which then covered the scarification site with a thin film. Side effects were found to increase proportionally with age and to occur less frequently with the scarification method; protection rate was about 90%.

Lloyd and colleagues adapted the pantropic "Asibi" strain to the chick embryo, terming it "17 D", and to mouse embryonic tissue as "17 E". Dry vaccine was produced from culture medium filtrates using the 17 D strain grown on chick embryo cells minus the brain and spinal cord. Prior to use, the vaccine was dissolved in normal human serum. However, the serum was omitted in 1942, after the discovery that it was often responsible for vaccine-related hepatitis; sterile distilled water or milk was used instead. Passaging the strain 17 D caused a loss in viscerotropic but not in neurotropic properties. The Vaccine was injected subcutaneously or by scarification; in the case of egg-allergy, the first dose was given intradermally, and the second with Dakar vaccine by scarification 1–2 months later. Consistency of the neurotropic and immunogenic properties was tested in the monkey, and the effectiveness of the vaccine, initially determined in a protection test in mice, was found to be 90% of those vaccinated for a minimum of 5 years. Issues addressed during the 1960's had less to do

Table 4.6. Anti-viral active vaccines in 1957

Vaccine	Vaccine type	
	Inactivated	attenuated
African horse sickness		+
Newcastle disease		+
Blue tongue		+
Equine encephalomyelitis	+	
Yellow fever	+	+
Hog distemper	+	+
Fowl pox		+
Influenza	+	+
Fowl plague	+	
Louping ill	+	
Measles	+	
FMD	+	
Smallpox	+	+
Poliomyelitis	+	+
Rinderpest	+	+
Swine fever	+	+
Rabies	(+)	+

(+) Partly inactivated

with the efficacy of the vaccines than with the problem of which particular vaccine to employ to achieve optimal results.

The subsequent development of anti-viral vaccines concerned itself with further significant problems, namely how to increase efficacy and safety while decreasing untoward side-effects. As for live vaccines, it was essential to reassess how each virus strain reacted during adaption to cell culture and how its pathogenicity could be suppressed without sacrificing immunogenic properties. Vaccine strains were expected to be genetically stable and to cause subclinical infection, which in turn would afford long-term immunity. The production of inactivated vaccine required a mastering of selected strong immunogenic virus strains which were generally also highly pathogenic. With time, supplementation with suitable adjuvants appeared feasible in order to increase the efficiency of the vaccine, not to prevent a subsequent infection with the same virus, but to diminish the severity of the disease. Table 4.6 shows the range of active anti-viral vaccines which were either in an experimental stage or already on the market in 1957.

4.1.2 Disease-Producing Insect Viruses

In 1957 approximately 200 diseases with or without cell inclusions were known in Lepidoptera, Hymenoptera, Diptera and possibly in Orthoptera. Butterflies, wasps, houseflies, gnats, crickets, locusts and cockroaches were among the virus-infected insects, and the majority of studies was concerned with inclusion diseases.

4.1.2.1 Inclusion Diseases

Virus-induced polyhedroses were the subject of more experimental studies than granuloses because their size range of 500–15000 nm facilitated staining and observation with a normal microscope. The primary victims of polyhedroses were Lepidoptera and Hymenoptera. The original concept that polyhedra and granula were independent organisms which contained viruses was disproved by Komarek and Breindl in 1924 with their dark-field microscopic studies. Research work on polyhedra by Paillot and Garcia in 1939 and on granula by the Bergold group since 1942, in which serological, biochemical and electron-optical methods were used, left no doubt that polyhedra and granula were the result of virus multiplication in the insect. Both crystal forms were shown to contain large amounts of virus.

It was primarily larvae which were infected, although pupae and adult insects were occasional victims. The typical symptoms of the disease became visible only 1–2 days before death, 5–20 days after infection. Death was not only infection-dose dependent, but was influenced also by the species, age and health status of the larvae. The course of infection for polyhedroses differed from that of granuloses, with various species of insects responding quite differently to the pathogen.

Nuclear polyhedroses were studied primarily in silk moths (Bombix mori), nun moths (Liparis monacha), and gypsy moths (Porthetria dispar). In these Lepidoptera, inclusion bodies were found in the nuclei of most organs, including the blood but not the Malpighian vessels or silk glands. In rare cases, inclusion bodies were also observed in the gonads and in nerve cells. About 4–5 days after infection, it was possible to recognize very small, mobile granulae in the nucleus. In silk moths particularly the granulae were found to form a wide ring around the denser, central chromatin mass. The nucleus became enlarged, often taking up the entire cell volume. Within 2 days, the granulae developed into full-fledged polyhedra, which increased in number in the cell – perhaps to 100 – in proportion to the disappearence of chromatin. In 1952–1953 Aizawa found that in the silk moth the number of inclusion bodies increased logarithmically in the hemolymph, leading to the death of the larvae. As a rule, the larvae and pupae of the moth contained rhombododecahedra, but occasionally cubes and tetrahedrons; tetrahedrons are normal for the nun moth and more irregular shapes in gypsy moths. In the final phase, the cells swell until the skin splits releasing a milky, white fluid containing polyhedra; crystals remaining inside discolour the larvae. Prior to death, the larvae move in circles, finally assuming the typical position of hanging by their forefeet.

As for the Hymenoptera, studies were carried out primarily with the European spruce sandfly (Diprion hercyniae) and the European pine sandfly (Neodiprion sertifer) by the Balch's group in 1952–1955. Infected larvae showed a discoloration between the third and fifth abdominal segment, became sluggish and rejected food. There was often a sticky, dark-brown excretion from the anus, or milky white in the case of the mouth, whilst the larvae hung from the trees. The terminal phase between the sixth and 11th day was similar to that in the Lepidoptera. Internal symptoms included swelling of the nucleus and nucleoli as well as chromatin clumping. The polyhedra in the chromatin were distributed over the entire nucleus. It is typical for these diseases that polyhedra develop only in the digestive cells of the midgut epithelium. Their growth is interrupted during the development of this midgut in the pupal stage, but as soon as the midgut functions in the pupa, polyhedra synthesis resumes. They attain a size of 500–2000 nm, and are always irregular in shape but never round. In 1949 Bird reported on tumour development in the midgut epithelium of D. hercyniae larvae. As soon as midgut development had taken place in the pre-pupal stage, growth ceased. Thereafter, new cells of this part of the intestine enclosed a hard, dark-brown pellet that remained. The full-blown tumour consisted of necrotic, pigmented epithelial cells surrounded by a tenuous tissue of small, growing cells.

With regard to Diptera, only the polyhedrosis observed in crane flies(Tipula palidosa) was confirmed by Rennie in 1923. The infected larvae changed from a clay coulor to white, and numerous, irregularly shaped polyhedra were formed mainly in the nuclei of hypertrophic fatty cells. At first, granules formed in clumps in the chromatin, each surrounded by a ring. Later they rested at the periphery, crowded together in onehalf of the nucleus.

In 1934 Ishimori found cytoplasmatic as well as nuclear polyhedra in the larvae of B.mori, with only the cytoplasmatic type in the cylindrical midgut cells. He concluded that these were two separate diseases. A similar discovery was made in 1941 by Lotmar in the nuclei of various tissues and in the cytoplasm of the midgut cells of the clothes moth (Tineola bisselliella), although he did not assume to two separate causes for the disease. There were many subsequent observations of cytoplasmic polyhedra in Lepidoptera, in Diptera and also in the potato leaf louse (Macrosiphum solani). In 1954 Bird and Whalen examined the larvae of spruce budworm (Choristoneura fumiferana), discovering cytoplasmic inclusions after 72 hours development and thereafter the nuclear inclusions. Using a micromanipulator, they isolated cubic polyhedra in the cytoplasm, ranging in size from 500–2000 nm and present als colonies enclosed in a membrane.

Granulosis, identified by the accumulation of granules 200–500 nm in size, had been found primarily in Lepidoptera before 1957. An exception was the granulosis Weiser had described in the midge (Camptochironomus tentants) in 1948. The following year, Bergold and the Steinhaus group described the course of this infection. Inclusion bodies developed in the fatty cells of the larvae which simultaneously underwent a colour change. The milky fluid which seeped out through the burst skin contained the granules. The inclusion bodies were found not only in the fatty cells, but also in the blood, hypodermis and trachea. In the blood cells they were often found amassed within a membrane in the vacuole. Fusion of the membranes often caused bubble- or capsule-shaped forms to develop, hence the name capsule disease.

To clarify the relationship of various insect viruses, rabbits were inoculated with purified inclusion bodies and viruses, and the immune sera produced were examined in precipitation, agglutination and neutralization tests. By 1957 MacGregor and Bergold were able to demonstrate the serological relationship of viruses in eight different species of Lepidoptera.

4.1.2.2 Non-Inclusion Diseases

In 1913 G.White described a condition in the honey bee which he named sacbrood disease. The small sacs which developed in the larval membrane contained small granules in a watery fluid. Since the larvae died shortly before the pupal stage, and the filtrate was known to be infectious, it was assumed to contain a virus. The virus was found to measure 6 nm in electron-microscope preparations of Wasser and Steinhaus in 1949.

In 1937 L'Heritier and coworkers discovered a disease in vineyard (fruit) flies (Drosophila melanogaster) and named the causative agent *Sigma virus*. The "virus" seemed to be neither infectious nor pathogenic, but was transmitted in the gametes, where it was able to survive in latent form. The only obvious sign of disease in the infected imagines was their carbon dioxide sensitivity. Further publications appeared concerning Lepidoptera and Diptera species, but the viral nature of the agents, though probable, lacked proof.

4.1.3 Virus Diseases in Higher Plants

The discovery that plant viruses were in essence nucleoproteins, represented enormous progress, and it was natural that a wealth of papers on viral plant diseases and their transmission was published before the first general definition of a virus in 1957. Since there was no method for isolating and cultivating single cells of higher plants prior to this time, knowledge of any interaction was limited to the virus and specific tissue. The model of infection in plant virology was the cuticule-covered leaf with its various tissue layers and opening and closing stomata.

4.1.3.1 First Contact With a Virus

The cuticule as such is impervious to microorganisms, but is perforated by innumerable stomata which are surrounded by a single-layered epidermis. The stomata be entry points for virus causing local infections, but from then on penetration of a cellulose cell wall and cell membrane is necessary to infect a cell, and thus the entire tissue via the plasmodesmata. Aided by the cell fluid, the infection may spread from cell to cell. However, this path of a natural local virus infection is merely theoretical, in practice, the way is via damaged outer tissue and cell membranes including those of epidermal hairs, or subsequent to microorganism penetration of the cell.

Natural systemic disease can ensue when a virus enters the vascular tissue pervading the entire plant. This can include the xylem, responsible for water transport, and/or the phloem, through which organic substances pass. Viruses can enter the vascular tissue through the root system or with the help of biting or stinging creatures. Minute

Table 4.7. Phytopathogenic insect viruses and their vectors

Virus	Vector	Detected by
Rice stunt	Nephotettix apicalis[b]	Fukushi 1933
Aster yellows	Macrosteles fascifronts[b]	Kunkel 1924
Clover club leaf	Agalliopsis novella[b]	L. Black 1943
Corn stunt	Dalbulus maidis	Kunkel 1946
Wound tumour	Agallia constricta[b]	Maramorosch 1950
Curly top	Circulifer tenellus[b]	R. E. Smith and Boncquet 1915
Potato leaf roll	Myzus persicae[a]	Stegwee and Ponsen 1958

[a] Aphid, [b] Leafhopper

carrier organisms, termed vectors, were deemed to be primarily responsible for local as well as systemic viral infection. Transmission may be passive, although the organisms, usually arthropods, might also serve as hosts to the virus. This kind of virus was termed a phytopathogenic insect virus. Passive transmission, however, does not simply imply virus attachment to the surface of a body. Between 1933 and 1939, Storey concentrated his studies on this type of vector, specifically on *Corn streak virus* and its vector, the cicada Cicadulina mobila. Storey was able to demonstrate the active function of the arthropod's suction organ in withdrawing the virus from plant tissue. Within one minute's time, the cicada was able to penetrate the parenchyma and phloem with its stylets, and 15 seconds later absorb the cell fluid containing viruses.

More recently discovered vectors include aphids (Aphidinae), leafhoppers (Cicadinae), white flies (Aleurodidae), mealy-bugs (Coccoideae), and thrips (Thysanoptera). The cicadas alone include 73 species known to transmit 30 different kinds of virus. All vectors of phytopathogenic insect viruses are small butterflies (Tortricidae), with just one exception, the peach leaf louse (Myzus persicae), which is an aphid, discovered to be the vector of the *Potato leaf roll virus* in 1958. Initially it was assumed that a symbiosis existed between phytopathogenic insect viruses (Table 4.7) and their vectors. However, in 1956 Virginia Littau and Maramorosch reported pathological changes caused by the *Aster yellows virus* in the fatty cells of its vector, Macrosteles fascifrons: whilst the cytoplasm decreased, the nucleus assumed the shape of a star and the cell membrane became less and less clearly defined. Later, in order to carry out vector-free experimental transmissions, virus suspensions were rubbed onto the plant, preferably the leaf.

4.1.3.2 Intracellular Virus Multiplication

Knowledge in this field stemmed from experimental infections with *TMV* published by Rappaport and Wildmann in 1957. The appearance of a visible focus some 40 hours after infection indicated the involvement of more than 1000 cells in the infective process. Constantly enlarging, the focus contained necrotic cells without demonstrable chlorophyll and was surrounded by a narrow yellow band to which a second similar band of degenerate cells attached itself. The authors calculated that approximately 100 new virus particles arose per cell. Since they found that the cell membranes had merely shrunk but were otherwise undamaged, indicating that they were impenetrable by the virus, it was assumed that the foci spread via the plasmodesmata in the form of infectious nucleic acid. Aside from the foci, the leaf showed no other visible defects. Nevertheless, Stanley had already proved in 1936 that symptoms of diseased leaves were recognizable as early as 4 days after systemic infection. The number of viruses per leaf increased by a factor of 10^7 in that period. Stanley established that only some of the newly synthesized virus particles were capable of infecting other cells.

Virus diseases in plants seemed to be insidious infections, and although some proved fatal, most ran a mild course. Even in the harmless cases, however, plants rarely fully recovered; they took up nutrients and continued to grow but the yield was often negligible or nil. Thus, normal plants were changed into stunted weaklings which, aside from crystalline and amorphous inclusion bodies, showed hardly any histopathological abnormalities.

4.1.3.3 The Spread of Viral Plant Diseases

In addition to the viral diseases transmitted by small organisms, some are caused by the spread of infected plant parts. Infection by seeds containing virus was known in the case of *Lettuce mosaic virus*. Large-scale dispersion by the wind transports parts of plants as well as the insect vectors. It was shown that *Rice stunt*, *Clover club leaf*, *Wound tumour-* and *New Jersey potato yellow dwarf viruses* were able to infect plants via the eggs of their vectors.

It is remarkable that different parts of plants contain significantly different virus concentrations, depending upon the stage of plant development. Thus, viruses were not found in the meristem (undifferentiated embryonic plant tissues) until all mitoses were completed, and tissue growth was due to stretching of the cells and formation of vacuoles. At this stage varying amounts of virus in different types of tissue were no longer found. Differences in the rate of virus synthesis in relation to the age of the leaves, despite decreased protein synthesis with age, were also negligible.

4.2 A Biochemical Approach to the Virus

By 1937 two fundamental facts had been established, namely M.Schlesinger (1934) had proved a nucleoproteid to be the generative substance of the bacteriophage, and Bawden and Pirie (1937) had also determined *TMV* to be a nucleoproteid. At that time, however, these findings on the nature of the virus and its reproduction had hardly any impact. Nor did the results of Breindl and Jirovec or L.Jansen attract even the slightest attention: in 1937, the former had demonstrated nucleic acid in polyhedras of L.dispar using the Feulgen reaction, whilst the latter had shown the nucleoproteid nature of *FMD virus*.

In the biochemical perspective of the 30's all cells of living organisms were in principle be the same: their main constituent, protein, exclusively determined their biological characteristics. Even the characteristic elements of self-reduplicating cells, i.e. nuclei, mitochondria, plastids and centrosomes, seemed to be proteinaceous. Finally, even the fermenting agents (enzymes) which triggered, maintained and completed chemical reactions in the cell were also proteins. In 1923, even though Avery and colleagues deemed carbohydrates to be meaningful for pneumococcal antigen specificity, Griffith linked them to the presence of specific proteins during the completion of his transformation trials in 1928.

The discovery of X-ray-induced mutations by Muller in 1927 served as a strong impulse to probe the molecular genetic structure of cells and the transfer of hereditary qualities. At that time, geneticists were already discussing the possibility that mutation was caused by an alteration in the thymonucleic acid molecule, whereas biochemists still considered mutations to be changes in proteins. Meanwhile, European scientists representing diverse disciplines were beginning to think about reduplication in terms of both proteins and nucleoproteins, which they had already accepted as being genetically decisive factors.

The uniform crystal-lattice structure underlying the homogeneity of *TMV* served to convince Stanley and others the virus was a protein molecule. The 5% of the virus mass containing polynucleotides, which was not discernible in the crystal-lattice structure, seemed of negligible importance and unworthy of discussion. Thus, it is not so astonishing that viruses, which had already been described as nucleoproteids 15 years earlier by biochemists, were still referred to in the literature as crystallizable and autocatalytic protein molecules. Biochemistry viewed the virus not as autonomous particle, but as a ferment, i.e. a product of the living cell. The chemistry, size, molecular weight, morphology, structure, and all of the functions were as yet to be determined, along with, if possible, the interactions with cells, tissues and organs. The results of investigations in these directions were published primarily in chemical and biochemical journals.

The interest in the biochemistry which Stanley had evoked with his *TMV* crystallization led Butenandt to initiate the creation of a German centre for virus research. He recruited the help of Berlin academic and industrial organizations, such as the Kaiser-Wilhelm Institut (KWI) für Biochemie, the Biologische Reichsanstalt, Siemens and Halske A.G., and the Allgemeine Elektricitäts-Gesellschaft, to work toward this goal. The Arbeitsstätte für Virusforschung of the Kaiser-Wilhelm Institut für Biochemie und Biologie became–after several metamorphic changes caused by the war–the Max-Planck Institute for Virus Research located in Tübingen. Another institute, that of colloid research in Frankfurt on Main, relinquished its reputation as a virology research centre when its leading virologist, the Hungarian-born physician M.Schlesinger, emigrated to London. He worked closely with Burnet at the National Institute for Medical Research until his early death in 1936, when viral research lost an imaginative and talented experimentalist.

Virus biochemistry was initially studied in those viruses which could be grown relatively easily and in large numbers, i.e. *TMV* and similar viruses, and some other plant viruses. These were easy to obtain in gramme amounts, whereas only milligrammes of animal viruses were available, with difficulty. Virology expanded rapidly, with new findings on the nucleic acids and other substances encroaching on the field of pure protein chemistry. After 1941 the discovery of HA resulted in an in vitro model, in the form of the influenzavirus, for investigating biochemical problems related to the interaction of a virus and its host cell. Many qualitative experimental results using a broad range of viruses suggested morphological diversity, and led to a distinctive picture of the biochemical structure of the virus in 1957.

4.2.1 The Morphology and Size of Viruses

The first electron-optical image of the rod-shaped *TMV* produced by Kausche and coworkers in 1938 ushered in an exciting new method for studying the morphology, size and structure of viruses. However, the high temperatures and vacuum required for EM were known to deform biological material, and often rendered inaccurate results. With time, the methods for isolating and purifying viruses, the dyes and ways of preparing micrographs were altered, and the techniques and resolving power of EM were enhanced. Several improvements on the early preparation techniques merit a mention. Metallic shadowing, which R.Williams and Wyckoff introduced in 1945, was supposed to prevent the slide preparation from drying out or shrinking, in addition to producing a 3-dimensional image. T.Anderson had been working towards the same goal, and he managed to convert the chemically fixed biological object into liquid CO_2 using high pressure and later remove the gas by increasing the temperature. As of 1948–1949, Heinmetz as well as I.Dawson and Elford developed a technique for examining hemagglutinating viruses electron-microscopically without the need for chemical purification. The use of saponin after HA gave rise to hemolysis, which made it possible to distinguish the virus adsorbed to a transparent membrane, and to count them.

Ultra-microtomy, introduced by Pease und Baker in 1948, was further developed by Newman and colleagues who embedded ultra-thin sections in polymerized butylmethacrylate. These advances enabled Magdalena Reissig and Melnick in 1955 to carry out EM studies on the reproduction of *Herpes virus B* grown in monkey kidney cells. The following year, C.Morgan and coworkers produced similar studies on the multiplication of an adenovirus in HeLa cells.

J.Johnson was able to obtain plant viruses for electron micrographs in 1948 without the hitherto required chemical purification. He applied water pressure to virus-infected stems and isolated virus free of plant material from the water droplets appearing at the edges of the leaves. Several of the small plant viruses, e.g. *Turnip yellow mosaic*, *Tomato bushy stunt*, and *Tobacco ringspot virus*, which had initially

Table 4.8. Examples of virus morphology and size

Virus[a]	Morphology	Size (nm)	Authors
Tob. mosaic	Rod-shaped	15 × 280	T. Anderson, Stanley 1941
Cuc. mosaic	Polygonal	30	Kaesberg 1956
Beet yellows	Filamentous	10 × 1250	Brandes, Zimmer 1955
Silkworm	Rod-shaped	40 × 279	Bergold 1947
Phage T2	Club-shaped		Luria, T. Anderson 1942
	head	80	
	tail	170	
Phage T3/T7	Club-shaped		R. Williams, Fraser 1953
	head	47 × 47	
	tail	10 × 15	
FMD	Spherical	20 - 30	Ardenne, Pyl 1940
Wee	Spherical	50	D. Sharp et al. 1943
Influenza	Spherical and Filamentous	80	A. Taylor et al. 1943
Vaccinia	Brick-shaped	210 × 260	I. Dawson, McFarlane 1948
Mumps	Spherical	180	I. Dawson, Elford 1949
Poliomyelitis	Spherical	27	Schwerdt et al. 1954

[a] Tob., tobacco; Cuc., cucumber; WEE, western equine encephalomyelitis.

been described as spherical using earlier preparation techniques, were then found to be hexagonal or polygonal. Stahmann and Kaesberg made this possible in 1955 by examining freeze-dried preparations, which Kaesberg then improved by using a light, metallic, shadowing technique in 1956. Steere developed a low-temperature method for frozen sections of plant virus crystals in 1957.

In order to clarify the morphology of *Vaccinia virus*, I.Dawson and McFarlane began in 1948 to combine electron microscopy and enzymatic treatment of the object. The micrographs showed an egg-shaped surface thickened in the centre. After applying pepsin, a pepsin-resistant, inner body about 90 nm in diameter was observed.

It is clear that electron-optical methods for determining the morphology and size of viruses varied greatly and often led to inconsistent results. This applied particularly to small, sperical viruses. Therefore, the data in Table 4.8 may be considered only as extract of the many viruses examined, and as indication of the diversity in virus morphology which had been established by 1957.

4.2.2 The Biochemistry of Viruses

Despite the fact that several hundred viruses were isolated in the two decades following Stanley's *TMV* crystallization, relatively few viruses were obtained in adequate numbers or in a form pure enough to permit the chemical analysis required for linking chemical structure to virus function. Most significant, without a doubt, was *TMV*, which was the focal point of biochemical interest in many institutes: At first considered a pure protein, it had within two short years been identified as a nucleoproteid; this earmarks the opening phase of specific, biochemical analysis of viruses.

4.2.2.1 Nucleic Acids

In 1939 Loring, while analyzing *TMV* nucleic acid, found it to be a ribonucleic acid which, though not identical with the RNA prototype, was indeed very similar. The results of inactivation experiments which Stanley published in 1936 and Schramm and H.Müller in 1940, were interpreted as effects on amino acids.

After the Swedish biochemist Caspersson had indicated the genetic importance of thymonucleic acid, Knapp and colleagues in 1939 investigated the influence of monochromatic UV irradiation on plant and animal germ cells at the KWI für Züchtungsforschung und Strahlenforschung. Their intention was to determine if either protein or thymonucleic acid alone produced a genetic effect, or only in conjunction with one another. They reported that the thymonucleic acid exercised a decisive effect, producing genetic as well as physiological changes. Furthermore, they assumed the presence of an additional "actual genetic substance", which in their estimation could only be a protein.

In 1940 Pfankuch had used alkali to split the nucleic acid from two different strains of *TMV* and had thereby obtained low molecular weight fragments which behaved similarly under electrophoresis. He concluded that *TMV* nucleic acid determines the biological *TMV* differences like the action of active enzyme groups. In order to learn more about the concurrence of proteins and nucleic acids, Schramm separated *TMV* nucleic acid using nucleotidase from the small intestine of calves. He obtained a nucleic acid-free protein which crystallized in the same manner as the untreated virus. He verified that the virus remained unchanged, indicating that loss of infectivity was due to removal of the nucleic acid, but he never maintained that the nucleic acid was the sole genetic agent. He was rather more eager to find out just how nucleic acid was linked to protein since "the presence of nucleic acid is absolutely essential for the reproduction of protein."

That nucleic acid itself was of paramount importance was to be repeatedly demonstrated in subsequent experiments with various microorganisms. Thus, Avery and colleagues reported that pneumococci were able to encapsulate after the transference of minute slight amounts of extracted DNA, a manipulation called transformation. Markham and K.Smith isolated and crystallized the *Turnip yellow mosaic virus* in 1949. Despite its homogeneity under electrophoresis, sedimentation studies revealed the presence of two different components: the heavier fraction comprising 70–80% of the material, contained 37% RNA, and the other fraction was just protein. The protein was the same in both sedimentation fractions, but only the part containing the RNA was infectious. Markham concluded in a lecture held at Oxford in April 1952, that the RNA was responsible for controlling virus multiplication. Proof of the reproduction function of DNA was provided in 1952 by Hershey and Martha Chase with their *Phage T2* studies. They infected E.coli with phages whose protein and DNA had been labelled, and found that about 80% of the phage protein remained outside the bacterial cell while at least 65% of the DNA entered. At least 30% of the parent DNA, but less than 1% of the parent protein, was present in the newly formed phage.

Similar support for virus RNA as the agent of reproduction was generated in 1956 by two groups working simultaneously on *TMV* using different methods: Fraenkel-Conrat in Berkeley, and Gierer and Schramm in Tübingen. Thus, the first crystallization of a virus prompted scientists from various fields to focus on the significance of nucleic acid as the carrier of genetic information in all living organisms, including as yet unclassified viruses. The more certain the role of nucleic acid became, the more frequently experiments emerged which proved that viruses were actually nucleoproteids. As of 1956, the experimental evidence supported the hypothesis that viral nucleic acid belonged either to the DNA or to the RNA type. By 1957 many plant viruses had been identified as nucleoproteids, but since only the ribose type of nucleic acid (RNA) had been found in the plants studied, it was assumed that RNA was characteristic of all plant viruses. In fact, the RNA content of ten tested viruses ranged from 5–35%. The apparently exclusive presence of RNA in plant viruses supported Caspersson's and Brachet's assumption that RNA was the determinant of protein synthesis. Working independently between 1942 and 1950 with histochemical techniques, they found a particularly high RNA content in cells which showed active protein synthesis for growth or secretion. Once the double-helix DNA structure had been established, J.D.Watson (1953) formulated the central dogma of biology which, after years of discussion, would finally define the control of protein synthesis. According to biochemical and virological findings, RNA was intricately involved in protein synthesis and thus DNA could no longer be considered the sole template for the protein. Hence, it was assumed that the path of protein synthesis led from DNA via RNA.

In 1942 Smadel and Hoagland as well as Beard established that the *Vaccinia virus* was a nucleoproteid of the deoxyribose type (DNA). Viruses of this type included the *Shope papilloma virus* (A.Taylor and coworkers, 1942), the *Gypsy moth polyhedra virus* (J.Smith and Wyatt, 1951), and *Phage T2* (Herriott and Barlow, 1952). The nucleic acid content was found to range from 1% RNA in the influenzavirus to 50% DNA in *Phage T2*. In 1952 Wyatt and S.Cohen discovered a peculiarity in the T-even phages of E.coli: 5-methyl-cytosine was replaced by 5-(hydroxymethyl)-cytosine. By 1953 this was no longer unusual in that *Phages T5* and *T7*, as well as the *Colias philodice*, *Vaccinia-*, and *Meningo-pneumonitis viruses* had been added to the list.

Publications began to appear in 1942 reporting the ratio of nucleic acids in virus particles as well as determination of particle and nucleic acid weight. Since various different methods were employed for purification, nucleic acid isolation and nucleic acid content determination often marked disparities between reports. For example, between 1942 and 1957 the molecular weight of *TMV*-RNA ranged from 0.2 to4×10^6 giving a mean of 2.4×10^6. The values given in Table 4.9 reflect this kind of deviation.

Interpretation of the structure of viral nucleic acid was derived from the observations on *TMV*-RNA which seemed to be rather simple: it was thought to be an unbranched polymer chain consisting of some 7000 mononucleotides. Based upon the known RNA content of *TMV*, it was assumed that nucleic acid of the small and middle-sized viruses was contained in a single molecule; large viruses were thought to divide their nucleic acid among a few molecules. Two other relevant observations were the multiple but similar *Vaccinia virus* DNA molecules, and the two dissimilar DNA molecules in *Phage T2*. Thus, the type and content of nucleic acid seemed to be characteristic for a given type of virus.

Table 4.9. Nucleic acid contents, particle and nucleic acid molecular weights

Virus	Nucleic acid content (%)	Particle weight (× 10^6 daltons)	Nucleic acid weight (× 10^6 daltons)
TMV	8.7 RNA		2.5
EEM[a)]	4 RNA		100
Influenza	1 RNA	280	2
Fowl plague	3 RNA	150	
Newcastle disease	4 RNA		
Poliomyelitis	26 RNA	6.7	
Rabbit papilloma	8.7 DNA	47	4
Vaccinia	5.6 DNA	3200	

a) EEM, equine encephalomyelitis.

4.2.2.2 Proteins

In 1936 Stanley as well as Wyckoff and colleagues investigated various *TMV* strains found in several different kinds of plant. They obtained virus proteins of molecular weights ranging from 30000 to several million – far higher than the known weights of plant proteins. The virus strains differed in their protein compositions. In 1937 Wyckoff found that *TMV* protein was a uniform substance which Schramm and H.Müller calculated to have a molecular weight of 23 million. They succeeded in splitting this macromolecular protein into pieces of differing weight. Later Knight (1947) identified 17 amino acids in the influenzavirus and discovered differences in the amino-acid composition of the A and B types. R.Williams and Fraser differentiated the *Phage T2* head and tail proteins in 1956. There were other indications the proteins might be bound to carbohydrates and lipids; furthermore, virus-specific proteins probably existed in soluble form, notably in the myxoviruses, and the *Vaccinia* and *Equine encephalomyelitis viruses*.

4.2.2.3 Virus Enzymes and Other Substances

Many publications claimed the existence of viral enzymes, but as these enzyme components were found exclusively in complicated viruses, it is not surprising that repeated experiments defined only three enzymes: the influenzavirus enzyme originally described by Hirst in

1941 and determined to be a mucinase by Burnet and colleagues from 1947–1951; adenosine triphosphatase (ATPase) isolated by the Mommaerts group in 1952 from the *Avian erythromyeloblastic leukosis virus*; and the lytic phage enzyme found by Sertic in 1929 and later termed phage lysozyme. Eager to assess the biochemistry and mode of action of mucinase, the three groups working with Eckert, Blix, Gottschalk and E.Klenk carried on the approach of Burnet and coworkers. Between 1946 and 1947, the latter had discovered receptor-destructive action in culture filtrates of the Vibrio cholerae similar to the enzymic activity of influenzaviruses, and termed the active viral substance receptor-destroying enzyme (RDE). In 1948 Brody demonstrated that heat-inactivated influenzavirus was able to bind to erythrocytes - called hemagglutination - but could not be eluted. Mucinase was identified to be neuraminidase, thus creating the corresponding term N-acetylneuraminic acid for the influenzavirus receptor. This, together with ON-diacetyl- and N-glycolyl-neuraminic acid, was generally named sialine acid.

Similar uncertainty was encountered with the lipid and carbohydrate components of the more complex animal viruses, which exhaustive research proved to be definite components of a virus. Lipid proportions ranged from 1,5% in the *Shope papilloma virus* to 54% in the *Eastern equine encephalomyelitis virus*, with mean values of 25% determined in the influenzavirus and the *Fowl plague virus*. The lipids included were phospholipids, neutral fat and cholesterol. A polysaccharide component was found in the influenzavirus by A.Taylor in 1944 and confirmed by Knight 3 years later. Besides the nucleotide ribose, mannose, galactose, fucose and glucoseamine were also present. Rather strangely, 0,05% copper as well as trace amounts of biotin and flavin were discovered in the *Vaccinia virus* by Smadel and Hoagland in 1942.

Generally speaking, in 1957 it was assumed that plant viruses and small animal viruses consisted of RNA and protein, whereas bacteriophages and larger, more complex animal viruses, including insect viruses, were of the DNA type.

4.2.3 The Structure of Viruses

At that time, the known structure of viruses was solely in terms of the analysis of the chemical components. And, by analyzing the arrangement of these substances, scientists attempted to explain the functions of the separate "structures". All the data available from chemical analysis and the pictures obtained by different methods were pooled in order to derive a concrete image. Microspectrophotometry using stained material, e.g. with the Feulgen reaction for DNA and Schiff's reagent for glycoproteins, greatly aided structure confirmation.

Since the morphology of virus types proved to be just as diverse as their chemical composition, it would seem advantageous to consider each type separately. As initially the only virus available for structural analysis was *TMV*, we will begin with the plant viruses.

4.2.3.1 Plant Viruses

In those early days, *TMV* was by far the most thoroughly investigated plant virus. In 1936 Wyckoff and Corey began using X-ray crystal lattice structure analysis for studying *TMV* protein crystals. They found them to be unaltered by even nine recrystallization cycles and complete inactivation by UV irradiation. With the help of X-rays, Bernal and Fankuchen (1941) showed that individual *TMV* rods were composed of regular subunits. In 1943 Schramm split *TMV* into pieces using a slightly alkaline solution, and found that as soon as the pH shifted back to acidity, the pieces recombined to assume their original form. However, both the pieces and the reaggregated forms were incapable of reproduction. During further experiments in 1955, Schramm, G.Schumacher and Zillig established that the rods really consisted of slices. These were 5–10 nm thick with a hole at the centre of diameter 3,4 nm, which corresponded to the diameter of the nucleic acid thread. They assumed, therefore, that the *TMV* slices were strung together on a thread of nucleic acid. In 1955 the arrangement of protein and nucleic acid components for reassembling infectious *TMV* was demonstrated by Fraenkel-Conrat and R.Williams at pH 6 using freshly isolated *TMV*-RNA under specific conditions.

Caspar's and Rosalind Franklin's EM images showed that the RNA thread was evidently not in the middle of the rod, but wound in a spiral around the central opening, surrounded and protected by the uniform pieces. Thus it was clear that *TMV*-RNA, and in parallel RNA infectivity could only be obtained by breaking the structure of the virus rod. Synthesis of the virus required both protein and RNA under defined conditions.

4.2.3.2 Enteroviruses

The term enteroviruses applied to the three types of *polioviruses*, 19 group A and 5 group B *Coxsackie viruses*, and 24 types of Echo viruses[1]. The poliovirus, which was the subject of the most intensive investigation, will serve as the representative of this group.

Although the first EM images of poliovirus were presented by Tiselius and Gard in 1942, it required an additional 12 years, i.e. until 1954, before an unequivocal image of the structure was obtained by the use of cell cultures. In the earlier years, organs from various sources were the only material available for analysis, and it was not possible to prove that the poliovirus was a pure nucleoproteid, or to produce consistent results on virus size and morphology. Interpretations ranged from threads measuring 15 nm to round particles of 50 nm diameter. It was not until 1954 that convincing results began to appear in the publications, for example of Bachrach, Schwerdt and Schaffer. This literature cited the size of an infectious, spherical, crystallizable poliovirus to be between 27 and 30 nm. According to A.Taylor and McCormick, the 24% RNA content of the virus was within the so-called core, surrounded by a protective layer of proteins.

4.2.3.3 Complex, Medium-Sized Animal Viruses

The influenzavirus and the classical *Fowlpox virus* are typical of this group. Proof of how closely these viruses were related began emerging from Schäfer's institute in 1950, with the help of Schramm, Munk, Sigrid Ullman and Zillig, with the use of biochemical and serological methods. The diameter of these myxoviruses measured about 100 nm. The composition of the influenzavirus was supposedly 1% RNA, 18–36% lipids, 5% carbohydrate and 70% protein. Part of the structural analysis was an enzyme, conspiciously adhering to the surface of the virus, which was able to destroy mucoprotein cell receptors. All of these components were presumed to be inside a round capsule, although there had been many reports of thread-like forms, as well as threads with sperical ends and other diverse arrangements.

---

[1] *e*nteric *c*ytopathogenic *h*uman *o*rphans

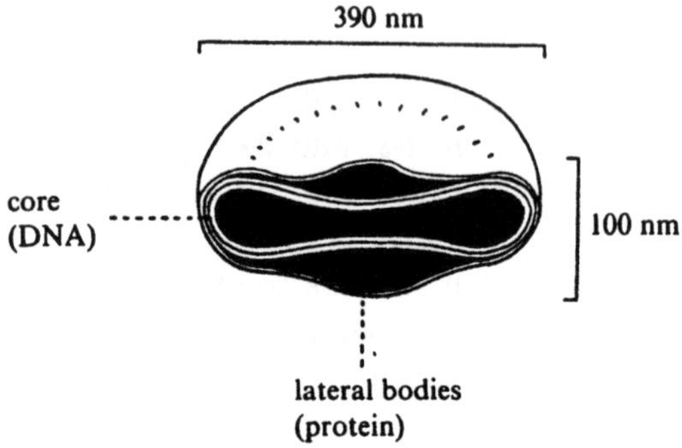

Fig. 2. Vertical transection of a Vaccinia virus

4.2.3.4 Complex, Larger Animal Viruses

The prototype for this group was the brick-shaped *Vaccinia virus*, with a composition of 5,6% DNA, 89% protein, 5% lipid, and 5,6% carbohydrate. In addition, F.Nagler found an HA component in 1942 which, 4 years later, J.Stone, identified to be a phospholipid not belonging to the inner part of the virus. In 1952 D.Peters continued the enzymatic analyses of the *Vaccinia virus* which I.Dawson and McFarlane had begun in 1948. His EM images showed a complex, pepsin-resistant, ring-shaped centre surrounded by several layers of nucleoproteids. Fig. 2 shows the structure of suspended *Vaccinia virus* constructed by Peters and Stoeckenius in 1954 using differential centrifugation.

4.2.3.5 Bacteriophages

As early as 1940, when H.Ruska made his first EM images and 2 years later, after Luria and T.Anderson had produced theirs, the first phage – E.coli *phage T2*[1] – was clearly recognizable as a complex particle. *T2* was then the most thoroughly investigated phage and a great part of our knowledge about phages relates to it. The cylindrical head could be differentiated from a thread-like tail, and the phage was found to contain 47% DNA, 51% protein and 2% lipid. At least 95% of the protein was in the form of the ghost (the sheath of the head), the remainder being soluble protein connected with the DNA in the core. In 1956 R.Williams and Fraser produced electron micrographs showing that the phage tail also had a protein-filled core.

[1] T for type

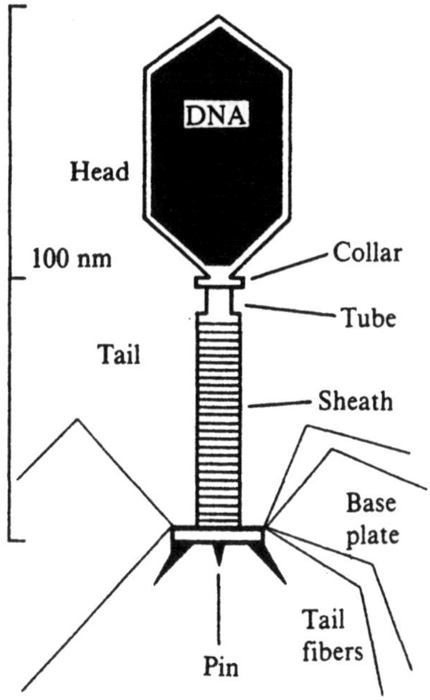

Fig. 3. Bacteriophage T2 model

In 1952 Hershey and Martha Chase were able to obtain evidence that during the infection of a bacterium, the protein shell of the virus remains outside the host cell, i.e. only the content of the head is transferred. After T. Anderson (1951) had demonstrated, with the help of 3-dimensional electron micrographs, that phages T2, T4 and T6 contacted target cells only with the tips of their tails, it was assumed that the adsorption mechanism consisted of an adhesive structure located at the end of a possibly contractile tail. W. Weidel and Kellenberger confirmed the presence of this mechanism in *Phage T5* in 1955. Kellenberger and Arber succeeded in separating *T2* and *T4* phages by oxidation to reveal their structural constituents. They also confirmed Hershey and Martha Chase's 1952 hypothesis whereby the release of DNA into the bacterium corresponded to an injection. This was accomplished because the tail of the phage included fibres for attachment and a thin internal tail pin which pierced the bacterium. Figure 3 shows the presumed structure of *Phage T2*.

Judging from the structures of various viruses proposed by 1957, it was clear that nucleic acid was contained in the central core, perhaps connected by proteins and surrounded by a protective shell. The surface of the virus was assumed to have certain chemical and morphological structures important for attachment to the surface of the bacterium.

4.3 Viruses as Genetic Macromolecules

The development of the rabies vaccine had already shown that virulence was alterable, and that it could be stabilized in the modified form and became heritable. The first attempts were undertaken in 1936 to adapt a pantropic strain of yellow fever, Asibi, to a stable neurotropic strain 17 D, to prove that viruses could undergo variation and mutation. At that time, however, only very few scientists regarded viruses as living organisms, and these few experimented with bacteriophages, which were considered to be neither viruses nor viable organisms. During the 1930's, there was a growing wave of certainty that viruses including bacteriophages, contained genes in their nucleic acid, as did the chromosomes in the more highly developed organisms. Thus, for those staunch supporters of the concept of viruses as microorganisms, it was then a matter of developing suitable experimental models in order to launch genetic or molecular genetic experiments to prove their point.

Prerequisite to this were defined and heritable traits which could be objectively assessed phenotypically, physico-chemically or immunologically. These demands were met for testing bacteriophages, but were only gradually fulfilled for plant and animal viruses. It was also essential to have pure virus preparations as well as suitable testing systems. The local lesion test had already been devised for plant viruses; genetic experiments did not present insurmountable difficulties for animal viruses, provided they could induce single lesions on the CAM or could be analyzed in the HA test.

4.3.1 The Genetics of Bacteriophages

Acute staphylococcus infection had been characterized by the English bacteriologist Twort in 1915, and the discovery of a similar "lytic principle" had been made by the Canadian bacteriologist d'Herelle in 1917 while investigating the dysentery bacillus at the Pasteur Institute in Paris. A third and new group of causative agents termed bacteriophages, or simply phages, was subsequently added. Soon phages were being found in Vibrio cholerae, typhoid bacilli, and E.coli, and hopes began to stir that they might also provide a therapeutic means of combating certain infectious diseases.

As for the nature of the pathogen, Twort and others still held to the concept of an enzyme. D'Herelle, who before 1926 obtained experimental results which served as a basis for his later investigations with the phage, decided rather early on that the phage was a corpuscular virus on or inside the bacterium. He described phages as being lytically active, i.e. during acute infection they kill their host cell by bacteriolysis in the final stages of their intra-bacterial multiplication. In 1929 Sertic was able to demonstrate how phage multiplication in bacterium led to synthesis of a lytic enzyme which partially dissolved the cell wall. The interaction of this phage lysozyme and the increased pressure caused by phage multiplication eventually burst the cell. Barrington and Kozloff (1954) showed that after phage adsorption to the cell wall during infection, the enzyme dissolved the wall in a punctated manner which facilitated or triggered the release of phage DNA into the host cell.

D'Herelle regarded the different types of phage to be variations of a single species *Protobios bacteriophagus*. As early as 1922, however, Bruynoghe and Appelmans had described two typhus phages which could be differentiated antigenically, and serological parameters later served as a basis for classifying phages. A second, non-infective phage, which may be carried by bacteria throughout generations, was discovered by Bordet and M.Ciuca as well as by Gildemeister in 1921. Since the cultures they used permitted the production of lytic phages, they were termed lysogenic. In 1922 a geneticist, Muller associated phage genes and chromosomes and concluded that the bacteriophage was the model for solving the gene problem. Nevertheless, it took almost 15 years before Burnet and Dora Lush were able to produce a phage with a mutated plaque morphology from a lysogenic culture. Meanwhile, d'Herelle had described plaques in 1926–the taches vierges– and proved that one plaque on the agar surface represented one bacteriophage, i.e. its offspring in the inoculated suspension. Later investigations showed, however, that multiple infections as well as mixed infections were possible, whereby "multiple" meant the invasion of several phages of the same type, and "mixed" applied to phages of different types multiplying within a given bacterium.

In 1929 Burnet confirmed d'Herelle's 1926 assumption that phage multiplication within a bacteria population is an intracellular step-by-step process which exhausts itself when all bacteria have been dissolved. Burnet succeeded in demonstrating this in his single-burst experiment, which enabled counting the actual number of offspring of a single bacterium. To execute this, he infected numerous small aliquots with, whenever possible, a single phage. In predetermined intervals, he transferred samples to culture plates, incubated them overnight and then ascertained the number of plaques per plate. In samples taken within the first hour, only 1 or 2 plaques formed, whereas more appeared in the later counts.

In 1929 Burnet and McKie proved that every cell in a lysogenic culture carried a phage in a non-infectious form, but that in such a culture few bacteria produced lytic phages. Later (1953) A.Lwoff termed these two different responses of bacteria to phage infection lysogenic and lytic (or lysogeny and lysis). Two years later, Den Dooren de Jong experimented with sporogenic strain 899 of B. megaterium from which he isolated a single phage. A non-sporogenic, non-lysogenic offspring of strain 899, termed *mutilat*, proved sensitive to the isolated phage, and Elisabeth and Eugéne Wollmann (1936–1937) used strain 899 to test the difference between lysis and lysogeny. They concluded that there was an infectious and a non-infectious alternation in the life-cycle of a phage. Whereas the non-infectious phage is part of the genetic structure of the bacterium, the infectious phage develops by means of, or from the non-infectious type. A phage which was potentially both was termed a temperate phage.

Interest in lysogeny diminished considerably when Delbrück and Luria, amongst others in the USA, initiated bacteriophage research based on genetic and molecular biological principles using lytic cultures. The year 1943 saw the termination of the research in lysogeny of Elisabeth and Eugéne Wollmann and, temporarily, of the Pasteur Institute in Paris. The successful work of the Wollmanns was brutally stopped off by their deportation to Auschwitz, where their lives were meaninglessly extinguished. A former colleague, A.Lwoff was able to continue their work from 1950.

Meanwhile, lytic phages began to provide new genetic and microbiological insights, increasingly and scientists began to see a closer relationship between bacteriophages and viruses. Thus, Ellis and Delbrück (1939) defined two further states of the bacteria-bacteriophage system in a standardized one-step-growth experiment. This made it possible to determine the burst-size as well as the latent period, i.e. the interval between adsorption and bacteriolysis. The duration of the latent period was found to be dependent upon the temperature as well as the host-cell system. Burst-size was independent of the temperature and the number of infecting phages, though dependent upon the phage strain, host cell and test conditions. Single counts of the progeny ranged from just a few to more than 1000 phages, while the mean burst-size was about 100. In 1957 Kellenberger and Arber devised a collodion-agar-membrane method which permitted assessment of the burst-size with the EM as easily as by counting the plaques.

The one-step-growth experiment was then regarded as the general, basic method for assessing phage multiplication, and later generally for virus multiplication. Ellis and Delbrück mixed growing bacteria with an excess of phages such that theoretically every bacterium in the culture would be infected by at least one phage. The phages adsorbed to the bacteria during an incubation period of a few minutes, then the mixture was diluted 1:40 in a medium which contained anti-phage serum to neutralize the non-adsorbed phages. After several minutes, a further dilution of 1:250 was made with fresh medium in order to discontinue the neutralizing action of the serum. This neutralized, 1:10000 dilution of infected bacteria, in which step-by-step multiplication had been eliminated, permitted only a single multiplication cycle. This method was the cornerstone of further plaque test experiments. In genetic experiments, attempts were made to combine phenotypic characteristics with genetic structures by simple mutation or by recombination with phage strains which differed in their phenotype. Both experiments required phage clones. Cloning was achieved by re-isolating single plaques in a double transfer.

Simple spontaneous or induced mutations were found on single plaques when the layer of phage-sensitive bacteria was covered with a layer of treated or untreated phage suspension. Non-adsorbed phages were removed by suction and the agar rinsed. The mutations related to plaque shape, host-cell spectrum, temperature sensitivity, sensitivity to irradiation and chemical substances, and required time for lysis. Other parameters measured were the mutation frequency, i.e. the number of mutants per replication, or the mutation index, i.e. the number of mutants in a phage stock.

Recombination experiments with the two related E.coli phages *T2* and *T4* in a single cell were begun in 1945–1946 both by Hershey and by Delbrück and Bailey. After mixed infection in a single cell, both phage characteristics occurred in a single burst, suggesting either multiple infection, or the possibility of phages exchanging their genetic material. In 1949 Dulbecco determined the maximum number of infective phages to be 10–30 per cell. Since M.Schlesinger had assessed the adsorption capacity of E.coli to be 300 for his phage strain 88, this meant that the adsorption capacity was far greater than the infection capacity.

The 1948 publication of Doermann aroused great interest in biochemistry and genetics circles. He discovered that due to a superinfection prior to the end of the latent period, lysis could be postponed for up to 60 minutes. As a result of this delay, the average burst-size, which usually remained constant in a normal mixed or multiple infection, turned out to be many times greater.

Luria (1951) unravelled the enigma of lytic phage genome reproduction by explaining the spontaneous occurrence of lysis mutants which altered the shape of the plaque. He had had to decide between one of three possibilities assumed in 1945: Firstly, there was geometric bacterial cell division; secondly, the follow-the-leader type in which only one of the offspring is triggered to duplicate; thirdly, the stamping-machine model whereby the parent genome supplies the matrix for the press to manufacture hundreds of offspring genomes by independent "stamping" acts. From the measured frequency distributions of mutant clones, Luria concluded that all three formerly-assumed multiplication possibilities should be discarded. It was true that phage reproduction occurred geometrically, though not by doubling but by increasing 100-fold! This finding was based on the 2874 agar plates and 1,850,000 plaques he assessed.

The task was still to determine exactly how multiplication occured. The earliest findings on the course of intra-bacterial reproduction had been presented by Doermann in 1952 with *Phage T4* (Fig.4). During the entire latent period, he had tested the intracellular condition of the phage by means of artificial lysis. Within the first 10 minutes post-infection, not a single infectious particle could be found. Thereafter, infectious phages gradually entered a 15-minute phase of exponential growth. The period lacking intra-bacterial infectious phages was termed the eclipse.[1] Doermann's experiments confirmed the findings of the Wollmann's in 1937 with the phage of B.megaterium. The unknown factor that now remained was the fate of the bacterium during the eclipse phase and the ensuing latent period.

Subsequent studies revealed that phage substances were already being synthesized during the eclipse phase, and that morphologically intact infectious phages were developing from prefabricated parts during the final or maturation phase. Using an immune-specific precipitation test in 1951, O.Maaløe and Symonds detected the synthesis of new phage protein several minutes before the eclipse phase had ended. About 20 minutes post-infection, the amount of protein sufficed to form 40 phages. The protein level was then

[1] Gr. ekleipsis: an abandoning of sun or moon

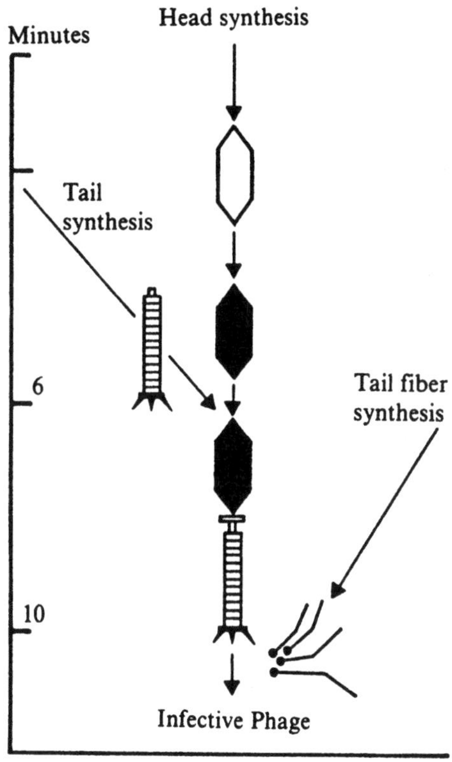

Fig. 4. Bacteriophage morphogenesis

maintained throughout the remaining latent period in a protein-pool for developing diverse incomplete forms of phage. Several of these forms had been observed with the EM by Wyckoff (1948) and Hercik (1950). Levinthal and H. Fisher had discovered tail-less heads or so-called "doughnuts" completely void of DNA. R. Williams and Fraser, who had succeeded in taking *T2 phages* apart after freezing and thawing, found intact heads, intact tails, head membranes, parts of tails, cores of tails, single fibres and fibres attached to tails. Whereas doughnuts appeared during the eclipse, tails were found afterwards. They remained in a fibreless form as a constant reservoir of 30 per bacterium ready for phage synthesis, according to Kellenberger and Janine Séchaud.

As for DNA, S. Cohen (1947) showed that its synthesis was blocked by the bacterium immediately after infection only to resume more intensively 5–6 minutes later. Neither DNA- nor incomplete protein-DNA forms were observed. Free, non-infectious DNA, however, existed prior to the end of the eclipse as a so-called vegetative phage, available in amounts which could constantly produce about 80 phages during a 40-minute period. In 1953 Stent and Maaløe discovered that the first mature phages synthesized consisted of 60% host-cell DNA. During the course of the latent period, the proportion of host DNA constantly decreased. This meant that there was a definite, phage-specific DNA and protein reserve in the bacterium before the latent period ended that could serve as a precursor pool for continuous reproduction of 50–100 phages.

In 1948 S.Cohen introduced radioactive marking to clarify the function of nucleic acid during phage reproduction. Additional insight was afforded in 1953 by Kozloff, as well as by Stent and Maaløe and, the following year, by Hershey's group. According to their findings, the bacterium contributed the most to the DNA pool, its DNA splitting into small fragments from which phage-specific polynucleotides were formed. Thus, 85% of the bacterium-DNA could be traced in the phage. On the other hand, the bacterium furnished little or nothing to the protein pool, as Kozloff and coworkers determined between 1951 and 1952. Hershey and colleagues confirmed in 1954 S.Cohen's 1947–1948 findings that the scope of bacterial protein synthesis was not influenced by infection. What did take place during the latent period, however, was that the rate of change to phage protein accelerated. Little happened in the first 5 minutes but, within the next 10 minutes, 60–70% of the synthesized protein became intensively engaged in phage reproduction. Exactly how precursor proteins were incorporated into the structure of a phage was not known in 1957. The sequence of steps involved in phage synthesis is shown in Fig.4. According to Burton's investigations in 1955, DNA and protein synthesis functioned independently following a phage infection.

The maturation process of a phage was also still a mystery. Microscopic investigations had convincingly shown that only a complete, mature phage could be infectious. Earlier it had been assumed that the head attached itself to the tail after the precursor DNA had been introduced. Findings by Maaløe and Stent in 1952 then showed that the head filled with DNA only in the final stage of phage synthesis. It was known that within 2 to 3 minutes of E.coli infection by a *T-even phage* its nucleoid disintegrated into small fragments which accumulated at the cell wall. Pardee and I.Williams (1952) presumed the fragmentation to be effected by an early phage enzyme. They had discovered that nucleoid disintegration did not occur in bacteria which did not sythesize this phage-specific enzyme.

Phage genetics was greatly enhanced in 1950 when A. Lwoff and Gutmann resumed their lysogenic experiments. At first they needed to determine whether only a few bacteria produced infectious phages in a lysogenic culture, or whether the infectious phages were secreted like enzymes from the lysogenic cells. The former standpoint was supported by Burnet and McKie, while Northrop held to the latter. Lwoff and Gutmann managed to observe microscopically 19 divisions of a lysogenic cell in a micro-droplet. Using a micromanipulator, they transferred a daughter cell from each division to agar and tested for lysogeny. In a parallel study they inoculated the strain mutilat with liquid taken from the microdroplet to test for free phages. Thus, they demonstrated lysogeny in the absence of free phages in 19 consecutive generations.

Subsequent experiments led them to conclude that the cells of a lysogenic culture carried a non-infectious phage structure called a prophage. This equipped the cell with the capacity to synthesize an infectious phage in the absence of any exogenous phage. A part of the growing lysogenic bacteria was spontaneously induced to produce infectious phages thereby causing the lytic death of the cell. Thus, the three possible explanations of a temperate phage were:

1. The non-infectious prophage undergoes synchronized divisions with the bacterial genome into which it was integrated.
2. The vegetative phage, which is non-infectious, multiplies independently of the bacteria genome.
3. The infectious phage develops from the vegetative phage.

Lysogenization, therefore, meant the conversion of an infectious, but resting phage to a prophage. The lysogenic character of a culture was generally stable, although this might be lost through the sudden occurrence of spontaneous or induced mutations.

E. Miller and Goebel (1954) showed that lysogenic bacteria did not contain the antigen of the homologous, i.e. the integrated phage. According to Stent and colleagues prophage DNA was identical to homologous phage DNA. Therefore, the prophage was, in reality, a DNA problem between the phage and a bacterium.

Further findings included:

1. A defined phage had a designated position in the genome of a bacterium.
2. Different types of phage had distinct positions in the genome of the bacterium.

3. A prophage did not displace a gene in the genome of the bacterium; it added its own gene to the genome.
4. Lysogenic bacteria were resistant to their homologous phages. Homologous genetic phage material did not multiply when it entered the bacterium; it gradually disappeared in the course of bacterial division.

When taking a closer look at lytic and lysogenic phages, it is clear that both are genetic systems. Lytic and temperate phages can multiply intracellularly and thereby kill the cells, as other viruses do. Prophages, which have been incorporated in the cell genome, contain a combination of the cell and phage genome. All the mentioned genome states can be altered spontaneously, i.e. without a perceptible cause, or induced by external factors, such as UV irradiation and chemical substances.

4.3.2 Plant Virus Mutants

Ever since Pasteur's time, attenuation has been a recognized method for changing hereditary characteristics in animal viruses. In 1927 Muller demonstrated the possibility of inducing mutations in plants and animals by chemical and physical means.[1]. Experiments with plant viruses and the same goals had been recorded since 1924–1925.

At that time, Carsner and Stahl reported attenuation of the *Curly-top disease virus*, a causative agent of disease in sugar-cane plants. Attenuation was achieved by passaging the virus in the nettle-leafed goose-foot (Chenopodium murale). In 1926 McKinney described yellow spots on the leaves of tobacco plants which had been infected by the causative agent of tobacco mosaic disease. When yellow-spotted leaves came into contact with healthy, green ones, these too developed the disease condition.

Experiments concerned with chemically induced, plant virus mutants were initiated by Ross and Stanley in 1938 using formaldehyde on *TMV*. On the one hand, scientists were interested in inducing mutants, whilst, on the other hand they were eager to know whether viruses which caused the same or similar symptoms in plants were somehow related. When viruses effected corresponding or identical

[1] Mutagenicity in bacteria caused by UV irradiation had been described by Henry as early as 1914.

symptoms in different test plants, these were referred to as strains, variants or mutants. When effects were macroscopically dissimilar, the tested viruses were considered to be different species. In 1947 Kunkel counted more than 400 strains of *TMV*, while about 300 plant viruses had been described as species by 1957. In order to differentiate strains from species, various biological as well as chemical methods were used. Kunkel examined the lesions which had developed on leaves to determine *TMV* mutation frequency in 1940. After retro-inoculation in healthy plants, he found that one mutation occurred in approximately 200 lesions. This was a considerably higher rate than Luria had calculated for bacteriophages, i.e. $1{:}10^3$–$1{:}10^9$. In 1941, F.Holmes evaluated the symptoms of disease to establish the difference between strains and species.

A less suitable method for distinguishing viral types was the sensitivity test to X-rays and the chemical substances which were also used for inducing mutations. In 1957 there was no substance or X-ray dose known with which in vitro experiments could have been carried out, the primary handicap being that the use of chemicals or X-rays meant inactivating the virus. There were also restraints on cross-immunity tests, and efforts were made to detect relationships by resistance and interference, although the basic mechanisms were as yet unknown. As of 1947, growing significance was allotted to biochemical criteria. These were initiated by Knight with his amino acid analysis, additional methods included the determination of nucleic acid content and nucleotide rates.

4.3.3 The Genetics of Animal Viruses

There were grounds enough to believe that influenzaviruses experienced spontaneous mutations: from reports on influenza epidemics; from antigen differences found in isolated strains during the A-epidemic in Great Britain from 1936–1937; and from the progress in the ability to change heritable properties of the virus experimentally. When Francis isolated an influenza strain which behaved differently from its predecessors in the neutralization test and in the CFR, researchers began to discriminate between strains A and B. In 1947 R.Taylor discovered type C. Type differentiation was determined from CFR results as set down by Fairbrother and Leslie Hoyle (1937), who had used antigenic material at first from infected mice lungs and later from egg membranes.

In addition to immunological criteria, numerous genetic markers were used to prove the mutability of the influenzavirus: adaptation of the infectivity to various species of animals as well as to the chorio allantois membrane and the chorio allantois sack of the chick embryo; the development of neurotropic strains; HA properties and HA inhibitors; and the assessment of cell-receptor destruction caused by the virus enzyme neuraminidase. The spontaneous appearence of new strains was explained by viral disappearence after epidemics and their persistence somewhere during inter-epidemic periods, and by the potential of renewed outbreak of the disease depending on the immunological defences of the population. The genetic differences between the strains had been earmarked in the laboratory. Thus, A-strains caused fever in ferrets, whereas B-strains ran an all but asymptomatic course. A-strains easily adapted to mice after passages in ferrets and chick embryos, while B-strains hardly adapted at all. Finally, Francis found requirements for B-strain adaptation which called for a long series of blind passages in the ferret and mouse. In 1939, Stuart-Harris developed a neurotropic strain by first passaging through chick embryos and later through the brain of young and then older mice.

HA differences were detected by Burnet and Bull in 1943. An isolated influenzavirus strain had a more favourable effect on human and guinea-pig erythrocytes than on those of the chicken. Natural human pathogens, the O-(original) form, which were bred in the ferret and the amnionic sack of the chick embryo, showed characteristics differing from the D(derivative) form which had been adapted to the allantois sack (Table 4.10).

In order to investigate antigenic structures more closely, van der Veen and Mulder tried using the cross-HI test in 1950. In the same

Table 4.10. Laboratory markers of influenza virus O and D forms

O-Form	D-Form
Agglutinates human and guinea-pig but not chicken erythrocytes	Agglutinates both chicken and mammalian erythrocytes
Does not infect the allantoic cavity	Multiplies in allantoic cavity
No enzymatic reaction with ovomucin	Reacts with ovomucin
Pathogenic for human beings	Pathogenic only in very high doses

year, Isaacs and Edney obtained a influenzavirus mutant with reduced enzyme activity by passaging highly diluted material through chick embryos. After 80 passages in the allantois sack, W.Smith and colleagues came up with an inhibitor-resistant A-PR8 mutant in 1951.

After Delbrück and Bailey had successfully undertaken recombination experiments with bacteriophages in 1946, Burnet and Patricia Lind (1951) began similar experiments on the reliably identified *Influenza virus-A* strains WSE and MEL, which differed from one another in six characteristics. In 1952 they set down the ground rules for performing influenza recombination experiments:

1. Only A-strains should be used.
2. The strains should have at least two differentiating genetic characteristics.
3. The majority of cells should have a multiple infection.
4. The strains must prove infective for the system.
5. Progeny must be taken from the first reproduction cycle.
6. Testing for recombinants must be undertaken with a suitable number of pure clones.

The problem of testing the purity of clones was not restricted to the influenzavirus, and arose from the fact that the given units for plant and animal viruses did not apply to one virus particle. Thus, the unit for plaque formation (PFU) of the poliovirus was equivalent to about 30 particles; the ID_{50}, i.e. the mean infectious dose, contained about 10 particles for influenzavirus and 5–7 for *NDV* in the chick embryo. In view of this inconsistency, reliable results in recombination experiments with these viruses could only be achieved when methods as well as results were critically analyzed.

In Burnet and Patricia Lind's recombination experiments, mice were given an intracutaneous virus mixture of a neurotropic and non-neurotropic strain. This, in turn, produced three new strains with altered enzyme activity, heat resistance and receptor properties. Hirst and Gotlieb criticised Burnet and Lind for not having adequately proved the stability of the newly found characteristics, and began recombination trials using the serological marker for HI. In the same year, W.Henle and Liu discovered the reactivation phenomenon, which required very stringent conditions. They reactivated influenzavirus, which had been partially inactivated by low doses of UV irradiation, by inoculation into the allantois sack. From their findings they concluded that reactivation was only possible if a cell had been infected by more than one particle; they therefore termed their process multiplicity reactivation.

Genetic experiments with influenzavirus assumed an increasingly important role in the following years. Recombination results using *HSV* in the chick embryo were reported in 1955 by Wildy. While searching for non-pathogenic poliovirus strains suitable for attenuated vaccines, several groups resumed genetic experiments in 1953–1954. Whereas Kolmer (1935), M.Theiler (1946) and Koprowski (1951) had attenuated poliovirus exclusively by animal passaging, Roca-Garcia, Cox and Koprowski managed to attenuate type 2 by adaption to the chick embryo. Simultaneously, Sabin and Melnick tried to alter the neuro-virulence of the poliovirus by passaging in monkey kidney cell cultures.

4.4 What Exactly is a Virus?

Experimental results on a broad range of viruses accumulated by 1957 encouraged A.Lwoff working at the Institute Pasteur, Paris, to define a virus at the 24th meeting of the Society for General Microbiology. This demanded critical discussion of the three divergent concepts concerning the nature of the virus. One, associated with names like Burnet and Andrewes, defined the virus as a microorganism. The diametrically opposite view was that of Stanley, who maintained that viruses were chemical molecules. Third, Bawden and Pirie simply substituted "molecules" for "small organisms".

To consider viruses to be simply chemical molecules was not justified, because there was no longer the slightest doubt that a significant component of the virus was its nucleic acid, which is associated, together with all its genetic implications, only with living organisms. Burnet himself had relinquished the "organism" idea when he wrote in 1957:

"A virus is not an individual organism in the ordinary sense of the term, but something which could almost be called a stream of biological patterns".

A.Lwoff advanced many arguments against the virus being a microorganism:

During the entire course of infectivity, a microorganism does not lose its identity, i.e. during its reproduction and while potentially developing pathogenic properties, it retains its structure as well as its functions. In contrast, the virus detaches itself from its structure and remains infectious and pathogenic exclusively by virtue of its genetic material which, in fact, comprises only a part of the virus. The elementary construction unit of an organism is the cell. Possessing a complete structure, the cell has at its disposal all the enzymes essential for effecting biological synthesis (Lipmann-system). The virus lacks such a system. The virus does not correspond to a cell, but rather to cell organellae such as chromosomes or plastids. On similar merits, therefore, the virus cannot be considered a living entity.

For A.Lwoff, the "molecule" was neither a microorganism nor a "filterable virus" nor an ultravirus" but simply a virus; hence his phrase, "viruses are viruses!".

Viruses could accordingly be defined as strictly intracellular and potentially pathogenic entities with an infectious phase, and

1. possessing only one type of nucleic acid,
2. multiplying in the form of their genetic material,
3. unable to grow and to undergo binary fission,
4. devoid of a Lipmann system".

This definition of a virus, originating in 1957, still applies today, despite heated debates on viruses being organisms or genetic macromolecules, kindled by newly emerging molecular genetic discoveries.

It is possible to sketch the virus as it was known in 1957 in terms of its characteristics, its chemistry, its physical-chemical and genetic structure, its mode of replication, and also the possibility of influencing its reproduction. Since the number of viruses described has meanwhile become so numerous, it would seem beneficial to classify them on the basis of recent findings.

The Chemistry of Viruses

Nucleic acid of the DNA or RNA type and proteins of various molecular weights are the major components. The number of proteins per particle ranges from one in simple viruses to many in complex ones. In addition, complex viruses often have a shell composed of carbohydrates, lipids, lipoproteins and glycoproteins. Enzymes can be virus constituents. Trace elements of copper and iron have also been found.

Physical-Chemical Structure

The morphology of viruses ranges from the simple sphere or rod shape (e.g. poliovirus and *TMV*) to the complex formation (e.g. *Vaccinia* and influenzavirus) and reflects their physical-chemical structure. The central part of the poliovirus and *TMV* contains the nucleic acid, which is enclosed in a simple, protective protein mantle. In complex viruses, the nucleic acid-protein core or nucleocapsid is additionally protected by a single or multiple-layered shell which does not consist of pure protein chains. In addition to the function of protecting the nucleic acid, parts of this shell serve as a link to the host cell. Thus, the glycoproteins found in the envelope of the influenzavirus link it to the receptor of the host cell. The entire envelope contains host cell-autologous lipids and carbohydrates. In 1943 Pfankuch and Piepenbrock were able to confirm the 1941 findings of the crystallographers Bernal and Fankuchen, according to which infectious *TMV* consists of hundreds of identical protein subunits. In 1955 Schaffer and Schwerdt demonstrated the presence of the same sperical subunits in the poliovirus. Crick and J.D.Watson (1956) hypothesized that simple viruses consisted of identical smaller structural units which were packed in a regular pattern.

Burnet (1956) and Schäfer (1957) split myxoviruses, i.e. medium-sized viruses, after pretreating them with ether at low temperatures. The ensuing nucleoprotein fractions were a source of 12 nm granula and 30 nm long hemagglutinins which contained proteins and carbohydrates.

The extracellular virus is cold-stable, losing its infectivity with increasing temperatures. Every type of virus has an optimal pH range. Chemical substances which can significantly change the optimal pH value have the same power to disinfect as do short-wave and, ultrasonic irradiation, or substances which denaturate proteins, split up nucleic acids, dissolve lipids, or have surface-activating or oxidizing properties.

Genetic Structure

Viral DNA or RNA carries the genetic information for virus reproduction into the host cell. Both forms reduplicate and both can regulate protein synthesis. Virus synthesis is initiated by the release of nucleic acid into the host cell. The beginning of identical reduplication for DNA viruses is the uncoiling of the double helix and the breaking of the hydrogen bond bridges between the nucleotides, according to the Watson-Crick model. The creation of virus protein was not clear, despite strong indications that it ensued via the RNA. That would also have explained the existence of RNA viruses as well as the demonstration that they contained genetic information for identical reduplication just like the DNA viruses.

Transfer of the viral genetic information to the host cell, set into motion potential cell suicide because synthesis was shifted to the needs of the virus. This might entail but a few minutes for phages, a number of days for other viruses. There was, however, also the possibility of the virus genome being instantly integrated into the host genome without any immediate consequences, similar to the prophage.

Multiplication of the Virus

In order to multiply, it is essential for the virus to penetrate the host cell. This is relatively simple in exposed tissue such as the mucosae of the respiratory or intestinal tracts. If tissue suitable for virus binding is enclosed by the epidermis, infection is possible through a wound or via a sucking or stinging insect. In this way, animal viruses can directly enter the blood, lymph or nerve pathways, and plant viruses can pass into the phloem-xylem system.

Binding to the host cell, or adsorption, requires suitable chemical structures both on the virus and host cell surfaces, Upon binding, the genetic material of the virus must be able to unite with the chromosomes of the host cell. Penetration of the cell occurs – as is the case with bacteriophages – by injection of its nucleic acid, or by ingestion of the complete virus particle in a phagocytic process with subsequent uncoating of the virus. This is known as pinocytosis when the virus is imbibed by a droplet. During this process of penetrating the cell, the virus begins to release its nucleic acid. If penetration and release do not occur simultaneously, the beginning of virus infection is termed viropexis or engulfment.

After it became apparent that some viruses contained only RNA, and the Belgian, Brachet, and the Swede, Caspersson, indicated the relationship of RNA to protein synthesis in 1955, it seemed justified to assume that protein synthesis occurred via RNA. This also explained the capacity of RNA viruses to carry the message for synthesizing their own proteins. Gierer and Schramm, as well as Fraenkel-Conrat, had found *TMV* to be of the RNA type in 1956; one year later Colter and colleagues determined *Mengo virus* to be an RNA virus, and Wecker and Schäfer proved the same for *Eastern equine encephalomyelitis virus*. In 1957, Crick incorporated these findings in his new version of the central dogma. Besides replication of DNA, it would neccessarily have to include an identical reduplication of RNA as well as protein synthesis in the absence of a DNA controlling mechanism.

After synthesis of the virus components, the assembly of the complete virus follows. Incomplete viruses which P.von Magnus had discovered in 1947 among influenzaviruses were subsequently found among other species. These incomplete viruses had been synthesized by quick sequential passaging of large amounts of virus through the allantois sack of chick embryos. They were not infectious but contained fully-developed hemagglutinating properties. The Henle's and coworkers also found similar forms in permanent HeLa cell cultures in 1955. During the assembly phase, synthesized structural units attach themselves to each other to form a virus particle. If, in the case of complex viruses, the formation of an envelope is needed, a maturation process follows in which, according to Burnet, host cell material such as lipids or mucoproteins can be utilized.

The adsorption phase can take up to several hours. During this time, the virus particle is very vulnerable to chemical substances and neutralizing antibodies. The duration of a complete viral synthesis cycle may be minutes for phages, or hours for other viruses. In order to indicate specific intervals, the terms lag-phase and eclipse were borrowed from studies of bacteriophages.

Virus and Host Cell Interactions

Virus infection can produce manifold effects in the host, ranging from affects on single organs to widespread systemic disease, either acute or chronic, including the development of cancer. On the other hand, virus infections can also exercise oncolytic action, namely by destroying degenerate tissue in tumours in which they multiply. The speed at which viruses multiply in the host or cause topical or systemic disease is expressed in specific terminology. Here, however, only the term "acute disease" is clear-cut, whereas adjectives such as "latent", "inapparent", "occult" or "persistent" demand further definition.

Viruses can be characterized as parasites with pathogenic, antigenic, pyrogenic, allergenic, teratogenic, mutagenic and/or carcinogenic properties.

Manipulating Virus Reproduction

It began to seem possible to combat certain destructive insects by promoting multiplication of viruses which had proved pathogenic to the target insects. Various institutes were encouraged for the first time to carry out research into biological methods of pest control.

In view of the totally different nature of a virus infection compared to that of a microorganism, it seemed most unlikely that the known anti-microbial substances would be suitable for combating virus infections. Experiments cited at the time were concerned with compounds influencing all phases of virus multiplication, including nucleic acid and protein synthesis. Some of the tested substances proved anti-viral in vitro, though none proved effective under natural conditions.

Classification

Virus infection can occur in all living organisms and affect all organs and tissues. The list of viruses known in 1957 comprised approximately 200 insect viruses, 300 plant viruses, and 200 animal, serologically differentiated types of virus. These latter also included viruses found in fish, e.g. the pathogen for lymphocystis disease of the orange file fish described by Weissenberg in 1938, which, for want of suitable methods, had not yet been identified.

After various attempts had been made to classify plant and animal viruses according to their filterability, their electron-micrographic image or their host and tissue affinity, a classification standard was finally settled upon – at first for vertebrates – at the 5th International Congress of Microbiology in Rio de Janeiro in 1950. Seven virus groups were formed on the basis of four fundamental criteria: nucleic acid type, size and morphology, nature of replication, and resistance to ether and bile salts. Thus it was possible to classify 80% of the then known viruses occurring in vertebrates. Additional criteria were defined for subgroups, the list of the seven primary groups being as follows:

1. Poxviruses
2. Herpesviruses
3. Adenoviruses (adenoidal agents)
4. Papilloma-polyomaviruses
5. Myxoviruses (a subgroup was later termed avian tumour viruses)
6. Arborviruses (arthopod-borne viruses)
7. Picornaviruses (small RNA viruses).

For medical purposes, clinical aspects determined the classification, e.g. viruses of the skin, respiratory tract, exanthema, venereal diseases, nerve tissue, and the eye.

5 The Period of Biochemistry and Molecular Genetics

The spectacular molecular genetic and molecular biological results of Delbrück and Luria in phage research after WorldWar II kindled the interest of biologists, biochemists, physicists and chrystallographers alike, to experiment along similar lines.

The discoveries in molecular biology which had such a momentous impact on experimental virology included the elucidation of the double-stranded DNA model, and the subsequent clarification of the genetic code and formulation of the central dogma. In addition there was the discovery of DNA polymerase, the oncogenes, reverse transcriptase in RNA tumour viruses, polyribonucleotide synthesis, as well as the development of hybridization techniques. The isolation of bacterial restriction endonucleases and their use in in vitro experiments offset an incipient stagnation in molecular biology by opening up new possibilities in gene technology or genetic engineering with the nucleic acids of bacteriophages and viruses, and ushered in an exiting new epoch in molecular genetics. A method for constructing biologically functioning bacterial plasmids was reported in 1973 by S.N.Cohen and colleagues. It entailed the isolation of a defined segment of a genome and ligating this to a plasmid, i.e. a small extrachromosomal self-replicating DNA molecule, in order to transfer it to another cell. The hybridoma technique for synthesizing monoclonal antibodies in vitro, in conjunction with the accumulated knowledge on the network of immunological defences, and the development of other improved methods, brought new insight into the antigenic mechanisms and virus-host interactions.

The model of the DNA double helix, arrived at by the American biologist J.D.Watson and the English physicist Crick in 1953,was the culmination of patient theoretical discussions and experiments which involved many other scientists. Besides the English crystallographers Rosalind Franklin and Wilkins, this group included Astbury, R.Hotchkiss, Chargaff and Pauling, among others. In 1945 Astbury established that the purine and pyrimidine bases were vertical to the DNA longitudinal axis, and that neighbouring bases maintained a distance of 0,34 nm. Hotchkiss and Chargaff (1948) separated and determined the nucleic acid components by paper chromatography and discovered that the four DNA bases, contrary to expectation, were not present in equal amounts. Chargaff (1950) found the ratio of purines to pyrimidines, i.e. adenines to thymines and guanines to cytosines, in the DNA to be almost 1:1. Pauling (1951) concluded that the backbone of polypeptide chains had an α helix structure.

In the early 1960's Kornberg and coworkers managed to shed light on the enzymatic processes involved in DNA synthesis. They purified an enzyme from E.coli, which was essential for DNA replication and called it DNA polymerase. Khorana (1964) created mRNA from synthesized polyribonucleotides and used it for protein synthesis.

The first DNA-RNA hybridization to determine the specific interaction of two polynucleotide strands with complementary base pairs was reported by Gillespie and Spiegelman in 1965. Hybridization exploits the separation (denaturation) of double strands on heating to 100°C and the renaturing (reforming) of the DNA that occurs on slowly cooling. Moreover, any two separate strands can be brought together that possess complementary sequences. By applying techniques for DNA-DNA, DNA-RNA and RNA-RNA hybridization, direct studies could be carried out of viral nucleic acid synthesis and its transcription, which requires only one DNA strand as a template. Embedding a denatured strand in agar, or fixing the strand on a nitrocellulose membrane filter to adsorb complex RNA but allow free RNA to pass through, using a fixation method developed by Nygaard and Hall (1963), facilitated the detection and quantitative determination of hybrid DNA.

The genetic code was finally elucidated in 1967, when Garen, Crick and Brenner discovered the nonsense codons which signalled the end of polypeptide chains and were correspondingly called terminators. The code enabled translation of the genetic "text" consisting of four different nucleotides into the protein "text" composed of about 20 amino acids. The debate stirred up by the idea of this code had lasted more than two decades. After the double helix was conceived in 1953, J.D.Watson and also the immigrant, Russian, theoretical physicist Gamow, teaching in USA, were amongst the most creative participants. First indications of the code's existence were presented by Nirenberg and Matthaei in 1961; they had managed to synthesize the amino acid poly-phenylalanine in a cell-free system using the codon poly-uracil.Thereafter, many teams joined in the search for the clarification which Crick provided with his first draft as carly as 1962. According to this concept, three sequential nucleic acid bases determined a specific amino acid. This confirmed Crick's sequential hypothesis of 1957, which maintained that the specificity of a nucleic acid was characterized by the sequence of its bases, this in turn denoting the order of the amino acids and structure of their related proteins. Using this knowledge, scientists could determine either the nucleic acid bases from their amino acid sequences, or the amino acid sequences from the order of the corresponding nucleic acid bases. Since analysis of the bases proved much easier, this remained the method of choice. As soon as restriction endonucleases were discovered, sequencing by exonucleases became redundant. The latter was a two-step process in which the 3′ or 5′ mononucleotides were digested, isolated and analyzed.

The central dogma of biology postulated by Watson in 1953, whereby the flow of genetic information proceeds from DNA to RNA to protein, and DNA serves as the pattern for replication, was modified again in 1974. In 1956 the discovery of the RNA viruses had demanded that the dogma include RNA replication in the absence of DNA. At that time, Volkin and Astrachan investigated the problem of RNA synthesis after bacteriophage infection, and found specific, divergent RNA activity in different host constituents. Besides a soluble intermediate varying in both intensity and activity, they observed two additional intermediates of which the smaller amount proved to be more active. The experiments carried out by Pardee and coworkers in 1957–1958 gave substance to the concept of genetic regulation whereby ribosomes functioned as the site of protein synthesis. Jacob and Monod's hypothesis clearly acknowledged the existence of an intermediate which transferred the information for protein synthesis from the genome to the ribosomes. Just as Volkin and Astrachan ventured into the problem of protein synthesis, Brenner and colleagues began the experiments which eventually provided the solution. While working with phage-infected bacteria in 1961, they identified mRNA as an unstable intermediate. Fig.5 depicts the changes which the central dogma had undergone up to 1970 when Crick published his diagram.

Knowledge about the flow of information from DNA to the synthesized protein was considerably expanded in 1977 and in 1979. Susan Berget and coworkers and Louise Chow gained new insight into the mechanism of mRNA synthesis whilst experimenting with *Adeno*

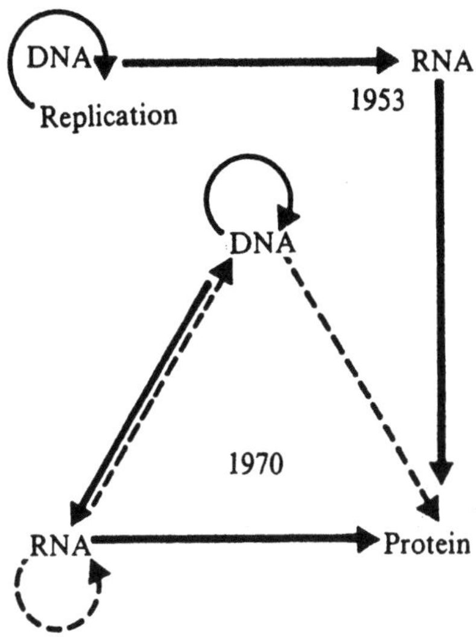

Fig. 5. The central dogma of biology --- Transfer that can occur

virus 2 in 1977. The enzyme-directed reactions in question became known as RNA processing, which entails the cleaving of the introns or non-coding, intervening sequences and splicing the coding regions or exons to form the mRNA. In 1979 several groups initiated experiments with various strains of the protozoa Ciliata tetrahymena to clarify the splicing process. Their findings 1981–1982 confirmed Crick and Leslie Orgel's 1968 assumption that there would most likely be a primitive tRNA which functioned as a sort of self-activating enzyme. Groups including Cech, Paula Grabowski and Kruger showed that ribosomal RNA possessed the capacity to splice itself in the course of autonomous self-processing. Instead of an enzymic protein, a very short piece of non-coding RNA, called a ribozyme, assumed catalytic endonuclease and ligase activity. Similar studies with yeast and fungi mitochondria by Burke and Raj Bhandary in 1982 indicated that self-splicing was not the exception but a normal process. With this discovery, the protein-free, infectious RNA first reported in 1967 by Diener and Raymer assumed a new significance; such RNA is known as a viroid since 1971.

Contemplating the activity of the RNA tumour viruses, Temin lined up a provirus hypothesis in 1964. He supposed that the RNA genome, because it was the converse of the structure envisaged in the central dogma, first required to be transcribed into DNA. Only upon integration into the host-cell genome was it transcribed back into viral

RNA. The enzyme required for this step was an RNA-dependent DNA polymerase or reverse transcriptase, which Temin and Mizutani, as well as Baltimore, discovered in RNA tumour viruses in 1970.

Equally important in the discussion on tumourgenesis were results on *Rous sarcoma virus (RSV)* obtained by several groups between 1969 and 1977. G.Martin and coworkers, as well as Toyoshima and P.Vogt, concluded from their genetic experiments with virus mutants that a viral gene product must in some way be involved in virus-induced cell transformation. Evidence for a sequence which might be responsible for this was found in 1971, by P.Vogt, and in 1972, by G.Martin and Duesberg, in the genome of the *(RSV)*. Duesberg proposed that this sequence be called onc or sarc gene with reference to the oncogene hypothesis put forth by Huebner and Todaro in 1969. The year following brought evidence of homologous sequences in normal cells, and the two types were differentiated by the prefix "v" for virus (v-sarc gene) and "c" for cell (c-sarc gene). Thus arose the discussions and experimental work in search of the significance of viral oncogenes and their corresponding cellular counterparts, the proto-oncogenes, and their roles in causing tumours. In 1977 Brugge and R.Erikson isolated and identified the gene product of the 10000 base-pair sarc region as a phospho-protein of 60000 dalton.

Luria and Human (1952), and Bertani and Weigle (1953), focused their attention on the bacterial defence system against invading foreign nucleic acids. This they found to consist of two enzyme types which acted at the same specific cleaving points of double-stranded DNA: restriction enzymes, or endodeoxyribose nucleases, recognize and cleave foreign DNA at the above-mentioned points, while modification enzymes protect the bacteria's own DNA at the same sequences, primarily by methylation. Arber and Daisy Dussoix (1962) recognized that synthesized bacteriophages carried a host-specifity for the bacteria in which they were produced, and Meselson and Yuan characterized the first restriction enzyme in strain K of E.coli in 1968. After H.Smith and coworkers discovered the cleaving point specifity in 1970, this idea was also applied to viral DNA. Adler und Nathans experimented with *SV 40*, whilst Kelly und H.Smith worked with *Phage T7*. Meanwhile, a large number of different restriction enzymes have become available that cleave DNA at palindromic sequences. These points are symmetrical in so far as they follow the same sequence when read from the same end of the opposing strands, as shown in the following model. It is also evident that their ends are identical and complementary. Thus, it is possible to cleave DNA fragments of different origin, and to recombine them provided they have been obtained by means of the same restriction enzyme.

```
---------→  5'  G A A / T T C  3'
            3'  C T T / A A G  5'  ←---------
            (/ marks the cleaving point)
```

Meanwhile, we have learned that there are also non-specific restriction enzymes.

Specific restriction endonucleases permit the plotting of restriction or physical maps of chromosomes which can be correlated with the genetic maps of the various viral gene functions. Furthermore, genes can be isolated, DNA molecules constructed, genome and DNA nucleotide sequences can be analyzed, viruses can be identified, and related interactions of different viruses can be detected. Sequencing by means of endonucleases facilitated the elucidation of a genome relatively quickly, supplementing or replacing methods for end group analysis or for the recognition of terminal sequences. The first completely sequenced genome of a virus was that of *Phage MS2*, which consisted of 3569 nucleotides and was published by Fiers and coworkers in 1976. A two-step procedure to cleave and to analyze a nucleic acid molecule composed of 1–10 nucleotides was described by Rushizky and Knight in 1960. One year later, the technique of nearest neighbour analysis to compare the frequencies of nucleic acid pairs was introduced by Josse and colleagues. It was applied first to bacteriophages, but also to animal DNA and plant RNA viruses soon thereafter. Fiers and Sinsheimer used an exopolynucleotidase to determine the terminal sequences of *Phage φ x174* and thereby discovered the first circular nucleic acid in 1962. Step-by-step sequencing of 5'-terminal RNA nucleotides was described by Steinschneider and Fraenkel-Conrat in 1966. Three years later, a method for base analysis of DNA was reported by Shatkin consisting of a two-step procedure to digest DNA with the help of chemicals or enzymes and quantify the resulting fragments.

With the help of restriction endonucleases the broad field of gene technology became the site of great activity, which led, among other things, to cloning – now an indispensable procedure in virology. Cloning entails the introduction of excised genes into microorganisms by means of plasmids, where they replicate with the genome and are expressed, i.e. the gene product, a polypeptide, is synthesized. The plasmids, found as small, circular DNA molecules in nature, can be compared in size to viral DNA. Both microorganisms and viruses serve as vectors or biological transmitters. Noreen and K.Murray as well as M.Thomas and colleagues used the DNA of E. *coli phage* λ as a cloning vector in 1974.

The usual steps in cloning are:

1. The plasmid is cleaved.
2. Foreign DNA which is to be cloned is specifically or non-specifically cleaved.
3. Plasmid and foreign DNA are ligated.
4. This recombinant DNA is added to specially cleansed cells which are transformable, i.e. they are able to take up information contained in DNA.
5. The bacterial enzyme system ensures that the ligated plasmid and the DNA fragment remain intact.
6. Selection pressure encourages the multiplication of the bacterial population carrying the plasmid.

Since 1985, we have seen studies – notably by Saiki and Erlich – which are primarily concerned with the polymerase chain reaction (PCR). This new in vitro cloning methods permits a million-fold amplification of an isolated DNA template of up to 2000 base-pairs within just a short time, even in the presence of extraneous material. This method is suitable to produce probes for diagnostic purposes and to detect a target DNA molecule present only once within a sample of 10^5 cells.

In 1975, the German biologist Köhler and the Argentine chemist Milstein successfully concluded their studies on cell fusion, and laid the groundwork for the hybridoma technique, which enabled them to synthesize monoclonal antibodies in vitro. The theoretical basis for this technique is to be found in the network theory of the immune system defined by Jerne in 1974. This system is extremely complex, in that all contributory factors are closely interrelated, and might perhaps be considered the final link in the train of thought which began with the side-chain theory by Ehrlich in 1900, and was enhanced by the postulation of antibody production by Burnet-Fenner in 1949, and ultimately leading to the clonal selection theory of acquired immunity by Burnet, Talmage, Jerne and Lederberg in 1959. According to this theory the embryo contains examples of all antibodies which an adult animal can produce, each antibody-producing cell being able to synthesize only one type of antibody. During a critical period in fetal development and early life, those cells producing antibodies against "self" antigen are killed or inactivated. At the end of this period all "itself" cells have been eliminated from the set of antibody-producing cells.

The experimental basis of the hybridoma method was the fusion technique using heterological cells which was introduced with HeLa and Ehrlich ascites tumour cells by J.Harris and Watkins in 1965. Using UV-treated Sendai viruses which contained the enzyme neuraminidase, they were able to induce cells to fuse. Köhler and Milstein fused B-cell myeloma and plasma cells of immunized mice, uniting two properties in the hybrid cell: the longevity of the myeloma cell and the capacity of the plasma cell to form specific or monoclonal antibodies. These, in turn, made it possible to diagnose and localize antigens, and to isolate pure antigens which, for example, could later prove valuable as passive vaccines. Parallel to these undertakings, considerable improvements were made in purification and concentration and quantification techniques for viruses, nucleic acids and proteins. In 1989 Huse and colleagues described a method for generating monoclonal antibodies for a large combinatorial library of the immunoglobulin repertoire in *Phage* λ, which facilitated the production of different specific antibody types within a short time.

When A.Lwoff defined the virus in 1957, the field of medically oriented, experimental virology had already attained such high laboratory standards that it seemed entirely possible that viruses would be shown to be causative agents of acute disease in most vertebrates, insects and plants. Where this had not already been established for specific acute infections (e.g. hepatitis was anything but unknown) the most likely explanation appeared to be that optimal conditions for isolating and propagating the pathogens had not yet been discovered. Since the assumption that viruses interacted with all forms of life seemed entirely justified, it was essential to develop new cell culture systems for the unexplored areas. For example, persistent infections and neoplasia, though well-known, had not been clarified by experimental studies.

The global, intense interest in the molecular genetics of bacteriophages and plant viruses shifted the emphasis of research from the virus as a pathogen to the virus as a macromolecule, capable of genetically directed, molecular reduplication with potential pathogenicity for the host cell or host organism. Initially, this meant redirecting endeavours towards biochemically oriented research, and later making the transition to molecular genetic experiments. Since this development continues at the time of writing, in fact since it has culminated in unforeseen progress in biotechnology, it would seem justified to designate the phase beginning in the 1960's as that of biochemistry and molecular genetics. The subsequent description of the period employs the terms "molecular genetics" and "molecular biology" according to Astbury's definitions. Thus, the morphological and structural problems concerning extracellular viruses, for example, are considered a part of molecular genetics, simply because of the opposition to the view of viruses as living entities. In contrast, molecular-genetic experiments with viruses which aim at solving problems of a biological nature belong to the field of molecular biology. These might include expressing the genetic code or formulation of the central dogma; to solve problems arising from virus-induced diseases such as neoplasia; persistent virus infections or "slow virus infections". More recently, virological problems concerned with self-processing RNAs or ribozymes have been placed in this category. They are helping us to understand infectious RNAs (the viroids) or infectious proteins (the prions), and generating new ideas on theory of evolution.

5.1 The Virion

Methods to recognize virion morphology and capsid symmetry were still in the inception stage in 1957. The biochemical structure of the virion was only very crudely explaned when the genetics of viruses very gradually began to appear. During the following two decades, questions about virions became a focal point for the investigations to identify and localize antigenic virus components. These studies called for improved methods which then followed in all their remarkable diversity: the purification and concentration of viruses; EM and cytochemical fluorochrome techniques (e.g. with acridine orange); immunological procedures including the present routinely used immunofluorescence; electrophoresis, chromatography, radiography and enzymology; hybridization techniques with nitrocellulose membranes developed in 1963; and density-gradient centrifugation.

5.1.1 Virion Morphology and Capsid Symmetry

Investigations in this field were considerably assisted not only by new methods of virus cultivation in cell cultures, but also by a spectacular advance in virus electron microscopy in 1958: the double shadow-casting technique. Applying this method, R.Williams and K.Smith were able to define the polyhedral form of the *Tipula iridescent virus* which had been discovered 4 years earlier. The virus suspension was applied to a collodion film coated with albumin, which would prevent aggregates from developing during the drying stage. The specimen was then freeze-dried and shadows cast from two different angles for microscopy. Since 1959, the technique of negative staining, introduced by Brenner and Horne, has also been available. They mixed a virus suspension with a neutral solution of potassium phosphotungstate, which allowed the formation of a strongly contrasting, electron-dense layer around the low-densitiy particle, rendering the morphology of surface structures visible. At first this method was used for extracellular viruses stained with various agents, and was later employed in investigations of virus-infected cells. In 1966 Caspar proposed the simultation method, as an alternative to negative staining, by which X-ray images could be obtained of viruses which had been embedded in X-ray absorbing material.

By using X-ray crystal lattice structure analysis, Bernal und Fankuchen with *TMV* (1941), Markham and J.Smith with *Bushy stunt virus* (1950), Markham with *Turnip yellow mosaic virus* (1951), and Schaffer and Schwerdt with poliovirus (1955) had all shown that the structures of these viruses, with arrangements of small protein units, resembled chemical molecules. In 1952 J.Harris and Knight found, by means of chemical end-group analysis, that the structural units of *TMV*, discovered by Bernal und Fankuchen in their X-ray structure analysis in 1941, adhered to a fixed pattern, which could be correlated to the nucleic acid-free fission products Schramm had analyzed physico-chemically between 1943 and 1947. Thus, it was no longer in doubt that the structure of viruses could be analysed, and indeed, by using various approaches.

The four species which had been investigated consisted of the two components nucleic acid and protein; *TMV* was rod-shaped, while the other three appeared as cubes. From this, Crick and J.D.Watson hypothesized that the highly specific nucleic acid of small viruses was protected by·a sheath of protein which was probably composed of small, identical subunits. The formationof these small subunits in the infected cell, could more easely be regulated by the virus than could one or two larger proteins. They further reasoned that since simple viruses were either rod- or cube-shaped, their potential structural arrangements would of necessity be limited. No rule was established for the rod-shaped, plant viruses but cube-shaped viruses were supposed to consist of asymmetrical protein units forming rotational symmetrical particles, which, in turn, might also contain subunits. The basic number of units for all cube-shaped viruses was thought to be 12, which allowed the formation of tetrahedrons; 24 units provided an octahedron, and 60 an icosahedron. It was in 1960 that Wildy und Horne demonstrated the possibility of measuring the structural elements of icosahedral viruses by EM; their numerical series followed the pattern 12-42-92-162-252. At the Cold Spring Harbor Symposium in 1962, crystallographers Caspar and Klug added an alternative numerical sequence, namely 32-122-272-supporting their alternative theory of quasi-equivalence of isometric viruses, due to icosahedral symmetry and capsomers totalling more than 60. Assuming a molecular weight of 20000 for the structural unit of an icosahedral virus with a 15 to 20-nm diameter, there would be room for only 60 rigidly arranged, identical units on its sheath. Therefore, it was necessary to explain how larger viruses, having different numbers of capsomers, could form dense capsids by means of self-assembly from identical structural units with minimal energy demands. Thus, an arrangement of quasi-equivalent lattice points supplanted the rigid equivalence of smaller icosahedral viruses. These points could be determined by triangulation, with a deviation of 5° on either side.

Since it seemed fair to assume that three structural units formed a morphological unit, it was not surprising that rules based on crystallographic evidence in time lost their significance, especially when conflicting values were published. On the other hand, X-ray diffraction analysis did not produce direct results; only after the interpretation of many pictures - obtained using complicated methods - could conclusions be drawn. One of these complicated methods was the isomorphic replacement method, developed by Cork at the W.Bragg Institute in Cambridge in 1927, which involved attaching an easily-detectable heavy atom like mercury to each capsomer. As early as 1967, the number of known capsomers ranged from 12 for *Phage* φ x174 to 812 cylindrical capsomers for the insect pathogen *Tipula irisdescent virus*. Whether there were 42 or 72 capsomers in the *Polyoma virus* was a matter of debate among both crystallographers and electron microscopists from 1960 to1975! The structure of 72 capsomers, of which 12 were pentamers and 60 hexamers, was accepted only until 1982. At that time, a group including Caspar published surprising results obtained by highly sensitive methods, and concluded that the ultrastructure of the

Polyoma virus determined by the quasi-equivalent method was untenable because all 72 capsomers were pentamers. Simultaneously, a discussion emerged on the packing arrangement of the structure units to form capsomers recognizable in electron microscopy.

Determination of the protein units was complicated in itself, and the fact that various investigators often came up with conflicting findings served only to make matters worse. Therefore another approach to the structure of the virus was established: analysis of the symmetrical arrangement of those elements which corresponded to the molecules of crystals. To ensure standardization, the structural elements were assigned names laid down in a proposal put forth at the Cold Spring Harbor Symposium in 1962, which later received approval in a slightly modified form by the Provisorial Committee for the Nomenclature of Viruses in 1965. Accordingly, the smallest units are the structural units, which can be detected by X-ray diffraction. Their molecular weights span a broad range: the approximately 18000 daltons for *TMV* may be considered the mean. The morphological units formed from structural units, can be detected by the EM, and build the capsomer. This serves as the basis for the capsid and can be differentiated from unlimited-growing crystals primarily by its genetically determined, restricted growth. In 1963 Caspar assumed that this genetic control was achieved by a recognition site on the *TMV*-RNA, where the development of rods, including their specific length, was initiated. Knowledge of capsid symmetry brought about firm concepts of capsid forms through the possibility of dividing a virus into equal, coinciding parts using symmetry axes or symmetry faces.

Rotation symmetry made it possible to ascertain how often coincidence occurred in the virus form during one complete turn on its own axis. Thus, 5-fold symmetry meant coincidence occurred 5 times during one rotation, i.e. coincidence was observed at 72° intervals. A virus has bilateral symmetry when it can be divided into complementary parts with two faces. A (5-3-2)-fold symmetry is frequently observed in viruses and characterizes the icosahedron, which consists of 20 triangular faces. This symmetry results when three rotation axes pass through the centre of the virus, as illustrated. (Fig.6).

A 3-fold and 2-fold rotation symmetry would indicate a tetrahedron which is defined by four equilateral triangles; a 4-fold, 3-fold and 2-fold symmetry, an octahedron. Rod-shaped viruses are bilaterally symmetrical, i.e. they can only be cut into symmetrical slices. As for bacteriophages, head symmetry and tail symmetry should be considered separately. Structural units can be linked in numerous

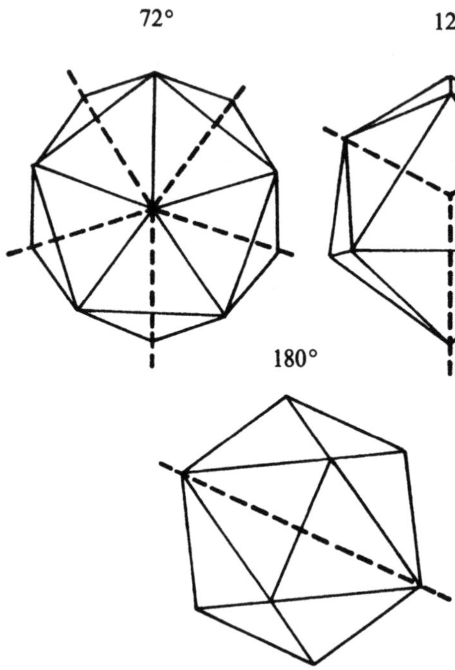

Fig. 6. Icosahedral 5-3-2 symmetry

ways – even inside a virus particle. P. Wildy and colleagues (1960) assumed the *Polyoma virus* to consist of capsomers composed of hexamers and pentamers. For example, the 252 structural units of adenovirus consist of 12 pentons at the vertices and 240 hexons on the faces. The two capsomer types are serologically distinguishable.

Capsid symmetry was found to be of both spiral and cube form. After the collection of data got by various methods, attempts were made to combine the results in a virion model; this required creating additional ultrastructural terms. The capsid, which in conjunction with its nucleic acid constitutes the nucleocapsid, seemed to determine the most common virus form, namely the polyeder, found in icosahedral, square, bullet and rod-shaped viruses. Under the guise of a polyeder there is often a difficult-to-detect icosahedron and, in the case of the bacteriophage, there may be a polyhedral head attached to a rod-shaped tail, often with spikes or tail fibres. Certain virus forms have characteristic surface structures such as a sheath and 7 to 10-nm spikes. All of these parts together constitute the infectious virus particle termed a virion. If a sheath is present, it is generally called an envelope. Should a part of the envelope be composed of virus-coded structural units analogous to the capsomers, then these units can be called peplomers. These form the peplos which, in contrast to the envelope, is free of host-cell material.

For the core, there exist two different definitions. The core as an recognizable electron-dense centre can be differently localized within the virion; being found in various RNA tumour virus types made them determinants for typing these viruses. Another type was the nucleocapsid-core which was isolated biochemically by digesting the lipid or protein envelope and employing various other chemicals to obtain the core composed of viral nucleic acid and bound proteins.

Morphology and capsid symmetry belong to the four fundamental criteria for the classification of virus families. The experimental basis of virus structure analysis was laid down in 1959, and subsequent improvements in optical techniques yielded a wealth of structural data on viruses and thus the classification criteria. However, it has also been shown that absolutely identical structures within members of a species or a family do not exist. Despite a considerable increase in the number of newly discovered viruses and the concomitant abundance of publications on their morphology and ultrastructure, there is still insufficient information to permit reliable models for all the various virus families or types. Thus, we know a great deal about *TMV*, but beyond that we can only say that plant viruses have been found to be round, branched, tubular or rod-shaped. Round viruses have a 30 to 50-nm diameter with definite or probable icosahedral symmetry. Other forms have been found to have a diameter of 10–20 nm and a length of 100–200 nm. Our knowledge of bacteriophages has hardly increased in this respect and information about insect viruses and the known viruses of fish, amphibia, mycoplasmas, spiroplasmas, green algae and protozoa has remained scanty.

The first and most thoroughly investigated virus was *TMV*. To establish a *TMV* model required the combined efforts during almost 30 years of numerous biochemists, crystallographers and virologists, including Bernal, Fankuchen, Rosalind Franklin, Caspar, Klug and Finch, among others, whose use of biochemical, immunological and other methods all contributed to the final model in 1959. *TMV* serves as an exemplary model of the spiral, tubular-shaped, simple virus forms. Table 5.1 presents a summary of the morphology and symmetry data which also served as the basis for classifying viruses from 1979 to 1985.

The number of capsomers and structural units relating to their size and shape is uncontested for relatively few groups. The 252 capsomers of the adenoviruses are 7–9 nm, those of the parvoviruses 3–4 nm, and

Table 5.1. Morphology and capsid symmetry of some virus species and families[a)]

Virus	Virion size (nm)	Form	Capsid symmetry	E	S	Host
Parvo-	18– 26	Icos.	Icos.			V
Picorna-	22– 30	Ikos.	Icos.			V
Toga-	40– 70	Spher.	Icos.	+	+	V
Papova-	45– 55	Spher.	Icos.			V
Reo-	60– 80	Spher.	Icos.	+	+	V
Adeno-	70– 90	Icos.	Icos.			V
Corona-	75–160	Pleom.	Helical	+	+	V
Retro-	80–100	Spher.	Icos.	+	+	V
Bunya-	90–100	Spher.	Helical	+	+	V
Orthomyxo-	90–110	Spher.	Helical	+	+	V
Paramyxo-	120–300	Spher.	Helical	+	+	V
Herpes	145–205	Spher.	Icos.	+	+	V
Rhabdo-	50– 95 × 130–380	Bullet-shaped	Helical	+	+	V
Vaccinia-	300 × 230 × 100	Brick-shaped		+		V
Potato y	11 × 840*	Filament.	Helical			P
Potato x	13 × 520*	Filament.	Helical			P
Carnation l.	13 × 660*	Filament.	Helical			P
TMV	18 × 300	Tube-sh.	Helical			P
Beet yellows	1600*	Filament.	Helical			P
Cuc. mosaic	28	Spher.	Helical			P
Turn.yel.mos.	29	Spher.	Helical			P
Baculo-	40–110 × 200–400	Bacillif.	Cylindric	+		IV
Micro-	27	Icos.	Icos.		+	B
Tecti-	65	Icos.	Icos.	+	+	B
Cysto-	75	Isometric	Cubic	+		B
Myo-	110 × 80	Cubic head				B

a) E, Envelope; P, plant viruses; IV, invertebrate viruses; S, spikes; B, bacteriophage; l, latent viruses; V, vertebrate viruses; Cuc., cucumber; Turn.yel.mos., Turnip yellow mosaic virus; Icos., icosahedral; Filament., filamentous; Bacillif., bacilliform.; *, mean values

those of the polioviruses 5–6 nm long. There is little reliable data concerning their surface spikes; they can be rod- or button-shaped, whereby the latter can serve as plugs at the end of the rods; they may be present on the surface of a virus in various forms and lengths, with different functions. For example, the 4 × 14-nm spikes on orthomyxoviruses carry the hemagglutinin while their 4×8.5-nm spikes contain the neuraminidase.

Considerably improved novel possibilities have come about through recent developments in high resolution electron microscopy, capable of resolving 0.29 nm, which approaches the miniscule dimensions of atoms. By employing this technique, the 3-dimensional structure of a virus can be ascertained, exact measurements carried out, viral synthesis can be tested, and the effects of external factors such as chemical substances or specific antibodies can be investigated during the entire multiplication cycle. According to a survey published by Rossmann and E.Johnson in 1989, new perspectives for topographical depiction of viruses have been uncovered. One of the ideas discussed in their paper entitled "Canyon Hypothesis" is concerned with the theoretical aspects of receptors and their interaction with antibodies and anti-viral substances.

5.1.2 Virion Structure

The term "virus structure" has undergone a transition as the use of biochemical methods have given way to molecular-genetic techniques. During the initial two decades nucleic acids and proteins were considered only to be the more important components in virus structure; but entering the molecular-genetic phase, viral nucleic acids became the favoured virus components. Therefore, now, viral structure is understood as molecular structure of the virus genome.

Scientific progress in both fields was made possible solely by the development of new separation and identification methods for virions, in addition to improved quantitative analyses of virus components. The abundance of new findings in biochemical or molecular-genetic virology led inevitably to the appearance of new journals, which, among other things, prompted the standardization of the virus classification. Here too, biochemical terminology served only in an intermediate phase, as a precursor to the presently acknowledged molecular genetic classification. Both, however, indicate the change in emphasis of the fundamental characteristics of the virus during various periods in the history of virology.

5.1.2.1 The Biochemical Structure of the Virion

The *TMV*-oriented concept of the virus, with all of its biochemical implications, conveyed the idea of a simple structure pertaining to all viruses. Even the insight gained from the Watson-Crick model of DNA and the central dogma of the genetic structure of the virus were in no way able to change the image of a simple virus structure, which was of a single, linear nucleic acid molecule plus one or a few additional proteins constituting the protective sheath. Other virus components such as carbohydrates and lipids seemed insignificant – apart from in a very few cases of animal viruses – until, in 1963, Best and Katekar demonstrated the 18,8% lipid content of the *Tomato spotted wilt virus*. The neuraminidase observed in the influenzavirus was considered an exception.

Perspectives changed rapidly, however, as laboratory methods improved. Analysis of the components of the virus nucleoproteid complex was soon possible qualitatively as well as quantitatively. Parallel to the advances in general molecular biology, virus research began to focus more strongly on various viral, virus-asociated, host, and virus-induced host enzymes.

Nucleic acids

Determination of the type of nucleic acid was indispensable to the definition of the virus. Smaller viruses were thought to have a single nucleic acid molecule; whilst in larger viruses the nucleic acid was divided amongst several molecules. Since the pattern of viral DNA was the open linear, double helix, the ratio of the base-pairs, i.e. adenine : thymine and guanine : cytosine was as well suited for defining the virus as was its molecular weight or its sedimentation constant. According to the central dogma that one DNA strand served as template for single-stranded mRNA, viral RNA was thought to be a single, linear strand. The presence of uracil indicated RNA, which could be further characterized by ultracentrifugation.

In order to define the nucleic acid type more precisely, it was essential to isolate it, purify it and prepare it for EM. Although the EM method offered the most reliable description of the nucleic acid form, results gained by other means soon met with equal recognition. These methods included sensitivity to radiation and to substances which inhibited DNA or RNA synthesis, resistance to DNase or RNase, alkali susceptibility, banding position in caesium chloride density gradients, or the determination of the melting points. With increased

Table 5.2. Nucleic acid structures

Year	Virus[a]	Nucleic acid type[b]	Authors
1959	Phage φX 174	ss–DNA	Sinsheimer
1962	EMC	ss–RNA	Montagnier, Sanders
1962	Vaccinia	ds–DNA	Joklik
1963	Reo	ds–RNA	Gomatos, Tamm
1963	TMV	ss–RNA	Sugiyama, Fraenkel-Conrat
1964	Cytopl. polyhedr.	ds–RNA	Hayashi, Kawase
1965	Murine leukemia	ss–RNA	Galibert et al.
1965	Murine sarcoma	ss–RNA	Mora, McFarland
1966	Mouse mammary tumour	ss–RNA	Duesberg, Phillis Blair
1962	Influenza	Several segments	Hirst
1968	Influenza	6 Segments	Pons, Hirst; Duesberg
1968	Cauliflower mosaic	ds–DNA	Shepherd et al.
1968	Blue tongue	ds-RNA	Verwoerd
1967	PST Viroid	naked ss–RNA	Diener, Raymer

[a] EMC, Encephalomyocarditis virus; PST, potato spindle tuber.
[b] ss, single-stranded; ds, double-stranded.

use of these more sophisticated methods, a startling number of publications began reporting viral nucleic acid forms which deviated from the expected linear double-stranded DNA and linear single-stranded RNA. Table 5.2 contains some of these results; they indicate how numerous the forms of virus nucleic acid can be, and make all too clear the virtual impossibility of deriving valid, simple and general rules.

It had been observed that a virion contained either DNA or RNA in one single molecule or segment, just as it had been established that nucleic acid was either single- or double-stranded. Thus, the nucleic acid of virions displayed a variety of structure surpassing even those of the higher organisms known at that time.

Proteins

The rapid development especially of electrophoretic and adsorption-chromatographic methods for protein analysis soon led to new findings on the second most important virus component, the protein.

The simple protein which Schramm had defined in *TMV* in 1943 was encountered again in plant viruses of similar structure. Animal viruses and bacteriophages displayed greater variation in protein structure, although the structure itself seemed simple. However, studies published on amino acid and protein determinations soon disclosed the multiplicity of viral proteins not only within the entire field of viruses, but within single virus species, as well. In 1973 Honess and Roizman identified 49 polypeptides involved in the synthesis of *HSV*, which consumed 75% of the coding capacity and contained 23 structural proteins. Esche and Siegmann (1982) quoted 19 early and 22 late proteins for *Adeno virus 19*. In 1982 Prusiner shattered the supposition that viral proteins were simply a protective mechanism for nucleic acid. This occurred when he found the prion, a tiny protein particle lacking nucleic acid but capable of inducing scrapie, a well-known latent disease in sheep.

The cautious surmise in the early 1980's that viruses also contained enzymes was modified shortly thereafter when it became clear that enzymes comprised a considerable proportion of virion components. Furthermore, it was found that virus-associated or virus-induced enzymes were essential in virus host-cell interaction. Thus, it was soon necessary to differentiate between viral structural proteins and non-structural proteins, the latter including virus-coded enzymes essential for distinct periods of the multiplication cycle in virtually all virus

Table 5.3. Detected viral structural proteins

Virus	Year	Number of proteins	Authors
Influenza	1961	3	Leslie Hoyle, Sheila Davies
	1972	6	H.Klenk et al.
φx 174	1963	1	Carusi, Sinsheimer
	1968	4	Hutchinson
Adeno	1968	9	Maizel et al.
	1973	10	Everitt et al.
Rous sarcoma	1968	3	Duesberg et al.
SV 40	1967	3	Anderer
	1971	6	Mary Estes et al.
Poliomyelitis	1963	4	Maizel
Herpes simplex	1965	4	Tokumaru
	1966	10	D.Watson et al.
	1973	23	Honess, Roizman

Table 5.4. Viral enzymes

Enzyme[a)]	Virus	Year and Authors
Neuraminidase	Influenza	1942 Hirst
Polys.depolymerase	Phage Kp	1956 M.Adams, Park
Virolysin	Staphyloc.phage	1961 Ralstone et al.
Capsule-depolym.	Phage A22	1962 Eklung, Wyss
DNA polymerase	Vaccinia	1962 M.Green, Magdalena Piña
Protein kinase	Smallpox	1963 Paoletti, Moss
Lysozyme	Phage T2	1963 Maas, W.Weidel
DNA-dep. RNA polym.	Rabbit pox	1967 Kates, McAuslan
Sealase (ligase)	Rabbit pox	1967 Kates, McAuslan
DNA-dep. RNA polym.	Vaccinia	1967 Kates, McAuslan; Munyon et al.
RNA-dep. RNA polym.	Reo	1968 Shatkin, Sipe; Borsa, Graham
Nucl.phosphoh.I,II	Smallpox	1968 P.Gold, Dales
Endo- and Exo-DNase	Smallpox	1969 Beatrize Pogo, Dales
RNA-dep.transcript.	Vesic.stomatitis	1970 Baltimore et al.
	Newcastle disease	1970 Alice Huang et al.
Rev. transcriptase	Rous sarcoma	1970 Temin, Mizutani; Baltimore
ATPase	Frog virus 3	1971 Villaginés; McAuslan
Poly A polymerase	Vaccinia	1973 M.Brown et al.
Proteinase	Rous sarcoma	1977 von der Helm

[a)] DNA-dep. RNA Polym., DNA-dependent RNA polymerase; Nucl. phosphoh., nucleotide phosphohydrolase; Polys., polysaccharide; Rev., reverse.

types. However, it was not only the number of pure proteins found, which increased, the number of newly discovered lipoproteins and glycoproteins rose in direct proportion to the increased complexity found in animal viruses. The majority of these proteins was observed in the envelope and exhibited a host cell-specific composition.

Tables 5.3–5.5 depict how the discovery of viral proteins and virus-induced proteins increasingly shifted the focus to more minute detail. Since information on the relationship between enzymes and plant viruses documented in the literature applies exclusively to virus-induced changes of plant enzymes, they have not been included in the Tables. Virus-induced proteins which attained importance in tumour development and as antigens have been described in the sections devoted to viral diagnosis (see Chapters 5.3.4 and 5.4). This compi-

Table 5.5. Virus-associated or virus-induced enzymes

Enzyme[a]	Inducing virus	Year and Authors
Deoxyribonuclease	Phage	1952 Pardee, I.Williams
Polysaccharide-DP	Phage	1956 M.Adams, Park
Deoxycytidylat-HM	Phage T2	1959 Flaks, S.S.Cohen
Thymidylat-S	Phage T2	1959 Flaks, S.S.Cohen
Arginase	Shope Papilloma	1959 S.Rogers
Deoxycytidine 5′-PA	Phage T2	1960 Keck et al.
Capsule-depolym.	Phage 22	1962 Eklund, Wyss
DNA polymerase	Phages T2 and T4	1962 Aposhian, Kornberg
Thymidine kinase	Vaccinia, HSV	1963 Kit et al.
RNA polymerase	Polio	1963 Baltimore et al.
RNA-dep. RNA polym.	Phage MS2	1963 Weißmann, Borst
Exonuclease	Phage λ	1964 Korn, Weißbach
RNA-dep. RNA polym.	Phage f2	1965 August et al.
DNA methylase	Phage T	1966 Hausmann, M.Gold
Sealase	Phage	1967 A.Becker et al.
Polynucl. ligasde	Phage T4	1967 Weiß, Richardson

[a] HM, hydroxymethylase; PA, phosphate deaminase; Polynucl., polynucleotide; polym., polymerase; depolym., depolymerase; dep., dependent.

lation shows that enzymes often comprise a part of virion structure and, after penetrating the host cell, they facilitate the initial step of multiplication, namely the transfer of information for viral nucleic acid synthesis.

By and large, proteins proved to be highly diversified, both as structural virus components and as virus-induced or virus-associated substances. They even seemed to offer suitable distinguishing characteristics for classifying viruses. With regard to the enzymes, it was obvious that viruses were more capable to induce and to associate enzymes via the host cell than to integrate them in the virion. Conflicting statements about the relationship of an enzyme to its host cell bear witness to the deficits in experimental investigation, and make pitifully clear that each virus behaves according to its own rules.

Lipids

The supposition that lipids were host cell-dependent and occurred only in complex viruses was given up in 1963 when a lipid fraction of 18.8% was found in *Tomato spotted wilt virus*. That this was actually the result of an interaction with plant cell metabolism and not simply an association with lipids within the host cell, was shown by Armbruster and Beiß in 1958, and confirmed by Best in 1968. The former had found phospholipids in *Influenza virus APR 8*, although these were missing completely in the host cells, namely CAM. Best, furthermore, had discovered a significantly higher fraction of lipids in the leaves of tomato plants following a virus infection.

Since the investigations of Burnet and Dora Lush (1940) and those by Andrewes and Dorothy Horstmann (1949), it has been known that myxoviruses can be dissociated and inactivated when treated with lipid solvents or surfactants. After this effect was successfully demonstrated for other viruses, and comprehensive comparative lipid analyses published, R. Franklin divided lipid-containing animal viruses into two groups in 1962:

1. *Viruses with peripheral structure lipids*: Myxoviruses, arboviruses (arthropod-transmitted viruses) *Vesicular stomatitis virus*, *Measles virus*, herpesviruses, avian tumour viruses, *Mammary carcinoma virus* in the mouse, leukemia viruses in the mouse, *Cytomegalo* and *Rabies virus*.
2. *Viruses without peripheral structure lipids*: Picornaviruses, Vesicular exanthema viruses, reoviruses, adenoviruses, *Polyoma* and *Shope papilloma virus*, and pox viruses.

The viruses of the first group were thought to assemble in cell membranes, whereas those of the second assembled independently of cell membranes. Since there is no uniform cell membrane and since it is possible for a single virus to multiply in various kinds of host cells, the analyses of viruses with peripheral structure lipids reflect the lipids of the host cells. The products of viral synthesis are neutral lipids, phospholipids and glycolipids.

Carbohydrates

Glycoproteids and glycolipids have been identified as the viral carbohydrates found in viral envelopes. Glycoproteids with special binding functions are known as spikes. Besides those carbohydrates known in 1957, galactosamine, glucose and neuraminic acid may be added. Viral glycoproteids are structural elements with a special ability to bind to viral lipids, to proteins and to cell membranes. As elements of viral surfaces, they are very important factors for determining the course of infection, for diagnosing diseases, for producing vaccines and for developing anti-viral chemotherapeutics. Table 5.6 shows the number of glycoproteids detected in some viruses.

Table 5.6. Viral glycoproteides (gp)

Year	Virus	Detected gp	Authors
1970	Rous sarcoma	2	Duesberg et al.
1972	HSV	4	Patricia Spear, Roizman
1972	Avian tumour	2	H.Bauer, Bolognesi
1972	Vaccinia	2	Sarov, Joklik
1972	Fowl plague	3	H.Klenk et al.
1976	Sindbis	2	Dalrymple et al.
1978	Vesic. stomatitis	1	Little, Alice Huang

5.1.2.2 The Molecular-Genetic Structure of Virions

The DNA model in 1953 was of a single long, polar, dextro-gyratory, double-stranded helical molecule with open ends, and with hydrogen bond bridges; the molecule was capable of precise reduplication, termed replication. One strand presented the template for an equally long, single-stranded RNA which functioned as an mRNA by transmitting the information for translation, i.e. for synthesizing proteins. Both nucleic acid types were found in higher organisms and, for want of more experimental findings on viruses, it was simply assumed that both types were typical also for viruses. It had been clearly established, however, that not only was mRNA transcribed from DNA able to carry the information required for protein synthesis, but an RNA capable of replication could also fulfil this function.This concept of viral nucleic acids soon proved to be an oversimplification. Table 5.7 which lists nucleic acids of various structure in the chronological order in which they were discovered, should be augmented by further information about genetic peculiarities.

In 1962, Kassanis discovered a small plant virus which could only be induced to multiply with the help of *Tobacco necrosis virus* (TNV). He termed this small, dependent virus satellite virus and the TNV a helper virus. There was no appreciable sequence homology between the genomes of these two viruses. Later, this phenomenon of mutual dependence was observed in animal viruses as well as in other plant viruses and led to the general term satellitism and to a special term virusoid for satellite plant RNA. Worthy of mention is Rous associated virus (RAV), originally discovered by Rubin und P.Vogt in

Table 5.7. Structural properties of viral nucleic acids

Year	Virus[a]	Nucleic acid[b]	Authors
1962	P. φX 174	c ss DNA	Fiers, Sinsheimer
1963	P. λ	Cohesive ends	Hershey et a.
1963	Fowlpox	Cross-linked strands	Szybalski et al.
1964	Phage T4	Terminal redundancy, Circular permutated DNA	Streisinger et al.
1967	Newcastle dis.	ss minus RNA	Nakajima, Obara
1968	Reo	10 ds segments	Shatkin et al.
1969	Phage PM2	Closed c ds DNA	Espejo et al.
1970	Phage λ	ds DNA in vitro c ds DNA in vivo	Lee, Davidson
1970	Cytopl polyh.	10 ds RNA segments	Fujii-Kawata et al.
1970	Papilloma	Superc c ss DNA	L. and E.Crawford
1971	Penic.stolon.	Two similar DNA's within one virion	Bozarth et al.
1973	Phage φ 6	segm hel ds RNA	Semancik et al.
1973	Densonucleolus	+ Strand and − strand in different capsids	Kurstak et al.
1975	Hepadnavirus	c Partial ds DNA with ss gaps	J. Summers et al.
1976	Influenza	Eight segments	Five working groups
1976	Viroids	c ss RNA	Sänger et al.
1986	Hepat.δ	c ss RNA	K.Wang et al.

[a] P., Phage; Penic.stolon., Penicillium stoloniferum; Cytopl Polyh, cytoplasmic polyhedrosis; Hepat., hepatitis;

[b] ss, single-stranded; ds, double-stranded; c, ciruclar; segm, segmented; hel, helical; d, disease.

1962. Two years later, Atchison and coworkers first observed the Adeno-associated virus (AAV) in the electron microscope. In 1986, K.Wang and colleagues described the Hepatitis δ virus as a satellite of *Hepatitis B virus*. Satellites do not possess a fully functioning nucleic acid to code for replication or protein synthesis, relying on helpers for these functions. Satellites may be deficient or defective. Deficient means insufficient to replicate without a helper *or* unable to synthesize protein; if the deficiency relates only to protein synthesis, this denotes that the virus exists only in its unstable nucleic acid form. A damaged or defective virus is capable of replication and protein synthesis but it produces an unsuitable protein which consequently cannot be used for virus assembly. Such a virus is incomplete. While investigating satellitism between *RAV* and *RSV*, Hidesaburo Hanafusa and colleagues in 1963 demonstrated that new and infectious RSV developed only if its helper RAV was present in the same cell. The helper function of RAV was needed only for constructing the sheath of the virus, and malignant transformation of RSV did not require synthesis of mature virus particles. *HBV* helps its satellites in the same manner. The phenomenon of activation of satellitism entails a helper virus or activator responsible mostly for protein synthesis. Instead of referring to this as a deficiency or a defectivity, the term unstable form is frequently used.

In 1963 Montagnier and Sanders demonstrated the existence of double-stranded RNA as replicative RNA in cells infected with the single-stranded RNA of *Mouse encephalomyocarditis virus*. Soon thereafter, when similar discoveries were reported with reo, polio, *MFDV*, *Silkworm polyhedrosis virus*, *TMV*, *Cowpea mosaic virus* and phages, it was essential to determine whether double-strandness was linked to virion structure or whether a replicative intermediate form was involved. In order to designate the RNA strands clearly, Weißman and colleagues (1964) termed the mRNA strand present in phages the plus strand; Baltimore and Girard (1966) called the two intermediate forms in the synthesis of polio virus, shown by Hattie Alexander and coworkers to possess infectious RNA in 1958, plus and minus; Weißman and Feix generally spoke of a replicative plus and minus strand complex in the context of viral RNA synthesis. The significance of mRNA for all viruses was compiled in Baltimore's six-class system of animal viruses in 1971. He labeled the infectious coding mRNA the plus strand and indicated its complementary minus strand.

During in vitro experiments a number of topological pecularities on circular DNA were discovered which led to phenomena termed cohesive ends, circular permutation and terminal redundancy. In1963 Hershey and colleagues found cohesive ends in double-stranded DNA of *Phage* λ in the form of 20 nucleotide, single-strand segments at both 5′-ends. These strands were complementary and capable of forming rings. A ring ensued not by directly closing the ends of the strands, but by H-bonds or by covalent or strong iony bonds in the single-strand regions. The other two phenomena were the subject of genetic experiments on *Phage T4* published by the Streisinger group in 1964. They explained the ring formation of linear double-stranded DNA after denaturation and renaturation by terminal redundancy of single strands with identical bases but unequal sequences. When the strands linked during renaturation to become double-stranded, single-stranded ends emerged lacking base partners. This caused a new, modified order, and changes of positions. In this changed organization, single-stranded regions remained outside the ring, which was closed by H-bonds.

5.1.3 The Taxonomy and Classification of Viruses

As cell culture techniques began to be used more widely in virology laboratories, there was a remarkable increase in the number of newly discovered viruses as well as corresponding details about them. Soon the known facts could no longer be accommodated in the above mentioned seven groups so that new biochemical and biophysical findings, e.g. acid-sensitivity and reaction to ether were included in the classification scheme. When regional and departmental committees were clearly unable to set up sub-classifications, not to mention the organization of the complete field of viruses, the IX International Congress of Microbiology in Moscow organized and delegated the virus section of the International Association of Microbiological Societies (IAMS) in 1966 "to find a universal taxonomic system for all viruses". In order to carry out this assigment, subcommittees were formed for bacteriophages, cryptograms, viruses occurring in invertebrates, viruses found in vertebrates, and plant viruses; in addition, study groups were assigned to each virus group. In 1973, the International Committee on Nomenclature of Viruses (INCV) was renamed International Committee on Taxonomy of Viruses (ICTV) with the following goals:

1. To define and to list groups (or genera) of viruses
2. to type members (species) of these groups
3. to propose names for the groups (genera)
4. to explore the use of a cryptogram as a taxonomic device

The Classification and Nomenclature of Viruses which was submitted in 1971 comprised 43 groups and acknowledged the following guidelines established in 1966:

1. The code of bacterial nomenclature shall not be applied to viruses.
2. Nomenclature shall be international and universally applied to all viruses.
3. An effort will be made towards a latinized binomial nomenclature.
4. The law of priority shall not be observed.

5. No person's name shall be used.
6. Sigla (acronyms) can be used but new sigla shall not be introduced.
7. The species is considered to be collections of viruses with like characteristics. The ending of the name is virus.
8. The genus is a group of species sharing certain common characters. The ending of the name is -virus.
9. A family is a group of genera with common characteristics, the name will end in -idae.

The resulting cryptogram comprised the following four pairs of symbols:

1st pair:Type of nucleic acid/strandedness of nucleic acid
2nd pair: Molecular weight of nucleic acid/percentage of nucleic acid in infective particles
3rd pair: Outline of particle/outline of nucleocapsid
4th pair: Kinds of hosts infected/kinds of vectors

The lack of consensus, even today, on whether a virus should be considered living or not is reflected in the nomenclature, which includes the alternative family = group; genus = subgroup; and species = type. The definition of species stirred up an interesting discussion which does not promise to be resolved in the near future. The concern is not only what a species (= type) is, but how it should be labelled, since endeavours toward binomial nomenclature have not yet been productive. Until this is the case, it seems reasonable to characterize the species by the separate word "virus".

The next report submitted in 1979 also contained subfamilies but ending in "virinae". It dispensed with cryptograms and classified virus families or groups according to their molecular genetics. Acceptable rules on nucleotide sequence homology, gene arrangement and gene function would doubtlessly prove useful for the classification and taxonomy of viruses.

The fourth report in 1982 contained additional rules for classification, for example, including an English vernacular name to standardize terminology. In addition to actually classified viruses, others were listed by name or number as possible or probable be members of a family, genus or group. Tables 5.8. and 5.9 attempt to give a survey of DNA and RNA viruses defined according to present guidelines.

About 30 additional virus groups were established for single-stranded non-enveloped RNA plant viruses named by a sigla. They

Table 5.8. Classified DNA viruses of animals and lower organisms

Family (group) ..viridae	Subfamily (subgroup) ..virinae	Genus ..virus	Prototype virus	Host range
Pox..	Chordopox..	Orthopox..	*Smallpox,*	Human, monkey, rabbit, mouse, buffalo, bovine, camel
		Parapox..	*Milkers node*	Bovine, milkers
		Avipox..	*Fowlpox*	Various fowls
		Capripox..	*Sheep pox*	Sheep, goat
		Leporipox..	*Rabbit fibroma*	Rabbit, hare, squirrel
		Suipox..	*Swinepox*	Swine
	Entomopoxvirinae	A, B, C	*Poxviruses of insects*	Coleoptera, Lepidoptera, Diptera
Herpes..	Alphaherpes..	Simplex..	*HSV 1*	Human
		Varicello..	*Varicella/Zoster*	Human, murine
	Betaherpes..	Cytomegalo..	*Human cytomegalo*	Human
		Muromegalo..	*Mouse cytomegalo*	Mouse, pig, rat
	Gammaherpes..	Lymphocrypto..	*EBV*	Human
		Thetalymphocrypto..	*Marek's disease*	Turkey, fowls
		Rhadino..	*Ateline herpes 2*	Rabbit
Baculo..		A	*Nuclear polyhedrosis*	Insects
		B	*Granulosis*	Lepidoptera
		C	*Oorytes rhinoceros*	Rhinoceros
Polydna..		Polydna..	*Polydna virus 1–4*	Insects

Table 5.8 (continued)

Family (group) ..viridae	Subfamily (subgroup) ..virinae	Genus ..virus	Prototype virus	Host range
Plasma..		Mycoplasm..	*Mycoplasma virus t*	Mycoplasm
TTV1 group			*Thermoproteus tenax 1*	Archaebacterium thermoproteus tenax
Irido..		Irido..	*Chilo iridescent*	Wide range of insects
		Chlorirido..		Wide range of insects
		Rana..	*Frog virus 3*	Frog
		Lymphocystis..	*Lymphocystis virus of fish*	Fish
Adeno.		Mastadeno..	*Human adenovirus 2*	Human, cattle, primates, sheep, horse, dog, goat, mouse
		Avian adeno..	*Fowl adenovirus 1*	Fowl, turkey, goose, pheasant, duck
		Rhizidio..	*Rhizidiomyces sp. F*	Rhizidiomyces
Papopa..		Papilloma..	*Rabbit papilloma*	Human, rabbit, cow, deer, dog, goat, hamster, horse sheep,
		Polyoma..	*Polyoma*	Human, mouse, rabbit, baboon, monkey
Caulimo..		Cauliflower..	*Cauliflower mosaic*	Cauliflower, dahlia, strawberry
Tecti..		Tecti..	*Phage PRD 1*	Enterobacteria, bacilli
Cortico		Cortico..	*Phage PM2*	Marine pseudomonads,

Myo..		Cyanomyophage	*Cyanophage AS-1*	Filamentous cyanobacteria
Stylo..		Cyanostylophage	*Cyanophage S-1*	Unicellular cyanobacteria
Podo..		Cyanopodophage	*Cyanophage LPP-1*	Filamentous cyanobacteria
Ino..		Ino..	*Phage fd*	Enterobacteria
		Plecto..	*MLV 51 phage*	Acholeplasma
Hepadna..		Hepadna..	*HBV*	Human, eastern woodschuck, squirrels, Peking duck
Parvo..		Parvo..	*Aleutian mink disease*	Cat, cattle, goose, rabbit, mink, rodents, feline, porcine
		Dependo..	*Adeno-associated*	Adeno virus
		Denso..	*Galleria mellonell*	Lepidoptera, diptera, orthoptera
Gemini..	A (vectored by leafhoppers)		*Maize streak*	Wheat, tobacco, maize
	B (vectored by whiteflies)		*African cassava mosaic*	Bean, tomato, tobacco, cassava, soybean
Micro..		φX phage group	*Phage* φX 174	Enterobacteria

Derivation of some names:
Pox: pustule, ulcer; Lepori: hare; Herpes: creeping, crawling; Irido: shining rainbow-like; Mast: breast; Polyoma: for many tumours; Myo: muscle; Parvo: small; Ino: muscle; Adeno: gland; Podo: foot;
Avi: bird; Sui: swine; Capri: goat; Rana: frog; Papilloma: nipple; Tecti: covered; Stylo: column; Denso: compact; Cysto: sack; Cortico: crust; Gemini: twins
Papova: Sigla from papilloma, polyoma and vacuolating agent; Baculo: stick, morphology-related; Caulimo: sigla from cauliflower mosaic

Table 5.9. Classified RNA viruses of animals and lower organisms

Family (group) ..viridae	Subfamily (subgroup) ..virinae	Genus ..virus	Prototype virus	Host range
Cysto..		Cysto..	*Phage* φ*6*	Pseudomas
Reo..		Reo..	*Reovirus 1*	Man, monkey, bird, cattle, bat
		Orbi..	*Bluetongue*	Insects, man, horse, monkey, rabbit, cattle, deer
		Rota..	*Human rotavirus*	Man, mice, calves, piglets, foal, lamb
	Plant subgroup 1	Phytoreo..	*Wound tumour*	Leafhoppers of gramineae
	Plant subgroup 2	Fiji..	*Rici dwarf*	Flowering plants, plant hoppers
Birna..		Birna..	*Infectious pancreatic necrosis*	Man, equine, bovine
Myco..	Helminthosporium maydis virus group			Fungi
	Penicillium stoloniferum virus group			Fungi
	Penicillium chrysogenum virus group			Fungi
Toti..		Toti..	*S. cerevisiae virus LA*	S. cerevisiae
Partiti..		Parti..	*Gaeumannomyces graminis*	Fungi
Toga..	Arboviruses A	Alpha..	*Sindbis*	Man, equine
		Rubi..	*Rubella*	Man
		Pesti..	*Mucosal disease*	Bovine, equine, hog
		Arteroi..	*Equine arteritis*	Equine
Flavi..	Arboviruses B	Flavi..	*Yellow fever*	Vertebrates, invertebrates

Corona..		Corona..	*Avian infectious bronchitis*	Man, feline, rat, cattle, porcine, fowls
Orthomyxo..		Orthomyxo..	*Types A, B, C*	Man, primates, pig, horse, bird, dog, cattle, bear
Paramyxo..		Paramyxo..	*Newcastle disease virus*	Man, avian, murine, bovine, simian, canine
		Morbilli..	*Measles*	Canine, bovine, ovine
		Pneumo..	*Respirator syncytial*	Man, bovine, mice
Crypto..	A and B		*White clover cryptic*	Plants (not associated with obvious symptoms)
Rhabdo..		Vesiculo..	*Vesicular stomatit*	Vertebrates, insects
		Lyssa..	*Rabies*	Vertebrates, insects
	Plant subgroup			Wheat, broccoli, lettuce, aphids, hoppers
	Plant subgroup B			A wide range of plants, aphids, hoppers, mites
	Mammal subgroup			Mammals
Filo..		Filo..	*Marburg*	Man, mouse, monkey, guinea-pig, hamster
Bunya..		Bunya..	*Bunyamwera*	Vertebrates, arthopods
		Phlebo..	*Sandfly fever Naples*	Vertebrates, arthopods
		Nairo..	*Crimean-Congo h.f.*	Sheep
		Uuku..	*Uukuniemi*	Vertebrates, ticks
		Hantavirus	*Hantaan*	Man, rodents
Arena		Arena..	*Lymphocytic choriomeningitis*	Man, mice, hamster

Table 5.9 (continued)

Family (group) ..viridae	Subfamily (subgroup) ..virinae	Genus ..virus	Prototype virus	Host range
Retro	Onco..	Type C, B, D tumor..		Mammals, avian, reptilia
	Spuma..	Spuma...	*Foamy*	Mammals
	Lenti..	Lenti..		Sheep, goat
Picorna..		Entero..	*Human poliovirus*	Man, murine, simian, porcine, bovine
		Cardio..	*Encephalomyocardis*	Murine
		Rhino..	*Human rhino*	Man, bovine
		Aphto..	*FMDV*	Mammals
Calici..		Calici..	*Ves. exanthema of swine*	Man, swine
Levi..		Levi..	*Phage MS2*	Bacteria
Tetra.		Nadaurelia group		Lepidoptera

Derivation of some names:
Cysto sack; reo: sigla from *r*eo *e*nteric *o*rphans; rubi: reddish; fiji: a county; toga: Roman gown; rabdo: rod; lyssa: rage; calix: cup-shaped in EM; arena: sandy appearance in EM micrographs
orbi: ring; rota: wheel; flavi: yellow; corona: crown; levi: light; vesiculo: bladder; spuma: foamy; h.f.: hemorrhagic fever; ves.: vesicular
Bunya: Bunyamwere, a place in Uganda; Phlebo: Refers to phlebotomine vectors, gr. "vein"; Uuku: From Uukuniemi, place in Finland; Nairo: From Nairobi

were not yet fully defined but according to their genome structure were separated into viruses with a monopartite, dipartite or tripartite genome, i.e. a genome contained within one, two or three molecules. Examples of these groups are: tobamoviruses, with the prototype *TMV*, comoviruses with *Cowpea mosaic virus*, and cucumoviruses with *Cucumber mosaic virus*, respectively. Further progress was also made in the field of the viroids with the main criterion for their classification being sequence similarity.

The available experimental data has to date aided in the classification of roughly 30% of the approximately 4000 species or types of viruses known. Greatest progress has been achieved in the animal viruses (45%), while, with only 6% of 2200 bacteriophages and about 56% of the 1000 plant viruses classified, there is still much to be done.

5.2 The Virus Multiplication Cycle

The principles of virus multiplication were adopted from bacteriophage research after it had been clearly established that phages were, indeed, viruses found in bacteria. In 1949, W.Henle proved only 1-3% of free, infectious particles in suspensions of influenzavirus shortly after beginning his experiments. Similar observations were reported by Schäfer and Munk while working with *Classical fowl pest virus*. Since the viruses disappeared within the cell and were, therefore, no longer visible under EM, Wecker and Schäfer employed biochemical methods for their intracellular detection. They found slower sedimenting components which they were able to identify by means of markers as virus fragments. Thus became the beginning of biochemical investigations of the multiplication cycle as an adjunct to conventional EM methods. The cycle, from the first contact of the virus with the host cell to the release of newly synthesized particles, is a continuous process, and can be understood more readily by dividing it into its various interacting phases.

In 1959, Brenner and coworkers proved that many proteins participated in the synthesis of *Phage T4*, and in 1963, R.Epstein and colleagues found that the morphogenesis of the same virus was directed by no less than 40 genes. According to the 1 gene = 1 enzyme hypothesis, a minimum of 40 enzymes was involved, and it began to appear more and more likely that the enzymes required for this process were greater than the number lent by the host cell. The virus multiplication cycle ultimately turned out to be an enzyme-dependent specific interaction of a defined viral parasite and its host cell, involving the virus itself and virus-associated, virus-induced and cellular enzymes in the genetically controlled course.

5.2.1 Adsorption

This term was generally used for the initial phase, although virus adsorption exhibited features common to physical-chemical adsorption only at the very beginning of the phase. During the search for a suitable partner – known as attachment – not only did the Brownian movement of the particle play a role but also specific physical-chemical factors influenced the process. In 1960, Allison and Valentine determined the actual adsorption rate of *Fowl plague* and *Vaccinia virus* in monolayers to be 50% of the theoretical values. The adsorption was not inhibited by protein or by increased temperature but it was clearly dependent upon the concentration of multivalent cations. By 1957, Bachthold and coworkers had described increased attachment of the poliovirus, similar to the observations of Ann Farnham and Anne Newton with *HSV*, in the presence of the inorganic salts $CaCl_2$ and $MgCl_2$.

Attachment entails the negatively charged virus surface finding a suitable binding site on the accessible, negatively-charged surface of the host cell membrane. Hirst called these binding sites "receptors" after observing them in hemagglutination in 1942, and while infecting respiratory tract cells of ferret lungs with influenzavirus in 1943. In 1952, Szollosy and colleagues carried out further trials with mice, rats, hamsters, guinea-pigs, rabbits and chickens, obtaining the same results. In 1952, Tolmach and Puck, and in 1957 Tolmach, again demonstrated the influence of pH on the phage/bacteria system, i.e. that the addition of cations expedites the partner-searching step. As far as temperature was concerned, there were conflicting results, perhaps explainable by the fact that the first three phases, attachment, adsorption and virus uptake, had not yet been separately investigated experimentally.

Whether attachment led to adsorption in every case was just as controversial as whether certain chemical substances influenced the search-for-attachment phase. Once a suitable receptor has been found, the adsorption phase begins and specific antibodies can no longer exercise an inhibiting influence. That receptors have a definite chemical structure had already been concluded from earlier investigations on the hemagglutination of influenza viruses. Experimental poliovirus infection studies, and findings on tissue tropism resulting from other animal viruses led to the same conclusion.

In vitro hemagglutination systems enabled Zilliken and coworkers, as well as E.Klenk and Lempfried in 1957, to demonstrate the reaction of virus neuraminidase with the neuraminic acid residue of the mucoprotein in the erythrocyte membrane. Since neuraminidase and hemagglutinin are localized as glycoproteins in spikes, the significance of both glycoproteins and spikes, as well as fibres and the baseplates for bacteriophages in adsorption, was finally recognized. McLaren, Holland and Syverton searched for the polio receptor in 1959 and discovered it only in sensitive primate tissue and identified it as a lipoprotein. Later, they were able to prove that isolated polio RNA could infect the cells of non-primates which were normally non-susceptible.The results of receptor experiments on hybrid human-mouse cells, in which human chromosomes had been discarded, led Dorothy Miller and coworkers (1974) to the conclusion that the polio receptor gene was linked to human chromosome 19. Investigations carried out by Abraham and Colonno in 1977 showed that many rhinoviruses shared the same cellular receptor, although four of the 24 tested serotypes had different receptors.

Binding to a tissue receptor does not automatically ensure multiplication, as both Krontiris and colleagues (1973) testing *Friend leukemia virus* and Menezes and colleagues (1977) working with *EBV*, were able to demonstrate. In these systems, lymphoblastoid T-cells proved suitable for adsorption but not for multiplication. There are, however, other obvious mechanisms to attach viruses to tissue suitable for multiplication. Fazekas de St.Groth came to this conclusion as early as 1948 in influenza experiments, when infection ensued despite prior destruction of the receptors. Consequently, neither chemically defined receptors nor a "combining area" were absolutely essential; uptake of the virus particle could proceed in a manner similar to that of a colloid dye. Fazekas termed this alternative process viropexis, although his results could also have been taken to show that occupied or destroyed receptors were replaced by newly formed ones within very short time.

Glycoproteids and lipoproteids constitute the prevailing virus receptors which occur in thousands on the surface of each and every cell, and offer possibilities for the adsorption of many particles of the same virus or of different viruses. In 1976 Lonberg-Holm and colleagues determined that the number of receptors on a single cell for a non-enveloped virus ranged from 10000 to 100000. Furthermore, it was found that several non-enveloped animal viruses shared specific receptors. These workers subsequently devised a receptor map for many viruses within four different receptor families, and deemed it possible that receptor-specifity influenced virus tropism in tissues and organs which might, eventually, evoke viral pathogenicity.

Despite the evidence afforded by numerous experiments, hypotheses on virus binding to host cell are manifold and often contradictory, even in the case of a single virus. Binding times are sometimes cited in minutes, although constant adsorption rates spanning hours have also been reported. Biochemical results on such a crude basis have never been exact enough to produce defined and safe anti-viral substances. The only statement that receptors belonged to viral surface substances such as carbohydrates, proteins and lipids has not produced functional results, even in concrete cases.

New perspectives of receptors have emerged in recent publications in which - apart from high-resolution electron microscopy and monoclonal antibody techniques - molecular genetic techniques have been applied, including hybridization, sequencing, and cloning. Relevant results on poliovirus, rhinovirus and human acquired-immunodeficiency virus published by Cathy Mendelsohn, Greve, Staunton and colleagues finally pointed to the immunoglobulins, and the subject of receptors has become closely associated with the field of immunity.

5.2.2 Uptake by the Cell

This phase is often termed penetration because it was patterned on the earlier concepts of influenza virus multiplication. According to that model, the virus bound itself to the receptor, dissolved it by means of its neuraminidase and thereby obtained entry into the cell. As early as 1949, Fazekas de St.Groth and D.Graham disproved this hypothesis, although it continued to be accepted and, in the course of time, led to the generally view that the virus alone assumed the active role in penetration. Between 1960 and 1962, Higashi and colleagues, together with Dales, Siminovitch as well as Choppin had demonstrated with *Smallpox virus*, *Vaccinia* and influenzavirus that a phagocytosis-dependent first contact between virus and host cell brought about intimate association between the virus and the host-cell membrane. In

1962, Leslie Hoyle interpreted electron micrographs of influenzavirus uptake as showing the dissolution of the lipid layer on the virus followed by fusion of the virus and cell the membrane.

The terms applied to virus uptake by the cell, phagocytosis, viropexis, pinocytosis and fusion, indicated that the cell was an active partner. As unequivocal as it might have seemed that the virus lost its structure upon uptake by the cell, and that in the eclipse phase virions were no longer detectable, there were initially other results. In 1962, Mussgay and J.Weidel found morphologically intact viruses in the cells of chick embryos 30 minutes after infecting them with *Newcastle disease virus* (NDV). They distinguished *Fowl plague virus* from *NDV* after finding that it was not necessary for *Fowl plague virus* to penetrate the cytoplasm as deeply as *NDV* for dissolution during fusion. In 1964, L.Crawford and Gesteland proved with their electron micrographs that really very few of the many bacteriophages which adsorbed to bacteria or their pili, were actually taken up by the cell.

Two fundamental different mechanisms have been suggested for virus uptake into the cell: endocytosis and fusion. According to Allison and Patricia Davies (1974) endocytosis is the uptake of soluble exogenous or particulate material from the cell environment by plasma membrane-derived vesicles or vacuoles. It is a temperature-dependent process. Virus particles bind to microvilli, i.e. the elongated membrane projections, like tufts of hair, on the cell surface, and move along these projections into the cell body, where they encounter special areas of the cell membrane, called pits, which are undergo continuously new development. On their inner surface, these pits may be covered with the protein clathrin. The complex of coated pit and attached virus is able to fold inwardly, be pinched off, and move into the cell as a coated vesicle or phagocytic vacuole. Once the clathrin, now on the outside, is removed, the vesicles or vacuoles fuse with lysosomes, allowing the lysosomal proteolytic enzymes to trigger the uncoating process near the nucleus. Steinmann and coworkers (1976) established that macrophages formed at least 125 pinocytic vesicles per minute, thus internalizing the equivalent of their cell surface area every 33 minutes. In addition to these results, they postulated that at the end of the uncoating process the internalized vacuole membrane is then recycled back to the cell surface. When endocytosis requires energy and can be inhibited by cytochalasin B, – an inhibitor of microfilament formation and phagocytosis – though not by colchicin, which also affects the mobility of the protein in the membrane, then instead of endocytosis, phagocytosis is said to occur.

There are two forms of the uptake of fluid known as pinocytosis: the fluid and the adsorptive form. During this process, the virus is rapidly internalized in coated vesicles. Macro-pinocytosis occurs when substances taken up lack a specific receptor on the cell surface. It involves fluid uptake in vesicles with a diameter of 300–2000 nm. Micro-pinocytosis, in contrast, involves the uptake of virus particles in a 70 to 100-nm diameter vesicle without the use of energy or the help of cellular microstructures. Considering these phenomena in relation to virus uptake by the cell, then endocytosis appears as a heterogeneous structurally, functionally and physiological process, yet for many viruses it is the definitive step to potential multiplication.

Cell fusion is an alternative possibility; the fusion of the virus membrane and the cytoskeleton under the influence of neuraminidase served as the in vitro model of virus uptake for paramyxoviruses. As early as 1958, Okada had found cell fusion with *Sendai virus* – a paramyxovirus identical to the *Hemagglutinating virus of Japan* (HVJ) – in Ehrlich tumour cells, although Compans and coworkers had described the uptake of *Parainfluenza virus SV 5* as phagocytosis in 1964. Fusion was further demonstrated by Hoggan and Roizman with *HSV* in 1959, Hidesaburo Hanafusa with the *Vaccinia virus* (1960), Heine and Schnaitman with *Vesicular stomatitis virus* (1971), and Susan Michelson-Fiske and colleagues with *Cytomegalo virus* (1975).

In 1970, Dourmashkin and Tyrrell showed in vivo that myxo- and paramyxovirus infections in the ciliated epithelial cells of guinea-pig trachea also resulted from fusion. Earlier fusion experiments with paramyxoviruses employed orthomyxoviruses as negative controls which explains why some researchers have difficulty in accepting fusion as a means of infection. Nevertheless, it is certain that many viruses enter the host cell in vivo by endocytosis as well as by fusion. Dales' contention (1965) that most viruses enter the cell by viropexis is no longer tenable today. In 1973, Granados proved cell uptake by fusion for an insect poxvirus in the intestinal epithelium of insect larvae. In 1981, Sunanda Banerjee and colleagues pointed out the probability of plant viruses also gaining entry into the cell by fusion, in their work with *TMV*, which is capable of fusing with liposome membranes.

The time required for uptake into the host cell may be expressed in minutes, the time range being very broad for different kinds of virus. In general, the first phases of the multiplication cycle are primarily interactions between viruses and cellular membranes.

5.2.3 Uncoating

This term covers all direct action involved in separating viral nucleic acid from the virus in preparation for replication. This requires, above all, the bringing together of all the cellular hydrolases contained in the lysosomes. In addition, synthesis of a virus-coded uncoating protein may be necessary, as Dales thought probable in his 1963 *Vaccinia virus* study. During uncoating, the virus loses essential properties: it becomes sensitive to nucleases and specific chemical reagents, in addition to partially or completely losing its antigenic capacity. The time and site of uncoating depend on many factors. Uncoating may begin at the moment of penetration if, as with negative-stranded RNA viruses, unencapsidated nucleocapsids manage to enter the vesicle. On the other hand, there are also double-stranded RNA viruses which simply split into smaller particles.

According to Yvette Chardonnet and Dales (1970), adenoviruses first lose their fibres and pentons and remain partially digested in the cytoplasm. As for reoviruses, parts of the envelope are digested and the newly formed subviral particles are released into the cytoplasm, where they remain unchanged during the process of multiplication; the RNA polymerase they contain is activated to transcribe the 10 nucleic acid segments. Encapsidated viruses may release their nucleocapsids during fusion due to the acidic environment of the vesicle and thereby immediately enter the cytoplasm.

The uncoating process of the various virus types leads to the three states: free nucleic acid, a subviral particle and a nucleocapsid. The site of uncoating is the cytoplasm. Upon virus uptake into a vesicle or vacuole, the uncoating process begins and with it the eclipse phase, i.e. the virion dissolves and loses its infectivity. Then originating in the cytoplasma, the next phase of the multiplication cycle begins: the synthesis or replication phase, which can occur in the cytoplasm or in the nucleus.

5.2.4 The Synthesis Phase

This phase encompasses all biochemical processes in the host cell directed or influenced by viral nucleic acid. Thus, it includes not only the synthesis of new virion components but also the reaction, participation and fate of the cell. For the virus, this means identical reduplication or replication of the genome with ensuing transcription and translation, i.e. the conversion of the information contained in the mRNA into viral proteins in which cellular ribosomes are involved. Each type of virus is a unique case of reduplication, related to the size of its genome and the coding capacity for structural and non-structural proteins, especially enzymes. Thus, there cannot be only one valid rule for virus replication which goes beyond of the model system based on results with the bacteriophages.

In so far as synthesis is not only initiated in the cytoplasm but runs its entire course there, the term viroplasm was created for the cytoplasmic site of virion formation. The term virosome is also relevant here, these being identical to the inclusion bodies and X-bodies described earlier for some viruses.

5.2.4.1 DNA Viruses

The genetic structure of the DNA viruses known at present is single or double-stranded with genomes ranging from 4 to approximately 300 kilobase(kb). This would mean 4–300 genes per virion, assuming an average of 1 kb per gene. Thus, the question of the virus to host-cell ratio in virus replication has multiple answers: the more genes a virion has, the more of its own enzymes are available either as structural proteins or as a capacity to code for necessary enzymes via the host cell.

The molecular genetic-model for DNA-directed replication proceeds from a double strand, requires a DNA polymerase for synthesizing the identical DNA molecule, a ligase for ligating the synthesized DNA strands, a DNA-dependent RNA polymerase for protein synthesis, and additional enzymes depending upon the complexity of the virion structure. Many of the required enzymes are known, although their role in specific viruses still demands extensive investigation in order to clarify whether they are viral structural components, virus-induced, or virus-associated. It has been established, however, that DNA-directed replication can be divided into early and late phases with possible interim steps. The late phase is understood to begin with the synthesis of newly virus DNA.

As early as 1947, S.N.Cohen proved that bacterial DNA synthesis stops for a few minutes immediately after phage infection. In 1956, Kornberg and coworkers showed that *Phage T4* induces synthesis of its own DNA within just a few minutes of infection, at a rate of 10-times faster than the host cells of E.coli. In 1962, Aposhian and Kornberg discovered that the polymerase induced by *Phage T4* differed from host-cell polymerase. Edgar and Wood (1966) described the various steps of the morphogenesis of *Phage T4* which, it was discovered, involved more than 40 genes.

The most thoroughly investigated animal viruses are *Vaccinia* and *HSV*, both having a genome as big as that of *Phage T4*. Since 1962, we have known that a DNA polymerase is a part of the *Vaccinia* and *HSV* structural proteins, and that thymidine kinase is induced by these proteins. *Phage T4*, in contrast, exploits the host-cell polymerase by a mechanism in which its own induced proteins suitably alter the host polymerase structure to suit its own purpose. In 1964, Joklik and Y.Becker investigated the time required for vaccinia DNA replication which commenced 1–1.5 hours after infection. They found that three hours later more than 90% of virus DNA had been synthesized, the rate then decreasing progressively.

In 1967, DNA-dependent RNA polymerase was found in vaccinia and, 10 years later, Nevins and Joklik distinguished virus-specific vaccinia DNA polymerase from host-cell polymerase. W.Bauer and colleagues (1977) reported the isolation of a vaccinia topoisomerase. This enzyme is involved in the synthesis of a protein which, with the help of other, called binding proteins, functions in forming secondary nucleic acid structures. This, for example, holds for uncoiling double-stranded DNA at the beginning of replication.

Between 1960 and 1978, the replication of the first single-stranded DNA virus to be discovered, namely *Phage φ X174*, was described. In 1962 the phage was found to have a circular genome and 6 years later, Sinsheimer and colleagues reported its 3-step multiplication cycle at Cold Spring Harbor: replication was initiated by formation of a double-strand; then semi-conservative reduplication of the replicative parent DNA ensued, producing 10–20 double-stranded progeny per cell. During the third phase, single-stranded DNA was formed for the new virions. Development of a circular double strand required not only a DNA polymerase and a DNA template, but also deoxynucleotide-triphosphates and an RNA polymerase. Immediately after its formation, cellular binding proteins encompassed the double-stranded DNA. This DNA-protein complex reacted, in turn with several additional E.coli gene products. The entire course of replication became clear when Sanger and coworkers (1977–1978) elucidated the sequence of the 5386 deoxynucleotides present in the genome. At the same time, the5226-nucleotide sequence of the *SV 40* genome was determined by V.Reddy and by Fiers and their respective colleagues, and subsequently correlated to the molecular-genetic results by numerous other groups.

The replication site of *SV40* is in the nucleus. Prior to DNA synthesis, half the strand is transcribed to give so-called early mRA and the other half to late mRNA. *SV40* codes for at least four major proteins and for a large early, so-called A-protein or T-antigen which is involved in virus-induced neoplastic transformation. Individual DNA regions could be identified coding for structural proteins, T-antigens, the origin of DNA replication, as well as DNA binding sites for proteins. However, it was not then possible to identify the recognition and binding site for RNA polymerase, called the promoter.

5.2.4.2 RNA Viruses

Since RNA polymerase can replicate just as well as DNA polymerase and, in the case of plus-stranded RNA viruses, the nucleic acid can act directly as mRNA, the replication process of RNA virus, is shorter than that of DNA viruses. The relative low size of 7–30 kb assessed at the present time would suggest that RNA viruses cannot produce the variety of replication processes which DNA viruses are capable of. As a result, there is also no clear-cut division between the early and late stages of the genetically directed functions - not even for double-stranded RNA viruses.

Some important milestones in RNA virus biology from the work of Kitamura and colleagues deserve to be mentioned beginning with clarification of RNA synthesis to the elucidation of the 7433 nucleotides in the poliovirus genome in 1981. Further progress came from T.Loeb's and Zinder's discovery of the first RNA phage, *Phage f2*, in 1961.

RNA infectivity has been known since 1957 in *Mengo virus*, poliovirus and *West Nile encephalitis virus*, demonstrated by Colter and coworkers, and in *Eastern equine encephalomyelitis virus*, as shown by Wecker and Schäfer. The following year, Hattie Alexander and colleagues proved that the RNA of poliovirus contained the complete information for its nucleic acid and protein synthesis. Darnell and coworkers (1961) found that the earliest starting point of poliovirus RNA synthesis was 30 minutes after HeLa cell infection. In the same year Reich's group was able to establish the dependence of both viral and cellular RNA on host-cell DNA by using the antibiotic actinomycin D. Simultaneously, Simon determined that DNA intervention was not required for the growth of either RNA-containing poliovirus or *NDV*. Reich and coworkers (1961) further found that viral RNA and cellular DNA synthesis and expression were completely independent phenomena. Since S.N.Cohen discovered virus-specific, newly synthesized enzymes in bacteria in the same year, it was thought possible this might hold true for all animal viruses, and it was assumed that *Mengo virus* was capable of cytoplasmic, RNA-dependent RNA synthesis. The results of further investigations on picornavirus RNA by Baltimore and R.Franklin, and the actinomycin D synthesis-inhibiting experiments carried out by Eggers and Tamm (1963) prompted the conclusion that a viral RNA-dependent RNA polymerase was present. This enzyme was found in reoviruses, i.e. in double-stranded RNA viruses, in 1968. Penman and colleagues (1964) demonstrated that poliovirus RNA functioned as mRNA, and Kingsbury (1966) showed base sequence homology between *Newcastle disease virus* RNA and the RNA synthesized by the cell after infection.

R.Franklin and Rosner (1962) recognized that RNA synthesis in the cell nucleus was quickly shut off during *Mengo virus* infection, and that the RNA which was later synthesized in the cytoplasm matched the viral RNA. One year later Rich and colleagues carried out EM investigations which clarified the synthesis of various structural proteins namely capsid protein, polymerase and inhibitors of host cell functions, in polysomes which required 40–60 ribosomes per RNA virus molecule. RNA and protein synthesis ran in parallel, beginning about 2 hours after infection, according to the findings of Darnell and coworkers (1961) and Wecker and Richter (1962); that the processes were linked in some way was clear from the reports of Scharff and colleagues (1963) and Baltimore and colleagues (1966). In 1970 Baltimore's group found that paramyxoviruses possessed an RNA polymerase capable of synthesizing a minus RNA which was complementary to its own virion plus RNA. The sequence of poliovirus RNA, elucidated in 1981, revealed that 89 of the nucleotides coded for a primary polyprotein. This sequence contained all of the information for the translation and the subsequent processing of the protein into 12 polypeptides with the help of a proteinase. At the 5′ terminus, the genome was linked to a small protein. What remained obscure were the signals and processes involved in the synthesis of proteins.

5.2.4.3 Retroviruses

All viruses which contain an RNA-dependent DNA polymerase, known as reverse transcriptase, and which, therefore, must undergo a DNA step in their multiplication cycle, belong to the family of retroviruses. Their single-stranded, linear virion-RNA is not infectious and has a very specific structure. This includes identical sequences of about 70 nucleotides at both ends called repeats. One gag gene codes for a polyprotein which is later cut into four proteins to be found as the main antigenic proteins inside the virion. Gag stands for *g*roup-specific *a*ntigen *g*ene. Toward the 3'end there is a pol gene responsible for the reverse transcriptase. The third, or env gene contains the sequence for a polyprotein which is split into two proteins to be stored in the lipid sheath. An additional gene, called src or onc gene is necessary for the induction and maintainance of cell transformation.

Reverse transcriptase performs the following three consecutive tasks:

- Storing the DNA strand complementary to the viral RNA
- Digestion of RNA contained in the DNA-RNA hybrid
- Formation of a DNA double strand

From the newly created double-stranded DNA, termed provirus, a viral RNA can be formed with the help of a host DNA-dependent RNA polymerase.

The structure of a member of the subfamily retrovirinae, Prague *RSV* strain, was determined in 1983 by Schwartz and coworkers. *HIV*, which was first included in the subfamily lentiviruses, lacked an onc gene but possessed a tat or transactivator gene as was shown in 1984/1985 by the Haseltine and the Muesing groups. Flügel (1987) and Maurer and colleagues (1988) described the genetic structure of the first human strain of the third subfamily, the spumavirinae, isolated in 1971 by Achong and coworkers and characterized in 1974 by the group of M.A.Epstein. This virus showed overlapping sequences and two additional genes in a genome of about 12 kb, and is thus the retrovirus with the longest sequence determined so far.

5.2.4.4 Hepadnaviruses

According to investigations by Robinson and colleagues (1974), viruses like the *Hepatitis B virus* (HBV) contain small circular double-stranded DNA and judging by the results of J. Summers (1975), also partly single-stranded DNA. This virus family is known as hepadnaviridae. They possess a full-length minus-strand and a somewhat shorter plus-strand. In 1973 and 1974 P.Kaplan and colleagues, as well as

Robinson and Greenman, found that shortly after infection virion-specified DNA polymerase augmented the shorter strand to about 3200 nucleotides, i.e. to the length of the longer one. J.Summers and coworkers (1975) observed that the number of nucleotides in the shorter strand varied between 1700–2800 in different molecules for the four presently known hepadnavirus types. In *HBV* DNA the portion of single-stranded DNA varies between 15 to 60% in length. There is some evidence that viral DNA is integrated in the host genome. Furthermore, minus-stranded DNA may be synthesized on an RNA template with the help of reverse transcriptase.

5.2.5 Virus Assembly

Assembly encompasses all events involved in forming a new virus in an infected cell after viral structural components have been synthesized. Schramm's experiments (1943) and those of Fraenkel-Conrat and R.Williams (1955) with *TMV* indicated that certain pH conditions caused the *TMV* to split up as well as to re-aggregate. This was conceivable provided the assumption was valid that the virus consisted of identical subunits arranged in a specific order – which was later proven. After examining the results of small plant and animal viruses, Crick and J.D.Watson formulated a hypothesis for small rod- and sphere-shaped viruses in 1956.

The rules for the assembly of virus structures set forth by Caspar and Klug (1962) turned out to be basically accurate for all types of viruses. According to them, genetically directed self-assemby was as regular as a crystallographic process. The accepted rule was that disk aggregates formed spiral-shaped viruses while cube-shaped subunits formed spherical ones. Formation was spontaneous requiring very little energy under appropriate conditions. Caspar (1963) stated that during the assembly of *TMV* the balance of electrostatic forces within the interacting structures was controlled by conformational changes in the coat protein. Under given physiological conditions, the coat protein alone associated to form small protein oligomers in equilibrium with the disk aggregates.

Step-by-step assembly of phages using prefabricated parts had been reported in many studies during the early 1950's (See Chapter 4.3.1). In 1966 Edgar and Wood described the morphogenesis of the very complex *Bacteriophage T4*. According to those findings, more than 40 genes programmed the highly organized intracellular assembly of the virion from preformed components. Three distinct ways were observed:

1. Formation of the head from immature precursors located on the cytoplasmic membrane of the bacteria. After separating from the cell wall, this newly-formed head was filled with DNA.
2. Formation of the tail.
3. Formation of the tail fibres.

The fully-developed tails aggregated with the heads before the fibres attached themselves to the base of the head-tail complex.

Klug (1971) working with Butler, A.Durham and Finch described the different steps of *TMV* assembly. The natural state of *TMV* protein aggregation, they reported, was the 20 S disk, which consisted of two paired rings of 17 subunits each. The subunits polymerized into the double-layered aggregates (Fig.7) at almost neutral pH, while the helical rod developed at acid pH. The mode of aggregation was controlled by the state of ionization. Assembly of *TMV* proceeded by the stepwise stacking of preformed disks on the rod and was initiated by the interaction of a disk with a special sequence on the *TMV* RNA. This sequence on the RNA specified the origin of assembly. The interpretation of their results was that assembly nucleates at a unique internal site on the RNA and subsequently grows in both directions but at unequal rates.During this process, only small protein aggregates are added along the 3′-tail of the RNA. Later, when the disk has been transformed into a helix, salt linkage is involved, as Bloomer (1978) and G.Stubbs (1979) and colleagues reported.

In their 1968 assembly experiments with type 1 *Poliomyelitis virus*, Jacobson und Baltimore as well as D.Summers and Maizel were able to show that at first the entire genome is translated into a single, giant polypeptide, which is then cleaved into smaller pieces for assembly. Although many aspects of self-assembly still remain unclear, there

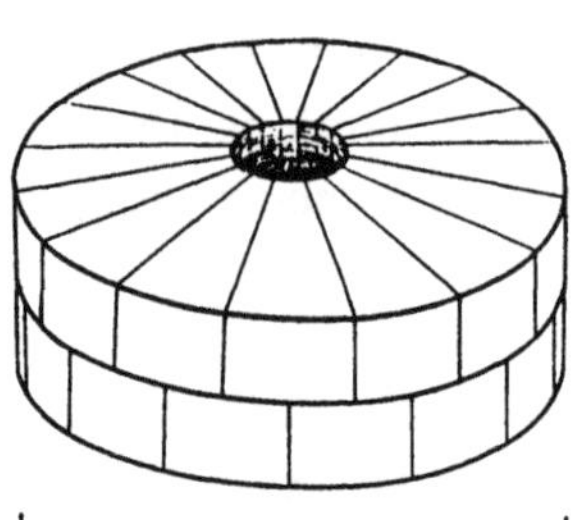

Fig. 7. 20s protein disk for TMV assembly

are, nevertheless, a number of features common to all types of viruses. For example, there is always step-by-step construction of the virus particle. Furthermore, virus assembly requires only nucleic acid and capsid proteins. That specific factors or substrates are indispensable in individual cases is well-known, however, generally applicable rules have not been found. This applies, among other things, to the specific packaging of nucleic acid within the capsid.

5.2.6 Virus Maturation

Maturation refers to the late period of the multiplication cycle in which virus morphogenesis occurs. This is true for all types of viruses which require a complete or partially virus-coded envelope to protect their nucleocapsids. Various cellular membranes may be involved in the maturation process. C.Morgan and colleagues (1959) were able to show that nuclear membranes of the host cell were involved in the maturation process of *HSV*. In 1972, several groups found that smooth as well as rough membranes of the endoplasmatic reticulum participated in arbovirus morphogenesis.

Investigations have thus far led us to believe that most viral envelopes are assembled on preformed cellular membranes. During envelope assembly, viral nucleic acid codes for protein components, while host cells provide the lipid elements. It is probable that viral glycoproteins, glycolipids and mucopolysaccharides are processed by cellular enzymes.

5.2.7 Virus Release

The earliest visualization of the release of newly formed viruses from infected cells was provided by studies on influenzaviruses in the early 1950's. C.Morgan and coworkers produced electron micrographs, and Leslie Hoyle offered dark-field micrographs, which showed viruses being released by detachment from the cells within cytoplasmic protrusions enclosed in cell membrane. This type of release was termed budding and proved typical for many viruses which are enveloped, including myxoviruses, rhabdoviruses and arborviruses, for example. Other ways out of the cell from the site of assembly - from the nucleus in the case of adenoviruses or out of the cytoplasm for poxviruses and retroviruses - include release aided by microvilli, cell disruption or lysis of the infected cell. Thus, according to present results, viruses of different families are released inversely to their mode of penetration.

5.3 Viruses as Pathogens

Rapid progress in laboratory techniques and the new results described in the previous chapters have promoted the further development of speciality fields associated with viral disease and its relevance for the individual and society as a whole. Therefore, it is not surprising – thanks to present culture and detection methods for virions and for viral components – that the number of detected viruses causing acute infections has increased so rapidly, and not only in the field of human medicine. Agreement has been reached on the definition of acute infection: after a short incubation period, specific, induced pathological-anatomical changes become manifest. The duration and degree of the disease are determined by the body's defences; thereafter, it was assumed that pathogens were no longer in the body.

New insight was gained into virus-host cell interaction during infections which remain imperceptible over a shorter or longer period of time. The various fields of virological research approached the subject of interaction in diverse ways, such that the terms inapparent, persistent and latent were used but not in a clearly defined and uniform sense. The latter term and phenomenon had been known and described for decades, despite little or nothing being known about the pathogen. The same holds true for the first detected latency of insect viruses described by Tubeuf (1892), a polyhedral disease in the nun moth, or for animal or plant viruses which caused inclusion bodies in the infected tissue. The terms persistent and non-persistent had been used by M.Watson and Roberts as early as 1938 to distinguish plant viruses of different types transmitted by aphids. The former survived inside their vectors for weeks or months, the latter only for days. In the context of insect viruses, the sudden appearance of viroses prompted the question of causative agents possibly originating endogenously, within and through the action of the cell. Thus, successful provocation experiments were interpreted with this premise by Yamafuji (1950) and Vago (1953). But Krieg (1955/56) proved that these were manifestations of latent contamination. There were five different ways known for activating latent insect viruses: physical conditions during plant growth; food quality; the application of chemicals; infection by insect viruses from other host species, and super-infection with the species-specific virus.

At the 1957 Madison, Wisconsin symposium "on latency and masking in viral and rickettsial infections", an attempt was made to standardize terminology for certain known, lasting virus-host cell interactions. One was inapparent infection, which ran an asymptomatic course, without recognizable damage to the host cell. Latent infection applied to a special type of inapparent infection with longer-lasting equilibrium between the virus and host-cell partners. This condition, through factors later termed trigger mechanisms, could cause apparent infection. Sub-clinical applied to an inapparent infection of limited duration. A condition lacking a demonstrable virus became known as an occult infection, essentially considered to be an inapparent infection. Persistent-latent was used to describe an infection of long duration which ran an asymptomatic course and might possibly be associated with continuous though weak virus multiplication. Since viruses are known which cause latent infections and induce acute disease, and there are slow virus infections which give no evidence of the time involved in virus multiplication, it was evident that there are no latent or slow viruses but rather latent or slow virus infections. In 1958 Bergold extended the term latency by indicating the possibility of arousing pathological activity by provoking latent conditions with certain treatments. The same year, Steinhaus reported on latency observed in insect viruses. Describing the external factors known to activate insect viruses from their latency, he called these factors stressors.

Since the Madison symposium concerned only animal viruses, the definitions which were established were not generally acceptable for other fields of virology. Thus, veterinary scientists found that inapparent infections existed for practically all acute diseases in fowl. However, since these were automatically linked to pathogenic virus strains, inapparency assumed a different meaning. Terms such as velogenic, mesogenic or lentogenic strains simply expressed various degrees of change caused by viruses and the key question became that of how to define the word "damage".

It wasn't long before the established definitions were inadequate for all the observed virus-host interactions. There were long-lasting occult infections which became apparent without recognizable trigger mechanisms to which the term slow virus infection was applied. Other virus types were capable of permanently linking their genome to that of the host cell, with detrimental consequences for the host organism, including benign and malignant tumours as well as diverse viral leukemias.The following definitions serve as a terminology reference for the ensuing text:

Acute infection: Proven virus infection; short incubation period; pathological-anatomical damage; immune response by the host.

Inapparent infection: Asymptomatic infection; lacks demonstrable virions; no observable immune response by the host; virus-specifity can be established by EM or by determination of the presence of nucleic acid and/or antigens.

Persistent infection: Asymptomatic infection; isolation of virus possible; induced immune response.

Latent infection: Persistent infection; a follow-up apparent infection can be provoked.

Slow virus infection: Apparent infection with a long incubation period of months or years.

5.3.1 Acute Virus Infections

The number of viruses discovered to be the etiological agents of acute disease increased mainly within the biological fields already studied. New areas were disclosed and in parallel the number of detected vectors or carriers of infective viruses increased. Thus, in 1981 K.Harris registered 381 species of plant virus vectors capable of transmitting at least one type of plant virus. Our progress in the knowledge of acute virus infections will be depicted by citing a few noteworthy examples.

5.3.1.1 Animal Viruses

Rhinoviruses

Since the beginning of this now dwindling 20th century, it has been easy to transmit the pathogen for the common cold, but its identification has eluded the unsparing efforts of even the most zealous scientists. Price (1956) as well as Pelon and colleagues (1957) cultivated rhinoviruses taken from the nasopharyngeal washings of infected humans, using rhesus monkey kidney cell cultures. Isolated strains 2060 and JH were designated *Echo virus* 28. Real progress in experiments with rhinoviruses, however, was achieved by Tyrrell and

Parsons (1960), who took conditions within the nose into account and lowered the temperature to 33°C, maintained a neutral pH 7 and gently rotated the virus growth – all of which proved to be significant factors. Additional improvements, which may possibly lead one day to routine measures, include the use of diploid cell strains which Hayflick und Moorehead developed with human embryonic fibroblast tissue in 1961.

Hepatitis viruses

Because of clinical observations, it has been assumed for decades that there are two kinds of hepatitis: infectious and serum hepatitis. The evidence for two distinct, clinical, epidemiological and immunological types was published by Krugman and colleagues in 1967. All of the time spent searching for their etiological agents has been in vain, and attempts to transmit the infectious type to animals met with negligible success, until in 1962 Deinhardt and coworkers described enzymatic and histologic changes in chimpanzees 25–40 days after inoculation with acute-phase serum or stool. These experiments were carried out one year after Hallis had reported an outbreak of infectious hepatitis among chimpanzee handlers at a U.S. air force base. In 1967 Deinhardt and colleagues transmitted and passaged the etiologic agent taken from patients to South American white-lipped marmoset monkeys (Saquinus species). The first convincing results in working with the infectious agent in cell cultures was published by Provost and Hilleman in 1979. They propagated the marmoset-adapted virus in primary cultures of adult Saquinus labialis livers and in a rhesus monkey kidney-cell line. In the absence of a cytopathic effect, they had to confirm the successful growth of the virus by reinfection of marmosets. The infectious or epidemic hepatitis agent was called *Hepatitis A virus* (HAV). As enterovirus 72 it was classified a member of the picornaviridae.

The etiologic agent of serum hepatitis was well on the way to being tracked down in 1963. Based on the epidemiologic differences to the infectious type, A.MacCallum (1947) had introduced the term hepatitis B for serum hepatitis or homologous serum jaundice. Blumberg (1963), cooperating with Alter and Visnich, found a precipitating antibody against an antigen obtained from an Australian aborigine in the serum of a hemophiliac patient who had received multiple transfusions, hence the name Australia antigen. After pursuing the geographical distribution of the antigen and comparing the corresponding antibodies in diseased and non-diseased groups, they detected a high antigen frequency in leukemia and multi-transfusion patients. Assuming that the agent was associated with a hypothetical "leukemia virus", they began testing the sera of Down's syndrome patients. In 1966, one of these patients developed hepatitis between two tests, and this, as a result of collaborative work by Sutnick, proved the hypothesis that the Australia antigen was related to hepatitis. Within the same year, the hypothesis was confirmed and it was suggested that presence of the antigen in thalassemia and hemophilia patients might be due to a virus introduced during blood transfusions. This idea was not accepted for two years until it was confirmed by Okochi, Vierucci, Prince and colleagues in 1968.

In 1970, Dane and coworkers described the hepatitis-B-associated virus which became known as the Dane particle. Brzosko and colleagues (1973) provided evidence of two distinct antigens in hepatocytes associated with hepatitis B infection, one was localized predominantly within the nuclei, the other in the cytoplasm. These two components were, therefore, termed hepatitis B core antigen (HBcAg) and hepatitis B surface antigen (HBsAg). In 1972 Maynard and colleagues reported the successful infection of a chimpanzee with *Hepatitis B virus* after Blumberg and coworkers had found HBsAg in the serum of other chimpanzees in 1968.

In 1978, three groups, namely those involving Alter, Hollinger and Tabor, found that when human hepatitis-infected blood was transmitted to chimpanzees, neither type A nor type B ensued. Moreover, in conjunction with blood transfusions, there were frequently non-A/non-B (NA/NB) carriers who closely resembled *HBV* infected cases. Early clinical observations with varying incubation periods, EM confimation of structural deviations, as well as results from cross-challenge tests in chimpanzees led scientists to conclude that there were at least two types of NA/NB hepatits pathogens.

One of these, a delta antigen had been discovered in serum and in the liver of *HBV* and HBsAg carriers by Rizetto and coworkers (1977) using immunofluorescence. They discribed it as a defective RNA virus which required *HBV* as a helper for replication. Between 1978 and 1980, three additional viruses of this type were discovered in animals and grouped with the human *delta virus* in the Hepadnaviridae family. These were the *Woodshuck hepatitis virus* (WHV) isolated from a marmata monax colony by J.Summers and coworkers at the Philadelphia Zoo; the *Ground squirrel hepatitis virus* found in Spermophilus beecheyi by Marion and colleagues in California; and *Duck hepatitis virus* found in Anas domesticus (Peking ducks) by J.Summers and colleagues in a region of China known to have a high incidence of hepatocellular carcinoma. Recently, two other types have been separated from the NA/NB complex with the help of a specific serological assay. It has been proposed to call them *Hepatitis C virus* (HCV) and *Hepatitis E virus* (HEV).

Gastroenteritis Viruses

Mild to severe epidemics of gastroenteritis in children and young adults have not been uncommon in the last decades, but despite intensive research, the specific etiologic agent has not been isolated. The successful use of human-embryo organ cultures in propagating common cold viruses since 1960, brought new impetus to work on the isolation of the suspected agent in acute intestinal disease. The earliest signs of possible success cropped up after an outbreak of gastroenteritis in Norwalk, Ohio in 1968 that attained epidemic proportions. Dolin and coworkers (1971) managed to transmit that infective agent but not to propagate it in vitro. The following year, Kapikian and colleagues observed a 27-nm virus-like particle in an infectious stool specimen by immune electron microscopy (IEM). Again, it was impossible to propagate this *Norwalk virus* or similar isolated viruses in cell cultures. Meeting with so little success, scientists attempted broad-range experiments in test animals, only to find again that they were unable to cultivate the agent. However, one glimmer of hope was that chimpanzees reacted with an immune response, which was confirmed by IEM and a radio immune assay (RIA); apes were also shown by RIA to excrete soluble Norwalk antigen. Further immunological studies made it clear, however, that the *Norwalk virus* played a negligible role in global intestinal disease epidemiology.

An intriguing phenomenon was the case of a 70-nm virus found in duodenal tissue obtained from pediatric diarrhoea patients in Australia by Bishop and coworkers in 1973, using the thin-section EM method. One year later, the same virus was isolated from stool extracts of patients with identical symptoms by Flewett and colleagues. Before it was officially named Rota virus because of its wheel-like structure, it had been called "orbivirus-like", "reovirus-like" and "duo-virus". Rotaviruses cause the highest mortality in cases of infant and childhood diarrhoea in developing countries.

Possibly to draw attention to especially severe diseases in this age of biochemistry and molecular genetics, some virus diseases were grouped together and termed exotic virus diseases. These diseases

Table 5.10. Exotic virus diseases

A.	*Mosquito-borne*
	Yellow fever
	Venezelian equine encephalitis
	Rift Valley fever
B.	*Tick-borne*
	Different types of encephalitis
	Crimean hemorrhagic fever
C.	*Rodent-borne*
	Lassa Fever
	Bolivian and Argentine hemorrhagic fevers
	Hemorrhagic fevers with renal syndroms:
	Nephropathia epidemica
	Korean hemorrhagic fever
D.	*Others*
	Marburg disease
	Rabies
	African swine fever

have occurred and are occurring all over the globe in defined geographic areas and pose a medical challenge for control and eradication. Table 5.10 contains some of the exotic virus diseases affecting man and domestic animals.

Fish Viruses

Although fish viruses were assumed to have existed for some time, Wolf and colleagues were the first to isolate a virus in trouts called *Infectious pancreatic necrosis virus* (IPNV) in 1957. Meanwhile, more than 50 other viruses and some virus-like agents have been discovered on all continents among fresh- and salt-water fish alike; about half of these have been characterized. They belong to 12 established DNA and RNA virus families and are considered etiologic in diverse fish diseases and plagues. Increased incidence of disease in fresh-water fish can be explained primarily by the highly intensive breeding in hatcheries, and in salt-water fish by the progressively deteriorating ecological conditions. Table 5.11 provides a survey of the most important known fish viruses. These cause systemic infections which are usually fatal early in the life of a fish. *IPNV* is perhaps the most prevalent piscine virus. Missing from the list of fish mentioned in the Table are: turbot, cod, river sheatfish, glass-eyed perch, butterfly

Table 5.11. Fish viruses as pathogens (After Ahne 1985)

Virus	Hosts	Geographic distribution
Channel catfish	Channel catfish Ictaturus punctatus)	USA
Infectious pancreatic necrosis (IPNV)	Predominant in trouts	Europe, Asia North America
Spring viraemia of carp	Carp (Cyprinus carpio) crucian (Carassius carassius) silver carp (Hypophtalmichthys molitrix) Sheatfish (Silurus glanis)	Europe
Pike fry rhabdo	Pike (Esox lucius) Grass carp (Ctenopharyngodon idella) Tench (Tinca vulgaris) Bleak (Blicca bjoerkna)	Europe
Hemorrhagic septicemia	Rainbow trout (Salmo gaidneri) River trout (Salmo trutta fario) Grayling (Thymallus vulgaris) Pike (Esox lucius)	Europe
Infectious hemapoetic necrose	Different salmon species (Onchorhynchus masou, O. nerka) Rainbow trout (salmo gaidneri)	North America, Japan

perch, goldfish, eel, whitefish, sole, flounder, golden shiner, tench, chub and steelhead trout.

Other aquatic animals subjected to virus infection are sea-lions, oysters, mussels, snails, giant and miniature shrimps and crabs, and seals.

5.3.1.2 Plant Viruses

In this context, viroids and killer systems present fields of special interest. The first will be treated under the subject of unconventional viruses in chapter 5.3.3.2. Killer viruses are double-stranded RNA viruses, found in various yeast strains and classified as mycoviridae. They excrete a thermolabile toxic substance which inhibits the growth of other strains by killing off the cells. These so-called killer systems have been recorded since 1968, when they were first detected in Ustilago maydis by Puhalla, and one year later in Saccharomyces cerevisiae by Somers and Bevan. Meanwhile, many more yeast species containing killer systems have been identified.

5.3.2 Inapparent Virus Infections

The development of tissue and cell culture methods has paved the way for experimental testing of inapparent virus infection in vitro and in vivo. Since 1956 there has been uninterrupted multiplication of viruses in subcultures of established cell lines without apparent destruction of the cells, as reported by several workers (see Table 5.12). Parallel to this, methods for virus determination have become much more sensitive, e.g. the co-cultivation as well as selective in vitro growth of T-lymphocytes which can be stimulated by phytohemagglutinin and/ or concanavalin A. Additional possibilities for detecting inapparent virus infections have been furnished by in vitro hybridization and by monoclonal antibodies. As a result, we have recently become aware of many more virus types which can cause inapparent infections. The possibility of inducing inapparency on an experimental basis revealed the factors which were essential for evoking the same effect in vivo. This meant, for example, being able to suppress pathogenicity by marked increases in antibodies, or using an apathogenic virus strain to induce interferon synthesis in a preceeding infection.

Inapparent infections in plants were discovered in 1968, when von Pullen detected small isometric viruses or virus-like particles which could not be transmitted to other plants. These particles, however, could not be eradicated in the carrier plants by heat treatment, which

Table 5.12. Inapparent virus infections in vitro

Year	Virus	Authors
1956	Polio	W.Ackermann, Hilda Kurtz
1956	Adeno	Ginsberg
1957	Western equine encephalomyelitis	Velma Chambers
1957	Newcastle disease	Cieciura
1957	Eastern equine encephalomyelitis	F. Bang et al.
1958	Mumps	Deinhardt et al.

was the usual method. Kassanis and coworkers (1977) purified these particles and called them *Beet cryptic virus* after the plant species involved. Today it is possible to identify many cryptic forms of viruses and carriers. Because they possess strong immunogenic properties, they can lead to spurious cross-reactions and to false serological identification.

5.3.2.1 Persistent Infections

Persistent infections can also be thought of as "regulated infections" because they are kept in check by immune mechanisms. In other words, during this type of infection, humoral and cell-mediated immunity as well as interferon induction result in slight though continuous multiplication of the virus. Persistent infections can be primary infections or may follow an acute infection. The list of viruses capable of persisting in experimentally infected cells in vitro is indeed long. Tables 5.13 and 5.14 summarize the early data obtained with different species of viruses. Evidently, even most acutely infections viruses can persist, and thus persistence will endure as long as virus multiplication is restricted to a small number of cells and cellular damage remains slight. According to Strohl and R. Schlesinger (1965), in vitro effects appear similar to those found in human and animal infections. Proof of in vivo persistence is of great practical significance in that it indicates healthy carriers of viruses that in turn might become sources of infection for their environment through their blood or secretions.

Table 5.13. Persistent virus infections in vitro

Year	Virus	Authors
1958	Newcastle disease	G.Henle et al.
1962	Mumps	Walker, Hinze
1962	Measles	Rustigian
1962	SV 40	Ashkenazi, Melnick
1964	Haemadsorption 2	Ishida et al.
1964	Rabies	Fernandes et al.
1965	Parainfluenza 3	Cole, Hetrich
1969	Western equine encephalitis	Simizu, Takayama
1969	Sindbis	Peleg

Table 5.14. Proven naturally persistent virus infections

Year	Virus	Authors
1953	Varicella/Zoster	Weller
1954	Cytomegalo	M. Smith
1957	African swine fever	DeTray
1959	Foot-and-mouth disease	Van Bekkum et al.
1962	Lymphocytic choriomeningitis	Hotchin
1969	Marek's disease	Calnek
1970	Herpes simplex	Plummer et al.
1971	Pseudorabies	Sabo, Grunert
1972	Rabies	Fekadu
1974	Meningoencephalomyelitis	Mayr, Danner
1975	Murine encephalomyelitis	Lipton
1975	Rubella	Cremer et al.
1976	Influenza A	Shortridge et al.
1977	Polio 2	L.Davis
1985	Border disease	Barbara Potts et al.

Deficiencies in the immune system or colonization of the virus/antibody aggregates in blood vessels or tissue can lead to chronic disease, with the central nervous system being the target organ for several persistent virus infections. Symptoms of chronic disease in the CNS include from disturbed development of the brain, progressive brain lesions, demyelination, spongiform changes and altered neuronal functions.

5.3.2.2 Latent Infections

Latency signifies the persistence of an unexpressed or partially expressed genome in a target organ; activation from latency to expression is known as induction. Among the several factors known to trigger induction are physical and mental stress, transplants and pregnancy. Animal viruses which can cause latent infection include *HSV*, *Cytomegalo virus*, *Varicella-zoster virus*, *Epstein-Barr virus*, and *HBV*.

The time it required to prove *HSV* latency in glia cells, already predicted by Goodpasture in 1929, was 42 years. In 1971 Stevens and Margary Cook detected latent *HSV* in spinal ganglia of mice, and F.Bastian and colleagues isolated *HSV* from the trigeminal ganglia of people who had died of trauma. R.Richter in 1944 concluded already that herpes labialis was caused by activated virus from trigenimal sensory-root sections. He was successful in isolating *HSV* from three patients who developed labialis vesicles postoperatively. There is now a general agreement on the course of *HSV* infection: mucocutaneous primary infection is followed by the colonization and acute infection of ganglia; thereafter latency is established in neurons.

5.3.3 Slow Virus Infections

A new class of infections was discovered in 1936 when Cuillé and Chelle transmitted the known chronic disease scrapie to an uninfected sheep. In 1954, Sigurdsson introduced the term "slow virus infections" for slowly developing infections of the central nervous system derived from his observations of chronic diseases in Iceland sheep. These infections were characterized by incubation periods ranging from months to years and a protracted clinical course which invariably proved fatal. Meanwhile pathogens have been isolated which comply with the definition of a virion, and others are called unconventional viruses because of their laboratory markers and other properties which differentiate them from known viruses. These cause subacute encephalopathy and subacute encephalomyelitis in humans and animals. The definition of a slow virus infection would justify the inclusion of tumours and leukemias of viral etiology, but for didactic purposes it may be advantageous to accord these diseases a separate section.

5.3.3.1 Slow Infections Caused by Viruses

Diseases in Animals

Visna/Maedi. Both diseases have been observed in sheep for more than half a century, and it was suspected that only one pathogen was involved, although the target organ differed; visna attacked the nervous system whereas maedi affected the lungs. Visna was identified as a sporadic disease manifesting subacute encephalomyelitis in Iceland between 1935 and 1951 after infected caracul sheep had been imported from Germany in 1933. In 1957 Sigurdsson and coworkers transmitted the infectious agent to young healthy sheep and isolated a cytopathogenic agent they suspected was a virus. From 1960–1966

Thormar and colleagues confirmed the virion morphology and the morphogenesis of an enveloped particle having a diameter of 80–120 nm. Stavrou and colleagues (1969) reported that the same pathogen caused infectious adenomatosis of the lung and called it granulomatous meningoencephalomyelitis; the observed incubation periods ranged from several months to 8 years or more. Haase and Varmus finally described the visna provirus in 1973, now classified as a retrovirus, the site of replication is in the lymphocytes.

Canine Distemper Demyelinating Encephalomyelitis. As early as 1962, publications by Innes, Saunders and Van Bogaert reported severe neurological disorders in dogs weeks or months after recovery from acute infections, or even without any preceding acute clinical disease. Using the co-cultivation method, Confer and coworkers (1975) succeeded in isolating infectious virus from the brain. Since there we know a serological relationship between *Canine distemper virus*, *Measles virus* and *Rinderpest virus*, these have been classified as morbilliviruses in the paramyxovirus family.

Chronic Progressive Myelitis in Mice. Joan Daniels and colleagues (1952) established that mice which survived acute infection caused by *Theiler's murine encephalomyelitis virus* could develop chronic, progressive demyelinating, and/or necrotizing myelitis. Lipton (1975) observed similar persistent CNS infection despite the presence of high levels of neutralizing antibodies.

Human Diseases

Subacute Sclerosing Panencephalitis (SSPE). The viral nature of this disease which usually affects children and adolescents was demonstrated by Bouteille and colleagues in 1965. Connolly and coworkers (1967) identified the infectious agent as *Measles virus* which belongs to the paramyxoviruses, and may be isolated from the brain and the lymph nodes. SSPE has been observed as long as 5 to 6 years after an acute measles infection. Tests employing various methods indicate a slight difference between the measles and the SSPE viruses, the latter being present in neurons and glial cells.

Progressive Multifocal Leucoencephalopathy (PML). Cavanagh and colleagues (1959) assumed a viral etiology in a demyelinating disease in the human called PML, and this was later supported by Gabriele

ZuRhein and coworkers (1965) in their electron-microscope observations. In 1971, two new papovaviruses were thought to be the etiologic agents: one, detected by EM in a renal allograft recipient B.K. suffering from ureteral obstruction, was named after the patient by Sylvia Gardner and Anne Field, the other, isolated from brain matter at autopsy, was similar in size to *SV 40* and called *JC virus* by Billie Padgett and colleagues. These viruses were also detected in renal transplant patients by Hogan and coworkers in 1980. Of the 61 patients studied, 11 excreted *JC virus* and 9 *BK virus*.

Progressive Rubella Panencephalitis (PRP). This is a very rare chronic infection of the CNS with symptoms similar in certain respects to those of SSPE. The disease is known from studies carried out by Lebon and Lyon (1974), Jeanette Townsend and colleagues (1975) and Weil and colleagues (1975), amongst others. Cremer and colleagues isolated *Rubella virus* from brain biopsy material in 1975, and Wolinsky and coworkers recovered it from peripheral blood lymphocytes in 1979. Patients can develop signs of this disease over a period of years after a childhood rubella infection. Symptoms may begin with mental deterioration, followed by neurological disturbances.

5.3.3.2 Slow Infections Caused by Unconventional Agents

This group primarily includes subacute spongiform virus encephalopathies in man and animals. Though these agents are called unconventional viruses, there is a real lack of evidence that they comply with the commonly accepted definition of a virus. Their physical, chemical and biological properties differ remarkably from conventional viruses, and, furthermore, it has not been possible to recognize them by EM. For a time, they had been considered to be related to free plant RNA's, i.e. to the viroids, whereas, now they tend to be grouped with the prions, i.e. the infectious proteins. It is for this reason that viroids and prions have been included in this chapter.

Diseases in Man

Kuru. Kuru, which was the first disease detected in this group of slow infections, has incubation periods of 30 years or more. The progressive, accumulative pathology, with signs of cerebellar and spinal ataxias as well as degeneration involving the corpus striatum, invariably leads to death. Originally Kuru was commonly observed only in the eastern highlands of Papua, New Guinea, where the Fore people lived. Infection was unrelated to sex or age, although the adult male showed a slightly lower infection rate. The practice of ritualistic cannabalism seemed to have a direct bearing on the incidence of Kuru, and with the cessation of this rite, the actual number of infections has decreased dramatically within the last two decades. Gajdusek and Zigas (1957) initiated a Kuru study and Gajdusek and colleagues were able to transmit and passage the agent in chimpanzees in 1966. In 1977, Gajdusek, cooperating with Clarence Gibbs, transmitted the Kuru agent and the Creutzfeldt-Jakob agent to New World monkeys. In 1980, Sotelo and colleagues detected auto-antibodies effective in axonal neurofilaments in 27% of the Kuru and 59% of the Creutzfeldt-Jakob patients studied.

Creutzfeldt-Jakob Disease (CJD). Common throughout the world, this progressively debilitating disease of the CNS presents a clinical syndrome of spongiform encephalopathy, myoclonus, ataxia and cerebral disfunction. It was originally described by the neurologist Creutzfeldt in 1920 and the neuropathologist Jakob in 1921. The incubation period is estimated to range from 4 to 47 months. The infectious agent was transmitted to the chimpanzee by Klatzo and colleagues in 1959.

Alzheimer's Disease. This disease was described by the German neurologist Alzheimer in 1907 and long considered to be a presenile dementia. It is now clear, however, that not only the post 65-year-olds are affected. Characteristic symptoms are brain alterations with neuronal loss in memory, cognition and thought-process centres. The cause of the disease is unknown, but among the various etiologic hypotheses, one links it to an infectious agent.

According to a compilation of Gajdusek (1975) some other chronic diseases in the human are suspected to be of slow viral etiology. Among these are multiple sclerosis, Parkinson's disease, ataxia telangiectasia, schizophrenia dementia, disseminated lupus erythematosis, chronic arthritis, dermatomyositis, scleroderma and juvenile diabetes.

Diseases in Animals

Aleutian disease. Here the virus belongs to the parvoviruses and is responsible for Aleutian diseae of mink, a chronic infection causing extensive plasmacytosis, glomerulonephritis, hypergammaglobulinemia, and lesions typical of auto-immune disease. In 1962 several groups, including Karstad and Pridham, Henson and colleagues, and Russell, presented evidence of the viral etiology of the disease and its horizontal as well as vertical transmission. Once a mink has been infected, the virus may be recovered from the serum, organs or urine during weeks or months. In 1967, Porters and Larsen reported that pathogenicity of glomerulonephritis was related to the formation of viral antigen-antibody complexes which had lodged in the glomerular capillaries. In 1969, they detected viral antigens in the cytoplasm of macrophages in the spleen, lymph nodes and Kupffer cells of the liver.

So far, no definitive virion has been isolated to account for unconventional agents, and, as a result, there has been a recent tendency to group the unconventional agents with the viroids or the prions. "Free nucleic acid" or "unconventional virus", or the unconventional "virus" which was termed viroid in 1971, was first detected by Diener and Raymer in their effort to characterize the infectious agent of potato spindle tuber disease in 1967. It was reported that they have a very low molecular weight and that virions were not detectable. In 1971–1972, the molecular weight, as determined by various methods, was found to be between 5×10^4 and $1{,}25 \times 10^5$ daltons. Other viroids which have been so far detected are shown in Table 5.15. Since RNAs of such low molecular weight contain a relatively small capacity for genetic information, it was astonishing that they were able to induce such severe disease in higher plants, with symptoms resembling those typical of virus infections. This new type of infectious agent for plant diseases seemed to comply with the characteristics of the scrapie agent, which, however, proved more highly resistant to UV irradiation than did the viroid. Until 1973, other unconventional agents were also thought to be of a viroid nature.

The molecular weights of four viroids with covalently closed circular RNAs were determined to be 110000–127000 daltons by Sänger and colleagues in 1976. Groß (1978) determined the nucleotide sequence and secondary structure of *Potato spindle tuber viroid* to be one molecule of a covalently closed ring of 359 ribonucleotides and serial arrangements of double-helical sections as a consequence of intramolecular base pairing. In 1981, Imperial and coworkers reported that infectivity in cadang-cadang disease in coconuts was associated with four RNAs. Meanwhile, several additional viroid sequences

Table 5.15. Viroid diseases

Disease	Described as viroid by
Potato spindle tuber	Diener (1971); Singh, Clark (1971)
Citrus exortis	Semancik, Weathers (1972); Sänger (1972)
Chrysanthemum stunt	Hollings, O.Stone (1973); Diener, Lawson (1973)
Cucumber pale fruit	Van Dorst, Peters (1974)
Avocado sun blotch	Desjardins et al. (1974)
Chrysanthemum chlorotic mottle	Romaine, R.Horst (1975)
Coconut cadang-cadang	Randles (1975)
Hop stunt	Sasaki, Shikata (1977)
Columnea latent (erythrophae)	Owens et al. (1978)
Tomato planta macho	J.Galindo, Rodriguez (1978)
Tomato apical stunt (identical to tomato bunchy top)	Walter (1981)
Burdock stunt	Chen et al. (1983)

that have been analyzed have convinced us that minor changes in viroid structure can cause drastic changes in biological properties.

Alper and colleagues used the term unconventional virus to describe the unusual features of the scrapie agent which seemed to make it fundamentally different to viruses. Publications dating from 1978 and 1979 claimed that DNA was an essential component of the infectious scrapie agent. Prusiner and colleagues (1980) were unable to confirm this, and in 1981 they reported that a protein molecule, not nucleic acid was necessary and responsible for scrapie infectivity. They used an improved analytic differential centrifugation method which enabled the analysis of subcellular organelles and a variety of macromolecules. In 1982, the infectious scrapie protein was declared a prion, defined as "a small *pr*oteinaceous, *in*fectious particle which resists inactivation by procedures that modify nucleic acids". Our present knowledge is that prions contain a modified isoform of a normal host protein but little, if any, nucleic acid. They are encoded by the host genome and do not invoke an immune response. Under these auspicious signs, it would seem more appropriate to consider the problem of slow virus infections caused by unconventional agents to be the "doings" of prions; the more so, now that ribozymes have been discovered. This concerns not only scrapie but transmissible mink encephalopathy, chronic wasting disease in deer and elk, as well as the three neurodegenerative diseases found in the human, namely kuru, Creutzfeldt-Jakob disease and Gerstmann-Sträußler syndrome, all of which are transmissible to experimental animals.

5.3.4 Viruses and Tumours

Although the term "tumour virus" does not exist in the classification of viruses, it is nevertheless used in connection with viruses which can cause tumours. According to the four fundamental criteria for virus classification, viruses thought to cause tumours were assigned to a papilloma group and to a sub-group of myxoviruses called avian tumour viruses in 1950. When it was established that RNA as well as DNA tumour viruses existed, the latter were designated papilloma and polyoma viruses. Later, both genera became members of the papovaviridae family. Other terms used for RNA tumour viruses were Avian leucosis-sarcoma complex und leukemia viruses.

Tumour viruses which at first had been observed only in animals, were shown to be transmissible cell-free as well as in organ suspensions and to cause both benign and malignant tumours. Subsequent experimental results can be considered under four aspects. The development of cell culture techniques enabled in vitro studies of the morphological changes in the cell induced by viruses. This step led naturally to that of immunology, which kindled keen interest among clinical virologists in the question of the antigens associated with tumour development. Almost simultaneously, biochemists interested in nucleic acids and proteins in virology made substantial strides in the analysis of tumour viruses. The discovery of the oncogenes acted as a hub for different disciplines in the approach to molecular oncology, which became an independent scientific branch. Experiments with nucleotide sequences and their gene products, i.e. specific proteins, aimed at clarification of the interaction of these gene products with the host organism.

The EM aided immeasurably in the discovery of leukemia-inducing viruses. Bernhard und M.Guérin (1958) observed typical structures in various mouse leukemias, which Bernhard proposed calling A, B and C-particles in 1960. In 1970, when reverse transcriptase was detected as a typical marker for C-particles, the EM differences served as a basis for classifying the RNA tumour viruses, which later became known as the retroviruses. A-type particles proved to be free of infectivity, possessing an electron-lucent centre which was surrounded by a double shell.

B-particles were characterized by their eccentrically located nucleoids. The majority of RNA tumour viruses possessed C-type morphology with a centrally located core; the immature virions have an electron-lucent, the mature forms an electron-dense core. D-particles were isolated from primate species; intracellular immature virions were ring-shaped and located near the plasma membrane, while their extracellular forms were located eccentrically and possessed an electron-dense nucleoid. In the course of time, several methods were to be developed for experimental work aimed specifically at different types of tumour viruses. This concerned replication and transformation assays, experiments with defective and pseudotype virus variants, susceptibility and resistance tests, tests for interference and receptor interaction, in addition to transfections and the use of genetic and molecular genetic methods. H.Rubin (1960) first reported that infection of fibroblasts with a non-transforming virus prevented later infection with a tranforming virus of the same subgroup. With this information, H.Rubin developed the Resistance-Inducing Factor Test (RIF) for leukemia viruses. By doing so, he made it possible to quantify non-cytopathic avian leukosis viruses in vivo.

5.3.4.1 In Vitro Tests

Stimulated by Griffith's transformation of pneumococcus in 1928, Berry and Dedrick (1936) initiated similar experiments with viruses. Using heat-killed *Myxoma virus*, they transformed live *Fibroma virus* into infectious *Myxoma virus*. Although confirmed by other investigators, their results were not completely consistent and other methods had to be found for studying viral malignancy.

Koegh (1938) had demonstrated that the CAM responded to inoculation of *RSV* by producing focal proliferative lesions which he termed "pocks". Halberstaedter and colleagues (1941) had also observed morphological changes of chick fibroblasts in explant cultures infected with *RSV*. Earle and Nettleship (1943) demonstrated that cultured fibroblasts which had been transformed by the carcinogen 20-methylcholanthrene caused sarcomas in seven different strains of C3H mice after intramuscular injection. Using the same carcinogen, Katherine Sanford and colleagues (1950) found that its ability to cause sarcomas was a reproducible response of C3H mouse fibroblasts to the culture conditions involved. Thus, even at that time, it was clear that normal tissue could not only be transformed into malignant tissue by biological material or cell-free filtrates of malignant tissue, as Rous had repeatedly been demonstrating since 1910, but also by chemicals and anaerobiosis during prolonged propagation in vitro as Goldblatt and Gladys Cameron demonstrated in 1952. Meanwhile, Avery and coworkers had used DNA to transform cells in vitro in 1944.

Proceeding on the assumption that a tumour-virus infected cell responded differently from a normal cell, Manaker and Groupé (1956) found that *RSV* induced discrete foci on chick embryo cells which had been propagated in a tryptose phosphate broth medium. This sharp observation then became the basis for the cell transformation test of tumor viruses. The following year, Kilham showed that *Fibroma-myxoma virus* transformation could occur in different types of cell cultures. At this point, it may prove helpful to distinguish between two terms. The effect of carcinogenically transformed cells to produce tumours in animals is known as oncogenic transformation. Another type, namely virus-transformed cells is known as neoplastic transformation. In the latter, injection of virus-transformed cells into the original host induces tumours. Neither oncogenic nor neoplastic transformation presents evidence of oncogenicity in the human.

In addition to these two phenomena, contact inhibition was described by Abercrombie and Heaysman in 1954. Investigating the "social behaviour" of cells in culture, they found that cells stopped growing in the direction that brought about contact with other cells; the result was a monolayer culture. During cell transformation, some of the infected monolayer cells did not react by growth inhibition on contact but by runaway uncontrolled growth to form toumor-like masses called foci. After Manaker and Groupé had proved that a tumour virus could be propagated in cell culture and that under suitable conditions cells of these cultures could be transformed during reproduction to easily recognizable foci, oncogenic transformation was to serve as the in vitro model for in vivo tumorigenesis. Sarah Stewart and Bernice Eddy successfully propagated lytic *Polyoma virus* on mouse embryo cells in 1957, making possible both virological and biochemical analysis. On the other hand, Stewart and colleagues induced multiple primary tumours and leukemias in newborn mice with supernatant fluids from cultured tumour preparations in the same year, but were unable to isolate virions from the tumours.

Svet-Moldavsky and Zilber (1957) reported that neoplasia of mammals could also be induced by variants of *RSV*. Two years later

Table 5.16. Cell transformation by DNA and RNA tumour viruses

Year	Transforming virus	Authors
1956	Rous sarcoma virus	Manaker, Groupé
1959/60	Polyoma virus	L.Sachs, Winocour
1961	Avian myeloblastosis virus	Baluda, Goetz
1962	SV40	Bernice Eddy; Koprowski et al.; Shein, Enders
1963	Epstein-Barr virus	M.Epstein et al.
1963	Bovine papilloma virus	P.Black et al.
1964	Adeno virus 12	McBride, Wiener
1965	Shope papilloma virus	Noyes
1965	Avian sarcoma virus	Klement, P.Vesely
1971	HSV 2-UV inactivated	Duff, Rapp
1971	Vaccinia virus	Jadwiga Koziorowska
1973	HSV 1	Darai, Munk
1973	CMV-UV inactivated	Albrecht, Rapp
1973	Marek disease virus	Akiyama
1976	Cytomegalo virus	Geder et al.

the first successful transformation experiments with DNA viruses took place; T.Hanafusa and coworkers transformed *Ectromelia* into *Vaccinia virus* in Earle's L cells. They assumed that these two viruses as well as *Shope fibroma* and *Myxoma virus* shared a common *n*ucleo*p*rotein named NP antigen. This term was introduced by Smadel and colleagues in 1942. In 1960, L.Sachs and Winocour as well as Marguerite Vogt and Dulbecco, proved that *Polyoma virus* evoked morphologically and functionally stable changes in mouse and hamster cell cultures; furthermore, these cells showed striking similarities to *Polyoma virus*-induced tumours. Thus, it was also proven that DNA viruses could develop in two ways. In permissive culture, virions developed by destroying host cells; in non-permissive culture, a virus-induced cancer-like state or transformation of the cells took place. Therefore, it seemed feasible to assess the oncogenic potentiality of a virus with the aid of the cell transformation test. Table 5.16 contains the first positive test results published.

Since transformation could be induced with active as well as with UV-inactivated virus, scientists reasoned that transformation and infective virus features were discrete, independent and, therefore, separable. The opinion that cell transformation tests were suitable for investigating cell-virus interactions was supported by quantitative studies on *RSV*. Prince published results of studies with the CAM, while Temin and H.Rubin described the features of an accurate tissue culture assay for *RSV* and Rous sarcoma cells. Foci were considered to be cell culture analogues of tumours found in chicks. Stoker and MacPherson (1961) exposed hamster cell cultures to *Polyoma virus* to observe the relationship between the number of foci produced in culture and the tumours which developed after subcutaneous inoculation of test animals.

Meaningful progress in cell culture experiments with the DNA-*Polyoma virus* and *RSV*, which Bather as well as M.Epstein and Holt (1957–1958) had found to be RNA virus, may be credited to various scientists. Temin and H.Rubin (1959) demonstrated that *RSV*-infected cells could divide and produce virus without lethal effects. Changes in behaviour of *RSV* and *Newcastle disease virus* in radiobiological experiments had led scientists to reason that an interaction between the virus and the host genome might be involved. Furthermore, the genome of *RSV* would have to integrate with that of the cell before virus production could begin. This was the starting point of Temin's extensive experiments with *RSV* which were summarized in his genetically implicated DNA provirus hypothesis initiated in 1960 and described in detail in 1976. At the Cold Spring Harbor Symposium (1962), he concluded that cell transformation did not depend on assessable virus in the supernatants of transformed cells, and that cell transformation and virus production were discrete phenomena in *RSV* infection.

Temin's conclusion that the provirus was composed of DNA gained support by several subsequent experiments. In 1963, he was unable to detect *RSV* or any substances that could interfere with the recovery or assay of *RSV* in *RSV*-transformed cultures, although the cells caused tumours when injected into newly hatched chicks. Then he showed that actinomycin D, a DNA complexing substance, inhibited *RSV* production, although substance-resistant *RSV* variants could not be found. This was contrary to findings on inhibitors of DNA synthesis in experiments carried out in 1961–1962 with other RNA viruses, namely *Fowl plague virus* by Scholtissek and Rott, *Reo virus 3* by Gomatos and coworkers, and influenzavirus by Barry and colleagues. Experiments with aminopterin, a DNA synthesis blocking agent, strongly led Temin to conclude that there was a DNA step in *RSV* production which was inhibited. In 1964 Temin explained his model in the following manner: "A virus enters a cell and directs formation of the virus. This new DNA, the provirus, then acts as a template for formation of new nucleic acid, RNA, for the virion." The same year, he was able to confirm his hypothesis of the homology between *RSV*-RNA and DNA in *RSV*-infected cells. A year later, further support was offered by Bader; while working with DNA antagonists, he determined that DNA synthesis was essential to the growth of *RSV* and *Rous-associated viruses*, reasoning that a viral DNA "primer" was synthesized in infected cells during the multiplication cycle of *RSV*. Thus, a mechanism common to tumour formation by RNA viruses was a suspected viral genome which had been integrated as a non-replicating or defective "endogenous" provirus at a specific site in the cellular DNA, causing, in turn, conversion and carcinogenesis. The term endogenous for a virus thus changed from an intracellular virus to an intrachromosomally defective virus.

Sabin and M.Koch (1963) published *SV 40* studies which indicated that the viral genome of *SV40* tumours existed in a non-infectious state in the majority of tumuor cells after 31 serial transplant passages of *SV 40* tumours in adult hamsters. In contrast, infectious virus was only synthesized in an occasional cell among several million. Aloni and colleagues (1968) described *SV40*-specific RNA which they had found in transformed cells, and in 1966, when T.Benjamin was able to demonstrate that *Polyoma virus* DNA was integrated into cellular DNA, the provirus hypothesis was firmly established for both RNA and DNA viruses. A further experiment in support of Temin's thesis was submitted by Duesberg and P.Vogt (1969), who showed that when an avian tumour virus infected cell was superinfected by a second virus, successful superinfection depended on DNA synthesis after the introduction of the second virus. Rokutanda and colleagues observed viral RNA-DNA hybrid molecule formation by DNA polymerase of sarcoma-leukemia viruses in 1970. In the same year, Baluda and Nayak detected DNA complementary to viral RNA in *Myeloblastosis virus*-infected leukemia cells. It was A.Galindo (1972) who demonstrated through RNA-DNA hybridization that *RSV* nucleotide sequences were present at very low frequency in DNA of cells infected. Simultaneously, Baluda observed the widespread presence of DNA in chickens complementary to the RNA genome of *Avian leukosis virus*.

In 1970, the enzyme required to transcribe viral RNA into provirus DNA was finally detected. It was an RNA-dependent DNA polymerase or reverse transcriptase, which served as the missing link for understanding the mechanisms of retrovirus integration and replication. With this feat, most scientists were convinced of the truth of the provirus hypothesis and acknowledged that the provirus was, indeed, an integral part of the retrovirus life cycle.

5.3.4.2 Virus-Specified Proteins in Tumours and Transformed Cells

In 1953, E.Foley detected specific antigens in methylcholanthrene-induced sarcomas of C3H mice. Thereafter, the search intensified for tumour-specific antigens in a variety of tumours known in test animals and humans. Zilber and coworkers (1958) reported an abnormal antigen in *RSV*-induced tumours. The protein primarily concerned was called tumour (T) antigen. It was first detected in *Adeno virus 12*-induced tumours of hamsters and described by Huebner and colleagues in 1963. Two years later, it was also demonstrated by Habel in "virus-free" polyoma tumours of hamsters as well as in primary hamster and mouse cell cultures exposed to *Polyoma virus*. This viral antigen did not influence tumour regression. Further results with different viruses showed that T antigens functioned in various ways, had a broad molecular weight range of 15000–110000 daltons, and were located in both the nucleus and the cell membrane.

One year later, Huebner and colleagues described a CFR for a group-specific antigen of avian RNA tumour viruses which corresponded to the internal structural proteins of a C-type virus grown in chick cells. This complement-fixation avian leucosis (COFAL) test facilitated the detection of avian RNA tumour viruses by making infectivity and transformation tests redundant.

In 1973 Reedman and G.Klein detected a complex of new antigens in the nuclei of cells having an integrated *EBV* genome. It was called EBNA for Epstein-Barr nuclear antigen. Presently, at least six different antigens encoded by *EBV* genes are known. They are expressed in vitro in infected B lymphocytes which have been latently transformed, and in vivo, but to a lesser extent. Meanwhile, a latent membrane protein (LMP) has been detected which assumedly is a component of the so-called lymphocyte-detected membrane antigen (lydma).

5.3.4.3 Genes and Their Products in Cell Transformation

As early as 1963 L.Crawford noted that *Polyoma virus* had a very small genome which would make it a suitable test virus in tumour research, and in subsequent transfection experiments (1964), L.Crawford's group demonstrated that different forms of DNA extracted from purified virus could evoke cell transformation. T.Benjamin (1966) was able to go a step further in his DNA–RNA hybridization studies by proving that transformed cells produced a small amount of mRNA that had been transcribed from *Polyomavirus* DNA.

Further significant results have emerged from experiments with *RSV* and especially with defective *RSV*-mutants. The earliest transformation-defect mutants were isolated from cloned *RSV* stock by Dougherty and R.Rasmussen in 1964. Robinson and colleagues (1965) isolated and purified *RSV* nucleic acid and Toyoshima and P.Vogt (1969) devised a method for obtaining temperature-sensitive mutants of an avian sarcoma virus. The virus was treated with the mutagen 5-azacytidine so that the mutants failed to produce foci and lacked the ability to replicate at non-permissive temperatures. Duesberg (1968) demonstrated that the RNA was a dimer composed of two single-stranded subunits. Amongst the results using different *RSV* strains published later, G.Martin (1970) concluded that a part of the viral genome was required for transformation but not for growth. The same year, Duesberg and P.Vogt observed electrophoretic differences between the RNA of transforming and non-transforming avian tumour viruses and added a class *a* subunit to the transforming potency. Three years later, Duesberg and P.Vogt collaborating with M.Lai and J.Horst, reasoned that only a part of the *a* subunit was involved in transformation, based on results of oligonucleotide fingerprinting and DNA-RNA hybridization.

Kawai and Hidesaburo Hanafusa (1971) suggested that a virus-coded, heat-labile protein, which was not a structural component of the virus, was needed for transformation, and the gene coding for this transforming protein was characterized by P.Vogt, Biquard and Vigier, by G.Martin and by G.Martin and Duesberg between 1970 and 1972. The four basic genetic elements of the RNA tumour viruses were defined at the Cold Spring Harbor Symposium in 1974: the env gene for coding the envelope glycoprotein, the pol gene for reverse transcriptase, the gag gene for structural proteins other than env, and the onc gene for a protein able to transform growth properties of the cells but not required for virus replication. In 1976, the Duesberg and the Joho group described the gene arrangement of the genome, whilst in 1977 Susan Weiss and coworkers confirmed this arrangement and determined the size and genetic composition of virus-specific RNAs of cells which were known to produce avian sarcoma-leukosis viruses. In

1976 Dominique Stehelin and colleagues determined the src or onc gene of different avian sarcoma viruses to be about 1600 nucleotides. And finally the complete sequence of 9312 nucleotides of *RSV*-Prague C strain was elucidated by Schwartz and coworkers in 1983.

The term onc gene corresponds to the oncogenes defined by Huebner and Todaro in 1969. Their oncogene hypothesis implied "that the cells of many, if not all, vertebrates carry vertically transmitted tumor virus information which serves as an indigenous source of oncogenic information which transform normal cells into tumour cells. Depending on the host genotype and various modifying environmental factors, either virus production or tumour formation, or both, may develop at some time in these animals and/or in their cells grown in culture. In addition, it implies that the occurrence of most cancers is a natural biological event determined by spontaneous and/or induced derepression of an endogenous specific viral oncogene(s)." An unexpressed viral genome, they assumed, was transmitted by a number of replicate RNA molecules or cellular genes in the form of a DNA provirus.

By means of the specific complementary DNA copy of the viral RNA template (cDNA probe), it was possible to detect src sequences in normal and transformed cells. Analysis of retroviruses after transformation made it possible to identify various viral structural components and to characterize virus-associated and virus-induced proteins which were suspected of participating in the regulatory mechanisms of cells.

The earliest indications that a retrovirus resulted from the recombination of a viral genome and an animal genome were published by Scolnick and colleagues (1973), and one year later by Scolnick and Parks as well as by Tsuchida and coworkers. Nucleic acid hybridization studies using murine *Kirsten* and *Harvey sarcoma viruses* enabled the Scolnick group to detect a set of nucleic acid sequences which had also been found in rat type C viruses. Tsuchida and coworkers discerned *Mammalian sarcoma virus* sequences which were related to RNA sequences in normal rat cells. The evidence that DNA sequences found in normal avian cells were related to the transforming genes of *Avian sarcoma virus* was published by Dominique Stehelin and colleagues in 1976. They demonstrated that the DNA of normal chicken cells contained nucleotide sequences which were unmistakably related to at least a portion of the transforming gene(s) of *Avian sarcoma virus*, defining these sequences as homologous to cellular genes. As a result of this publication, similar sequences were detected in a wide variety of vertebrates. Non-transforming cellular homologues of retrovirus transforming genes were later termed proto-oncogenes or cellular (c-) oncogenes. These specific sequences have a fixed position on the chromosome, including those in humans. Meanwhile, about 30 different viral oncogenes with transforming properties have been identified, each with a distinct structure and some with different activities. Sequences related to the onc gene have also been detected in Drosophila and in the worm Caenorhabditis elegans.

That protein homology between viral onc's and their cellular counterpart c-onc's is high was demonstrated by Van Beveren and colleagues in 1982: 1157 corresponding nucleotides of v-mos and c-mos differed in only 25 positions. But this small number of different positions in quantity mark also such in quality. Since it had been established that avian and mammalian DNAs possessed sequences closely related to the onc gene of *RSV*, it was thought that a viral oncogene might be the result of recombination with its cellular counterpart. The close relationship between v-oncogenes of highly oncogenic retroviruses and c-oncogenes was explained as the evolutionary conservation of sequences. The small deviations between the two types of oncogenes can be helpful in explaining the multiple stages of carcinogenesis, the causes of which can be manifold and include chemicals, radiation and other physical as well as enviromental factors, including infective viruses.

On the whole, experimental results with v-oncogenes support the biochemical view that cancer is the result of a multi-step process which still needs clarification. The role of tumour viruses, their oncogenes and their cellular counterparts are being questioned as stable components of the cell genome involved in the genesis of human cancers. With the tumour suppressor genes or anti-oncogens new ideas arose which further stimulated the oncogene discussion. Now it is supposed that both activation of oncogenes and inactivation of tumour suppressor genes are involved in oncogenesis, the whole being the result of induced genetical disturbances in the cell cycle.

5.3.4.4 Results on Some DNA (Tumour) Viruses

DNA tumour viruses also possess oncogenes and are able to confer these transforming genes on cells. Just as with RNA viruses, the presence of one or two of these viral oncogenes in an activated form, i.e. having been transcribed and translated, in an appropriate normal cell can result in cancer either immediately or after additional genetic changes. The known oncogenes of DNA viruses encode proteins which act only in early viral functions and basically regulate cell growth. How they participate in oncogenic cell transformation is still an enigma.

Papillomaviruses

Microscopical studies of *Shope papilloma virus* with proliferating tissue by Noyes and Mellors (1957) proved the absence of viral particles within papillomas but the presence of fluorescent antibodies in the wild and domestic animals tested. During the two subsequent years, Noyes was able to describe the morphology of *Shope papilloma* and human wart viruses. Further studies by Noyes' (1965) on primary human embryonic skin and muscle cultures, using extracts obtained from pools of human wart specimens, found dense focal areas but no other morphological entities. Pellets of wart extract revealed typical *papilloma virus* particles in electron micrographs.

Further Investigations indicated that in the otherwise successful transformation experiments with intact virus and in transfections with viral DNA efficiency remained low. An additional problem was the very low cytopathogenicity exhibited by the virus which hampered development of a reliable cell culture system. Zur Hausen and colleagues (1974) demonstrated a variety of papillomaviruses in different human infections and in malignant genital tumours using nucleic acid hybridizations. Of the known 48 human papillomaviruses, two types are associated with benign genital condylomas and six are correlated to genital and pharyngeal carcinomas.

Polyoma Virus

This virus has been known to cause tumours since it was detected by L.Gross in 1953; J.Smith and colleagues (1960) found that it contained DNA. Hamster cells, which are semi-permissive and do not support virus replication but can be transformed at greater frequency than mouse cells, were deemed a suitable system for this transforming and tumour-inducing *Polyoma virus.*

Herpesviruses

Beside *Locke's frog virus,* other viruses in this family are known to possess cell transforming and oncogenic properties in animals and in man.

Marek disease virus: this causative agent in avian neuro-lymphoblastosis was propagated by Churchill and Biggs as well as by Van

Nazerian and coworkers in 1968. It infected T-lymphocytes and was cell-associated, according to Van Nazerian (1972). Akiyama (1973) demonstrated that Marek's tumour cells could be established as a transformed permanent T-lymphoid cell line.

Epstein-Barr virus: In 1958 Burkitt described an unusual tumour of the jaw occurring in children in Uganda. Since the geographic distribution of the disease coincided with that of its arthropod vector, he firmly assumed that the tumour was induced by an insect-transmitted oncogenic virus (1962). Meanwhile, the tumour has been named Burkitt's lymphoma. In 1963, M.Epstein, Y.Barr and Achong established several cell lines with lymphoblastic characteristics taken from tumour tissue specimens.

Simultaneously, M.Epstein and colleagues and Pulvertaft (1964) reported encountering virus particles which possessed herpesvirus morphology, namely the *Epstein-Barr virus (EBV)*, in lymphoblasts that had been cultured from Burkitt's lymphoma (BL) explants. In the following 3 years, human lymphoblastoid cell lines were established using specimens from patients manifesting different forms of leukemia by Itakawa and Grace (1964), G.Foley (1965), Moore (1966) and colleagues. E.Jensen (1967) and Levy (1968) and coworkers procured additional cell lines from lymph node biopsies of reticulum cell sarcoma, as well as carcinomas of the lung and spleen removed because of spherocytosis. In 1966, W. and Gertrude Henle, Moore, Levy and colleagues concluded that all patients with BL possessed antibodies for *EBV*. One year later, W. Henle and coworkers reported that normal lymphocytes obtained from an infant girl could be made to grow indefinitely in culture provided they were co-cultivated with X-irradiated BL cells. This phenomenon, later called immortalization, was shown to be dependent upon *EBV*, and is central to the pathogenicity of lymphoproliferative syndromes associated with this virus. It was the first virus known to convert immunoglobulin-producing B cells into established cell lines, and has thus far been the most effective transforming virus.

Because different geographical areas with endemic or sporadic BL showed large fluctuations in the presence of EBV-DNA and expression of EBNA, and because some BL do not contain any *EBV* sequences, there have been various interpretations of the association between *EBV* and BL. As all malignant cells unlike hematopoietic stem cells contain c-myc oncogenes, one explanation for the cause of BL is that c-myc is the breaking point for chromosome translocation and deregulation.

EBV is also the causal agent of infectious mononucleosis (IM), a polyclonal transformation of B cells. Successfully cultured cell lines from the blood of IM-infected patients have been reported by Pope (1967), Glade (1968), Gertrude Henle (1968) and colleagues. *EBV* was isolated by Diel and coworkers (1968) and by H.Moses and colleagues (1968) from cell lines derived from IM-patients, but also from persons not showing signs of the disease. Infectious mononucleosis is usually the primary *EBV* infection; prolonged primary infection syndromes have been described in children with immune defects or in patients undergoing immune-suppression for organ transplantation. It is known that the virus can persist latently after a primary infection in any human population regardless of geographical location.

That *EBV* is associated with nasopharyngeal carcinoma (NPC) was established by Gertrude Henle and colleagues, by Niederman and colleagues (1968), and by various other groups through hybridization experiments between 1970 and 1975. In 1970 Zur Hausen and Schulte-Holthausen reported *EBV* nucleic acid homology in a "virus-free" cell line isolated from a Burkitt lymphoma.

Hepatitis B virus (HBV)

During the 1950's, some internists of the medical school at Dakar suggested a relationship between the consumption of alcohol and the incidence of primary hepatocellular carcinoma (HC). Test series initiated by Sherlock and colleagues (1970), and carried out in various parts of the world, proved that *HBV* surface or core antigen was associated with the carcinoma. In 1978, Kew reported that *HBV* infection preceded tumour development. This conclusion was supported by J.Summers and coworkers and by London in the same year, and by the Goudeau and Thung groups in 1979. They detected *HBV* antigen as well as viral DNA sequences in the tumour tissue. Bréchot and coworkers (1980) reported integrated *HBV*-DNA sequences in the cellular DNA of human hepatocellular carcinoma. Confirmative results followed in 1981 when Shafritz and colleagues reported integrated *HBV*-DNA sequences, this time in chronic hepatitis and HC patients. Mills and coworkers as well as Goudeau and coworkers (1981) finally confirmed a conspicuously high rate of previous exposure to *HBV* in alcohol-dependent patients. Bréchot and colleagues (1982) then compared the serological markers of *HBV* with *HBV*-DNA found in the livers of alcoholic liver-disease patients with or without HC. Their results indicated a direct link between *HBV* and the pathogenesis of alcoholic primary liver-cell carcinoma.

5.3.4.5 Results with RNA Tumour Viruses

At first the RNA tumour virus group comprised only avian, murine, and mice mammary tumour viruses. They were known primarily to cause leukemias, lymphomas, sarcomas, mammary and other tumours in rodents, chickens, cats, cattle and primates. The best known C-type viruses had an expanded host range including fish, snakes, worms, pheasant, quail, partridge, turkey, tree mouse and agouti. They were transmitted either as proviruses or as exogenous particles. RNA tumour viruses, now classified as retroviruses, are the only RNA viruses known to induce tumour growth.

For viruses which are transmitted vertically in a proviral state, there may be factors which change this state or which help to bring about oncogenicity. Horizontally transmitted viruses can cause tumors if they are exogenous. Since RNA tumour viruses seemed to be ubiquitous, some groups began searching for them in human tumours. In 1977 Uchiyama and colleagues as well as Takatsuki and colleagues reported on adult T-cell leukemia (ATL) in endemic areas in Japan. Based on clinical, hematological, cytogenetic and other studies, they came to the conclusion that ATL was of viral etiology. Poiesz and coworkers (1980) isolated a type C retrovirus particle from freshly cultured lymphocytes taken from a cutaneous T-cell lymphoma patient. This had been made possible because of the tremendous improvements in the cultivation of human hemapoietic cells and long term T-cell cultures. Prival and coworkers (1974) had cultured myeloid cells on agar plates using human peripheral blood lymphocytes treated with phytohemagglutinin. This substance stimulates the synthesis of a T-cell growth factor which since 1979 has been known as interleukin-2. In 1976 Doris Morgan and colleagues were able to propagate phytohemagglutinin-stimulated human T-lymphocytes selectively and permanently, paving the way for Poiesz and coworkers' use of this cell culture system.

The particles isolated by the Poiesz group were visible in both mature and immature form in thin-section electron micrographs and called *Human T cell leukemia virus (HTLV)*. The following year, they reported isolating a new C type virus from a Sézary T-cell leukemia patient, and found that it closely resembled the virus characterized in 1980. In the same year, Reitz and colleagues published results of hybridization experiments indicating that *HTLV* was a member of a new class of C type viruses transmitted horizontally as exogenous infectious agent.

In December 1981, *The New England Journal of Medicine* published three papers about a novel lethal disease, which had first been observed in New York, Los Angeles and San Francisco in mid-1981. It was simply called AIDS for acquired immune deficiency syndrome. The authors, Gottlieb and colleagues, Masur and colleagues, and Siegal and colleagues described drug abusers and homosexuals with intact humoral but depressed or acquired cell-mediated immunity. The immune deficiency was linked to reduced helper lymphocytes as well as to Kaposi's sarcoma and to unusual opportunistic infections. Until that time, such infections, normally caused by harmless microbes were known but thought to be pathogenic only in weakened people. In 1982, Kalyanaraman and coworkers isolated a second *HTLV* from a patient with hairy cell leukemia and called it *HTLV-II*. That year serological surveys were begun involving laboratories from all parts of the world. The results published by the Gallo group (1983) indicated that HTLV was associated with Japanese adult T-cell leukemia and T-cell lymphosarcoma cell leukemia. Other endemic regions of *HTLV*-associated malignancies were detected in the Caribbean, the south-eastern part of the USA, Central and South America, and equatorial Africa.

From papers published by Trainin and colleagues (1983), it was known that permanent secondary immunodeficiency were caused by *Feline leukemia virus*, which in turn prompted serological studies with sera from AIDS patients. These results published by Essex and colleagues (1983) showed that antibodies reacted with a surface antigen of a T-lymphoid cell line which had been *HTLV infected*. Thereafter, Barré-Senoussi and coworkers detected a different retrovirus in a patient with lymphadenopathy syndrome (LAS) which had also been observed in AIDS patients. The same virus type was repeatedly isolated by independent groups, and named lymphadenopathy-associated virus (LAV), *HTLV-III* or AIDS-associated retrovirus (ARV). In 1984, Beatrice Hahn and colleagues concluded, after extensive sero-epidemiological studies and molecular-genetic characterization of isolated viruses, that HTLV-III virus was associated with AIDS. In 1986, the name for all these types of virus was changed to human immunodeficiency virus (HIV). Between 1980 and 1986, viruses of this type but related to different diseases were isolated in various geographical areas, not only from humans but also from monkeys. Table 5.17 summarizes the viruses thus far isolated and typed: LAV, HTLV and H(human)IV or S(simian)IV.

Table 5.17. Isolated human and simian immunodeficiency viruses

1980	HTLV-I	Poiesz et al.
1982	HTLV-II	Kalyanaraman et al.
1983	LAV	Barré-Senoussi et al.
1984	LAV, HTLV-III, ARV, HIV-1	Gallo et al.
1985	SIV mac	Daniel et al.
1986	HIV-II (LAV-II, HTLV-IV)	Clavel et al.
1986	HIV-III	Gallo et al.; Popovic et al.
1986	SIV mangabey	Fultz et al.; Murphey-Corb et al.

5.4 Viruses and Immunity

When scientists first attempted to define "the virus", no clear-cut difference was drawn between the organism's natural defence - called resistance-- and acquired immune responses - called immunity - to various types of antigens. The degree of immune response to known causative agents, as well as the duration of immunity, which had proved to be disease-dependent, were firmly entrenched scientific results in immunology. It had been suggested that the thymus was the key organ in immunity, while the spleen, lymph nodes and bone marrow played important roles.

In 1939, the synthesized, circulating, humoral defence substances called antibodies, had been determined electrophoretically to be gamma globulins which exhibited slow anodic mobility. When serum proteins showing even greater electrophoretic mobility were disclosed to be linked to antibody activity and to possess γ globulin-like antigenic properties, they were called immunoglobulins (IG). Deutsch and coworkers (1946) distinguished between the low and high molecular weight components 7 S and 19 S by means of ultracentrifugal analysis. In addition, the IG were defined according to their serological activity as agglutinins, precipitins or lysins. Based on previous experience gained through serological testing during the development of inactive influenza and poliomyelitis vaccines, it was assumed that the amount of IG and its duration of circulation in the serum indicated to some extent the organism's defence against re-infection. It was Astrid Fagraeus (1948) who demonstrated that plasma cells were the effector cells in the immune response afforded by these circulating antibodies. For a long time, these antibodies were considered to be the sole protective elements against viruses which infected vertebrates. This applied to induced as well as to congenital immunity by the passive transfer of antibodies to the fetus in pregnancy. At that time it was impossible to prove the synthesis of demonstrable IG in fetal life. S.Anderson and Hamilton first described this for primary herpes simplex infection in 1949.

Between 1935 and 1939, Traub detected an immunological phenomenon which Burnet and Fenner a decade later were to regard as immunologic tolerance. It was identical to the "horror autotoxicus" which P.Ehrlich had postulated. This was defined as the inability of an individual later in life to respond immunologically to a specific antigen stemming from his fetal life. Traub's model was *LCMV* in mice. Later it was shown that induced tolerance was not restricted to the newborn, and that susceptibility to tolerance induction depended on a number of differing factors. It is as specific as immunity; broken tolerance is termed autoimmunity. The theory of humoral antibodies being the organism's main and only defence against virus infections was gradually discredited. For example, Bruton (1952) and Good and Zak (1956) detected hypogammaglobulinemia in measles patients which indicated that the bodies response to viral infection differed from its response to bacterial infection. Moreover, an immune response to some congenital human virus infections could not be detected, and the role of humoral immunity was uncertain in experiments with tumour viruses in animals.

The introduction of the term delayed-type hypersensitivity, which had been called tuberculin-type immunity in bacteriology, conveyed the idea that an additional factor independent of humoral immunity was a necessary element. An example of this kind of immunity was reported by Fenner (1948) with *Ectromelia virus* in mice, and by J.Brown (1953) in patients recovering from primary herpes simplex infection. Cellular transfer of delayed-type hypersensitivity in guinea-pigs was demonstrated by Landsteiner and M.Chase in 1942. At that time, it was assumed that lymphocytes and macrophages were the operative cells in delayed-type hypersensitivity. When Burnet set up his Clonal Selection Theory of Acquired Immunity in 1959, he postulated that there might be multiple clones of globulin-producing cells. He regarded the thymus as the "first level" immunological organ in which lymphoid cells originated. They subsequently proliferated to produce functionally active descendant or collateral cells in "second level" organs, namely spleen, lymph nodes, bone marrow and local lymphoid-cell aggregations.

By clarifying that specific and non-specific as well as humoral and cellular immunity were parts of a genetically-directed immunological network, in which many differentiated cells participated, a gate was thrown open to completely new possibilities in experimental virology. This involved proving the presence of viruses and virus fragments in cells and tissue, immunological diagnosis, and also development of active and passive vaccines. Significant milestones in immunological research which were to have an impact on experimental virology included greatly enhanced knowledge of the complement system, heterogeneity of the immunoglobulins, the significance of the T-lymphocytes and their activation, and interferon.

The Complement System

In 1958 only four components of the complement system (C'system) were known. The C' system is the primary humoral mediator of antigen-antibody reactions. In time, 17 different serum proteins were added, making a total of 21. Of these, 13 were considered to be true components, 8 of which were regulator proteins which could be activated sequentially. They were synthesized by macrophages, fibroblasts, intestinal epithelial cells as well as by peripheral monocytes and lymphocytes. There are three ways of activating the C'system which entail many factors. These include the IG and their aggregates, IG-containing immune complexes, all types of antigens, virus infected cells, different types of mitogens, polyanions, trypsin-like enzymes; the properdin-system (described by Pillemer and colleagues in 1954); and interferon. Disorders in this system, whether they be aquired or congenital, result in an entire array of symptoms which lead to the conclusion that immune response to a virus can cause disease.

Immunoglobulins (Ig)

The nomenclature for immunoglobulins which was adopted by the WHO in 1964 referred to only three IGs which had been characterized on the basis of their electrophoretic behaviour and their specific physical, chemical and antigenic properties. These were IgG (γ G), IgA (γ A) and IgM (γ M); in subsequent years, IgE and IgD were added. They were produced under genetic control by B lymphocytes which circulated while carrying specific Ig molecules on their surface membrane for recognizing a corresponding antigen. Ig molecules were found to possess extensive heterogeneity in their antigenic and electrophoretic properties.

In 1963, Uhr and colleagues and Finkelstein and colleagues published results on the formation of rapidly and slowly sedimenting antibodies after administration of various antigens to humans, rabbits, guinea-pigs, sheep embryos and several non-mammalian vertebrates. Results showed that 19 S Ig first appeared, replaced by primary 7 S antibodies one week after infection, followed by 7 S memory antibodies. These papers confirmed the results with poliovirus in rabbits obtained by Svehag and Mandel (1962) and those of Stelos in 1957 with antibodies to red cells.

Important Characteristics of Ig

IgA $(7S)_n$, (150000)n daltons: This type occurs predominantly in extravascular secretions, namely saliva, tears, digestive tract fluid, nasal and mammary gland secretions. About 10–15% of serum is IgA with a variable sedimentation coefficient range of 7S–11S. IgA exists as dimers, trimers, and even higher multimers of 7S monomers; usually they circulate in monomeric and dimeric forms.

IgD: Occurs predominantly on the surface of human lymphocytes, particularly in cases of lymphatic leukemia. Besides IgM, it is the second major B cell receptor for antigens.

IgE: This is a very small component of serum. Since human antibodies of this type are involved in allergic reactions, they are known as allergy-mediated antibodies.

IgG (7S, 150000 daltons): Occurs predominantly as late antibody in human serum. It can be transferred from the circulatory system of the mother to that of the fetus through the placenta. About 75% of the Ig of higher vertebrates is IgG. Its sedimentation coefficient is between 6,5 and 6,6 with a molecular weight close to 150000 dalton, and it is synthesized as the secondary response in memory B lymphocytes.

IgM (19 S, 900000 dalton): About 10% of human serum Ig consists of this pentameric *M*acroglobulin. This is the first antibody formed in an immune response and the first produced by the neonate. It is involved in the pathogenesis of some auto-immune diseases. IgM functions as a major antigen receptor on the surface of the mature B cell. It has high electrophoretic mobility and S values of 18–23. Normally IgM consists of five identical subunits and cannot pass the placenta; however, as membrane Ig, it is monomeric.

All Igs have a common polypeptide chain, but smaller structural variations differentiate the various molecules from one another and characterize their heterogeneity. These variations exhibit specific antigenic and electrophoretic behaviour.

The Role of T Lymphocytes in Cell-Mediated Immunity

By 1961, it had been proved that the thymus was part of the immune system. The heterogeneous group of lymphocytes synthesized in this organ cannot produce antibodies but they do acquire various special properties to recognize antigens. After leaving the thymus, they begin circulating via the paracortical areas of the lymph nodes, the periarterial sheets of the lymphoid follicles of the spleen, the efferent lymphatics and possibly via the blood vessels of the spleen back to the lymphoid compartment. They cooperate with antibody-secreting cells.

An immune response involves a cascade of intracellular events. Thus, antibody formation by B cells happens only if an antigen hits a B cell. This requires the help of both accessory cells and T cells. Here, T cells called CD4 or T4-positive helper/inducer T cells are very important. They cooperate by stimulating B cells to proliferate and differentiate into antibody-secreting cells. Their support in B cell response to most protein antigens, i.e. thymus-dependent antigens, is absolutely essential. Helper cells carry a CD4 molecule which can be recognized by suitable monoclonal antibodies. The function of this marker, as well as that of the CD8 molecule on the other main suppressor/cytotoxic subsets, has not been clarified. Suppressor cells restrain immune activity while cytotoxic cells are the most important effectors in immune response; they lyse cells bearing antigens for which they are specific, e.g. tumour antigens or virus-specific antigens on virus-infected cells. Cytotoxic cells, responsible for cellular cytotoxicity without prior sensitization, are known as natural killer cells (NK cells). A T cell carries either CD4 or CD8, never both. The assumption is, that CD4 or CD8 molecules somehow stabilize or strengthen the bond of a T cell receptor to the antigen.

In the various steps necessary to induce immunity, T4 helper cells are involved which, for example, release interleukin-2. The different factors produced by T helper cells further stimulate T cell and B cell growth. Some products synthesized by T cells encourage a shift in production from IgM to IgG, IgA, or IgE. Certain activated B cell clones produce B cell progeny which gradually return to the G_0 growth state. These lymphocytes display mutated receptors of higher affinity as so-called "memory" cells.

Delayed hypersensitivity may include inflammatory reactions in tissues characterized by activated macrophages infiltrating specific sites where antigen is present. This infiltration is dependent on specifically sensitized T cells and takes 24–48 hours to reach a maximum after the antigen has been introduced. Sensitized T cells produce a number of soluble mediators responsible for recruiting and activating macrophages. One of these released by T cells is the macrophage migration inhibition factor (MIF). An activating factor released by macrophages is called interleukin-1.

Antibody-forming B cells are the best understood effector cells. Nevertheless, the same broad rules of activation apply to other immunocytes. Thus, a type of CD8-positive T cell, known as a cytotoxic T cell, can cause lytic damage to a target cell, such as a virus-infected cell or an allogeneic transplant, to which the host is immune. We now know that helper-inducer CD4 positive T cells and accessory cells are present in all tissues, but that very high concentrations, predominantly in lymph nodes and the spleen, are usually required for activating cytotoxic T cells.

Secondary immunodeficiency, which is preferentially associated with defects in cell-mediated immunity, can result in viral infections, and viral infections can in turn result in secondary immunodeficiency. This has been proven during the last three decades. That secondary immunodeficiency induced by immunosuppression for organ transplants is followed by viral disease was reported by Merselis and coworkers for varicella-zoster (1964), H.Anderson for *Cytomegalo virus (CMV)* (1966), Underwood and Weed for herpes (1974), and Yokken and colleagues for rotavirus infection (1982). Viral infections resulting from developed lymphomas have been reported by Shanbrom and colleagues (1960) for varicella-zoster, Casazza and colleagues (1966) for herpes, and Lynfield and colleagues (1969) for *CMV* infection. That symptomatic primary *CMV* infection can be the cause of a disorder in the ratio of OKT4 to OKT8 lymphocytes was reported by Rinaldo and coworkers (1977). During the course of the disease, activated cytotoxic OKT8 cells increased strikingly, while helper OKT4 cells decreased only slightly. The immunodeficiency viruses first detected and characterized as *Feline leukemia virus* in 1983 and thereafter as *HIV* opened a new field in virology and also in viral immunology.

All theories and hypotheses in immunology propounded to explain investigational results as well as to stimulate further experimental undertakings culminated in Jerne's Network Theory and in the hybridoma technique, completed in 1975. The theory comprises the factors known to regulate the highly complex system of immune response, and it links biological molecules to the cells participating in immunity. Thus, there are partly synergetic and partly antagonistic interactions of T and B lymphocytes; antibodies can recognize or be recognized; antibodies may occur as free entities or as receptors on lymphocytes; antigen-sensitive lymphocytes can respond either positively by cell proliferation or activation and antibody secretion, or negatively by tolerance and suppression.

The immune system also includes interferon (IF), which was discovered by Isaacs and Lindenmann in 1957 and was thought to be the substance responsible for the interference phenomenon known from various viral test systems even since 1935. Because IF did not seem to be related either to antibodies or to phagocytes but was identified as being activ also between inactivated and infectious virus by the Henle's in 1944 and in co-operation with Pauker in 1958, it was considered to be a special agent with antiviral activity induced immediately after virus infection. Meanwhile, we have learned that interferons are low molecular weight proteins induced in cells by viruses and also by a number of different biological and chemical substances. In vitro and in vivo experiments have demonstrated that IF acts upon all types of viruses but only in a host-specific manner, which indicates that it can be effective only in that system in which it was produced. This can happen by active induction of endogenous IF through viruses or IF-inducing substances, as well as by administering exogenous IF doses. Thus, it was thought that a potentially severe epidemic could be warded off by prophylactically administering harmless, attenuated virus vaccine which possessed high IF-indution capacity. To administer exogenous IF to humans meant that it had to be produced in human cells, e.g. in the leucocytes. Now, it is known that there exist some IF types with different properties which have to be produced in diverse cells, e.g. lymphoblasts, fibroblasts, etc.

As time passed, IF was gradually recognized to be a non-specific factor in the immune response. Thus, it might activate natural killer cells, or it might modulate the immune response by curbing lymphocyte proliferation or it might even affect the growth of certain tumours including those of non-viral origin. The synthesis of IF is the first short-term protective response by virus-infected animals and man. In the early stages of biotechnology it was thought that at least eight cloned human IF genes within bacteria would suffice to produce the huge amounts of various pure interferons necessary for prophylactic or therapeutic purposes in the human. The effective and beneficial role of IF has been documented in numerous experimental investigations with test animals. Administration of exogenous IF to humans, however, can produce fever, fatigue, malaise and lymphopenia. We are now quite certain that similar symptoms observed during many common acute viral infections are due, in part, to the production of endogenous IF. Nasal spray administration of IF against rhinoviruses produced a local inflammatory reaction not unlike those observed during viral infections of the upper respiratory tract.

5.4.1 Humoral and Cell-Mediated Immunity as Parameters of Protection

Viruses use the vascular and lymphatic systems of an organism not only as a means for their own distribution; many cells required for the development of immune defences are also extremely important as host cells for viral replication or integration. Besides the plasma cells, these would include the cells of the aforementioned organs plus the lymphocytes, monocytes, macrophages, polymorphonuclear leukocytes and Kuppfer cells. When these become infected, the immune response is weakened; despite this, however, an induced immunity can as a rule be serologically demonstrated. This humoral type of measurable specific immunity subsequent to infection elicits at least two fundamental questions with respect to complete immunological defence: Can humoral antibodies provide instant protection against a severe course of infection? Can they prevent reinfection?

Serological tests do provide a relatively easy way of proving the presence of a specific viral infection, although it is far more complicated to make a reliable prediction about desirable immediate and permanent protection on the basis of in vitro test results. This applies particularly to neutralizing antibodies, and also to those tested in other systems such as CFR and hemagglutination-inhibition. When present in systemic circulation, neutralizing antibodies can block infectivity and promote clearence of those viruses which spread viremically, namely polio, measles, hepatitis A, rubella, smallpox and varicella. However, if cytotoxic T cells are required for protection as with pox- and herpesviruses, protection will only be half of that required. Mucosal immunity by IgA plays a role in reinfection by viruses that replicate exclusively in the mucous membranes, namely rhinoviruses, influenzaviruses, parainfluenzaviruses and *Respiratory syncythial virus*. Serum antibodies appear not to protect the upper but the lower respiratory tract.

In 1940, Francis demonstrated the type-specifity of the influenzavirus soluble antigen (S antigen) in CFR. The results were different, however, when the chick embryo was used for virus replication. Friedewald (1943) and W.Henle and Wiener (1944) showed that infected allantois fluid contained two components of complement-fixing antigens; one was type-specific while the other was strain-specific. There was no obvious relationship between the two antibody types. The CFR antibody was only short-lived whereas the HI-type was long-lived.

After Hirst (1942) had shown that hemagglutination of influenzavirus could be inhibited quantitatively by antisera, test methods were developed to determine the hemagglutination-inhibiting antibodies in the so-called strain-specific HI-Test. Hirst's conclusion that there was a close relationship between the neutralizing and the HI antibodies became a controversial issue in the following decades. Factors entailed in the testing methods were not assessed critically enough, i.e. the presence of inhibitors, the amount of virus used in the test, and HA measuring techniques. On the whole, it was conceded, a qualitative relationship did exist in the results of both tests, which permitted drawing conclusions on strain specific immunity.

In 1957, a new influenzavirus strain was held responsible for a global epidemic indicating the importance of immunity to surface viral glycoproteins. While the nucleoproteins and matrix-proteins were closely related to those of previous strains, the HA and neuramindase glycoproteins were novel. The assumed reason was that a previously unexperienced population had become infected. Askonas and colleagues (1982) demonstrated that monoclonal antibodies targeting HA or neuraminidase protected mice from the challenge by virulent wild-type virus, whereas those targeting internal proteins offered no protection. Thus, there was proof of a direct relationship between measurable HI-titers and their protective properties.

Antigenic variations of influenza viruses have been known since it was possible to classify strains by HI tests and to detect and to assess HI-antibodies in individuals and populations. These variations were divided into drifts and shifts. Drift described a relatively minor change in the surface glycoproteins which occurred in type *A* as well as in type *B* influenzaviruses. The shift had been observed only with type *A* viruses and involved far more dramatic antigenic changes in "novel" viruses as a cause of pandemics. First described in 1957, and later in 1977, the deviation was characterized by a change in the antigenic properties of the hemagglutinin and the neuraminidase gene. The origin of these novel virus strains was assumed to be the resurgence of old or enduring genes, or recombination with strains from different animal species, or by spontaneous or induced mutation.

Another phenomenon in virus-antibody interaction was detected by Hultin and McKee (1952) with influenzavirus and confirmed by H.Rubin and R.Franklin (1957) for *Newcastle disease virus*. They demonstrated the aggregation of a virus-antibody complex with susceptible cells. The following year, Mandel concluded that there was a very low probability of reassociation and spontaneous virus-reactivation from such complexes or even a dissociation of the binding complex from the cell under specific physiological conditions.

When Isabel Morgan (1949) showed that the amount of humoral poliovirus antibodies had a direct bearing on the degree of protection against an infection, this seemed to prove that neutralizing humoral antibodies afforded some measure of specific immunity. Since experimental neutralizing tests with animals were long-term, expensive and rarely possible, the option was to analyze the significance of neutralizing antibodies assessed in serological tests. Howe and coworkers (1950) discussed the different implications of antibody-dependent immunity for reinfection of the digestive tract and invasion of the CNS after infection trials in chimpanzees. Burnet's (1955) former assumption that an antigen-neutralizing antibody reaction was completely reversible in animal viruses was refuted by Dulbecco and colleagues (1956) in neutralization studies with plaque assays of *WEEV* and poliovirus. Their results, however, also showed a small portion of viruses which resisted neutralization despite exposure to excessive amounts of neutralizing antibodies. In addition, they stated that this process was a first order reaction, i.e. proportional to the concentration of antibodies. In the same year, this was confirmed for *Newcastle disease virus* by Rubin and R.Franklin.

Epidemiological and serological data for *Measles virus* led Babott and J.Gordon (1954) as well as Gisela Ruckle and D.Rogers (1957) to consider that immunity acquired through infection probably persisted throughout life. In 1958, F.Black and Rosen proved that the presence of measles antibodies was closely related to measles infection, and that neutralizing and HI-antibodies indicated the immune status against measles virus, while CFR antibodies probably did not. The antibodies which are induced by primary infection have a sedimentation coefficient of 19, as R.T.Smith ascertained in 1960. Subsequently, IgM is replaced by 7 S antibodies, namely IgG, within 2–3 weeks, as was shown for several viruses. Working with *Measles virus*, Ann Schluederberg (1965) undertook the determination of IgM for recent infections, and assessing the difference in IgM titres during primary and secondary infections in humans. She found that IgM was synthesized after primary antigen stimulus; within a few weeks this disappeared and was replaced by long-term 7 S IgG. The transient nature of IgM antibodies was subsequently demonstrated for most primary viral infections. Although numerous methods were gradually developed for their determination, the situation nevertheless remained complicated because of the occasional absence of IgM response as well as by false-positive reactions to related viruses.

The role of T lymphocytes in cell-mediated immunity is of prime importance for this protective mechanism in the organism. Because there are many factors involved in cell-mediated immunity, a single in vitro test cannot validly answer questions about protection in the case of a serious course of infection or reinfection; an in vivo protection test is the only way to quantify this type of immunity. In 1970, Blanden suggested that cell-mediated immunity was essential for recovery from mousepox and demonstrated the transfer of highly efficient specific anti-viral activity by immune spleen cells.

Human immunodeficiency viruses, first isolated in 1980, are known to infect specific cells of the immune system. Susceptible cells have a specific receptor on their surface, called CD4. This receptor reacts with glycoprotein gp 120, which is a membrane protein on the virus surface. The most important CD4-bearing cells are T-helper lymphocytes and macrophages. Other susceptible cells are Langerhans', glial and denditric cells, as well as distinct B lymphocytes. Beatrice Hahn and colleagues determined the genomic heterogeneity of different *HIV I* isolates; divergences were found exclusively in that part of the genome which encoded the envelope gene. Unfortunately, as with other virus infections, there is no single immunological test which would permit valid conclusions about possible protection.

Humoral antibodies are significant for diagnostic purposes as well as for protection against severe infections and reinfections. Their protection value is definitely curtailed in those virus diseases which spread from cell to cell, namely *HSV* and *Smallpox virus*, and thereby circumvent neutralization through antibodies. These humoral antibodies display unrestricted protective capacity during infections in which newly formed viruses leave their host cells to reach other cells and organs via vascular or lymph passages, namely polio- and influenzaviruses. In different forms of active and passive immunization, as well as the application of immunostimulators are useful for strengthening the network of immunity.

5.4.2 The Diagnosis of Virus Infection

The introduction of new cell culture techniques for virus propagation, as well as laboratory biochemical methods for splitting virus particles, greatly enhanced the diagnostic possibilities for identifying viruses. This applied just as much to direct proof of the particle or individual components as it did to indirect evidence of the immunological traces left in the host organism.

Constantly endeavouring to discover ever smaller intracellular and extracellular virus fragments or miniscule traces of immunological action, scientists were led to design more sophisticated and more sensitive methods for obtaining virological and immunological proof. The basic methods, besides microscopy, were the techniques already known for virus purification and virus concentration, e.g. diffusion techniques, chromatography, ion exchange, electrophoresis, as well as antigen–antibody reactions, precipitation, agglutination and lysis. The virological methodoly were greatly enriched by new techniques in molecular genetics, and the field of immunology advanced greatly by adopting hybridization techniques for producing monoclonal antibodies. The fundamental framework for all these methods was the knowledge that each type of virus possessed a fixed genome and a genome-fixed specific morphology, structure and immunogenicity.

Deviations from the basic genetic structure result from recombination, mutation and selection. Immunological deviations such as those observed in *Influenza A viruses* and in *HIV* should be regarded as important as immunological similarities, e.g. those found in the poxvirus group. Woodroofe and Fenner (1962) confirmed the close relationship between vaccinia, cowpox, rabbitpox, and ectromelia, as well as between myxoma and fibroma, which share a nucleoprotein antigen (NP antigen) detectable in serological tests. A description of the most important diagnostic developments is presented in the following text, augmented by a collection of data presented in Table 5.18.

Diffusion Tests: In a simple technique, antigen and antibodies are placed on agar in separate wells and, during the diffusion process

Table 5.18. Developments in diagnostic methods

Year	Method	Authors
1950	Direct immune fluorescent antibody technique	Coons, M. Kaplan
1953	Immune electrophoresis	Grabar, C.Williams
1954	Indirect immune fluorescent antibody technique	Coons, Weller
1959	Molecular sieve chromatography	Porath, Flodin
1959	Immune electron microscopy	Singer
1959	Radioimmunoassay	Berson, Rosalyn Yalov
1962	Radioisotope precipitation	Gerloff et al.
1962	Radial double immunodiffusion	Ouchterlony
1964	Immuno-Osmophoresis	Ragetli, Weintraub
1966	Latex agglutination	Bercks
1966	Radioimmunosorbent technique	Wide, Porath
1971	Enzyme linked immunosorbent assay (ELISA)	Eva Engvall, Perlmann
1972	RNA fingerprinting	J.Horst et al.
1975	DNA transfer technique	Southern
1975	Monoclonal antibody technique	Köhler, Milstein
1977	Northern blotting	Alwine et al.
1979	Western blotting	Towbin et al.

which follows, precipitin lines are formed. The method has been improved by labelling the antigen radioactively and using autoradiography. Alternatively, the antigens can first be separated by electrophoresis. Other techniques are the Radial Diffusion Test and the Immuno-Osmophoresis Test (also known as Counterimmuno-electrophoresis) in agar gel. In the first test the antiserum is incorporated in an agar gel layer in a Petri dish. Wells are then cut into the gel and filled with antigen preparation; precipitates form around the wells. In the second test, antigen and antibodies are brought together by electric current.

Aggregation Tests

Agglutination: In addition to agglutination of erythrocytes which can be inhibited by antibodies, virus or antibodies can be adsorbed to latex particles, then used in agglutination experiments. In tests with plant viruses, chloroplast agglutination is the simplest serological test. In an alternative method, red cells are treated with tannic acid to adsorb protein antigens and subsequently this complex is mixed with antiserum and incubated.

Coombs'Test: This technique is used for detecting cell-bound Ig. In the direct test, red blood cells taken directly from a sensitized individual are agglutinated by anti-gammaglobulin antibodies. In the indirect test, a patient's serum is incubated with red blood test cells; the sensitized cells are then agglutinated with the anti-immunoglobulin or with Coombs' reagent.

Immune Precipitation: Indirect immune precipitation with radioactively labelled protein permits the detection of small amounts of antibodies.

Molecular Sieve Chromatography

These are filtration techniques using column gradients for analytical and preparation problems. Sephadex seems to be the matrix of choice for the column.

Acrylamide Gel Electrophoresis

The organic substances acrylamide or polyacrylamide are both available in pure form and can substitute for biological gelling agents such as agar, agarose or partially hydrolized starch. Since these substances form gels ranging from 2%–20% and can be mixed with other gels to give different pore sizes, molecules between 2 million to 1000 daltons can be separated by a molecular sieving technique. After electrophoresis, the material can be analyzed by staining, by fractionating the gels in a number of ways, by radioautography or by immunotechniques.

Labelling techniques

Fluorescence Antibody Techniques: Coon's original technique described by Coons and M.Kaplan (1950) and by Coons (1958) has, after modifications by numerous laboratories, became routine. In a direct staining technique, viral antibody is conjugated to a fluorescent marker; substances usually used are fluorescein-isothiocyanate (FITC) and Rhodamine (Tetramethylrhodamine-isothiocyanate, TRITC). The specifically stained complex combines with antigens in virus-infected cells. In the first step of the indirect staining technique, viral antigens within cells combine with specific, non-conjugated antibodies. Thereafter, the antibodies of the antigen-antibody complex are stained with marker-conjugated anti-gammaglobulin. In a so-called sandwich-technique, antibodies within tissues are confirmed by the indirect method.

Radioimmunoassay (RIA): This assay permits the measurement of antigens on the basis of competition between the unlabelled and the radioactively labelled antigen to form an immune complex with the specific antibody, which is linked covalently to an insoluble matrix, e.g. Sephadex or an antibody complex.

Enzyme-Linked Immunosorbent Assay (ELISA): This assay is analogous to the direct or indirect immunofluorescence technique. Instead of using a fluorescent dye, an enzyme is conjugated to an antibody. The commonly-used enzyme is horseradish peroxidase. To measure the amount of antigen, antibody is bound to a solid support. After a test solution containing antigen, a second enzyme-labelled antibody is added.

Immune Electron Microscopy (IEM):
This method can be used to detect viruses which cannot be propagated by standard tissue techniques or those which are present in negligible amounts. The virus particles are aggregated by mixing with homologous antibody, sedimented to concentrate the aggregates, then negatively stained for visualization.

Molecular Genetic Tests
Fingerprinting: Nucleic acid molecules are digested completely to oligonucleotides by specific nucleases. After separation by a two-dimensional method, the digestion products are labelled and autoradiographed. Such genetic markers are as unique to each individual as the whorls on his or her fingers.

Disease Diagnosis by Recombinant DNA Methods: These methods for detecting DNA mutations or different species are highly sensitive and discriminating. Thus, they are proving especially helpful in virus classification.

Immune Complexes and Their Detection
Immune complexes can bind complement and activate them. They are able to react with erythrocytes, leucocytes, macrophages, thrombocytes and lymphocytes via various receptors on their surfaces. Moreover, they can bind to other cells such as Langerhans cells, hepatocytes, tumour cells and virus-infected cells. Particularly meaningful from an immunological standpoint, is their capacity to bind to monocytes, granulocytes and lymphocytes. Immune complexes rep-

resent a major pathway for pathological expression of the immune response. They can activate the complement system and can be detected by serological methods as well as by tissue analysis using histological techniques and electron microscopy. Within fluids, they can be detected by antigen-specific or by antigen-nonspecific methods like ultracentrifugation and filtration, adsorption and precipitation by polyethyleneglycol, electrophoresis, etc.

Western Blotting

Here, the components of a protein are separated by gel electrophoresis, transferred to a solid support and probed with specific reagents, e.g. antibodies that react specifically with an antigen attached to the solid support. This technique is particularly useful for identifying specific proteins in protein mixtures which have not been radiolabelled.

Hybridization Techniques

In Situ Hybridization: In an early step in virus multiplication, viral gene products can be localized in cells by using immunofluorescence and immunochemistry. In situ hybridization makes it possible to detect, quantitate and locate viral genes in particular types of cells and defined anatomical sites. For this method, only few viral genomes and transcripts are required. Preparations of cells or tissue sections are annealed to a virus-specific probe with a radioactive or non-radioactive reporter group to achieve in situ hybridization.

DNA Transfer Technique (Southern Blotting): After digestion using restriction enzymes and separation of the fragments by gel electrophoresis, the DNA is denatured and transferred from the gel to a solid support, e.g. a nitrocellulose filter or nylon membrane. DNA fragments which are attached to the support are hybridized to radiolabeled DNA or RNA. The positions of the bands complementary to the probe are determined by autoradiography.

RNA–cDNA Hybridization (Northern Blotting): Subsequent to electrophoresis on a gel such as agarose, RNA species are transferred to a support, i.e. diazotized paper, activated cellulose, nitrocellulose, glass or nylon membranes. After checking the efficiency of the transfer, RNA is hybridized with radiolabelled DNA or RNA and autoradiographed.

Monoclonal Antibodies

This technique makes use of the genetically determined specificity of antibodies and overcomes their heterogeneity. The important benefits include:

- An identical antibody can be produced in mass culture and used for a prolonged period in various laboratories.
- A monoclonal antibody recognizes only one antigenic determinant, thus selecting exclusively its specific antigen from many different ones.
- With the help of monoclonal antibodies, it is possible to select the most immunogenic antigen for immunization purposes.
- With the help of monoclonal antibodies, minute changes in a macromolecule or organism can be detected.
- Monoclonal antibodies can define antigenic structures

5.4.3 Immunoprophylaxis in Virus Infections

The development, production and administration of inactivated polio vaccine have taught us how interdependent safety, efficacy and tolerability really are. Simultaneously, new knowledge obtained in immunology clarified the fact that inactivated vaccine conferred only a limited partial immunity. This in turn stimulated a four-fold development in prophylactic immunology: the use of adjuvants to increase the efficacy of inactivated vaccines, concerted efforts to develop live vaccines, in vitro synthesis of antigens, and the administration of immuno-stimulators.

The number of adjuvants was augmented by inorganic γ and highly dispersed δ aluminium oxide by Drescher (1958) and by Grafe (1962), as well as by organic substances, as shown in Table 5.19. Non-metabolizable, particulate inorganic adjuvants can form intramuscular depots and act as foreign bodies. The justified assumption that these substances might be potential carcinogens motivated scientists to look into the possibility of using instead soluble substances with an assessable mode of action and simultaneous immune-stimulating activity; these might be added to an antigen, or they could carry the antigen to a desired site. Both of these functions would be filled by natural or synthetic liposomes, which could incorporate or attach to the antigens, and protect them from rapid metabolic degradation during transport to the immuno-competent organ.

Table 5.19. Organic substances as adjuvants or carriers in virus vaccine production

Year	Authors	Virus	Substance
1975	Mooney et al.	Sindbis	Lipos. from sheep erythrocytes
1975	Almeida et al.	Influenza	Lecithin-dicetylphosphate lipo.
1977	Webster et al.	Influenza sub.	Muramyl dipeptide adjuvant (MDP)[c]
1978	Morein et al.	Semliki forest	Vesicles of egg lecithin
1980	Sanchez et al.	HBsAG	Synthetic multilamellar liposome
1981	Lawman et al.	HSV-1	Small unilamellar synthetic lipo.
1981	Naylor et al.	HSV-1 sub.	Antigen + lipid A incorporated into synthetic liposomes
1982	Bittle et al.	FMDV[a]	keyhole limpet haemocyanin coupled
1982	North et al.	EBV gp 340	incorporated into synthetic lipo.
1985	M.Epstein et al.	EBV gp 340[b]	incorporated into synthetic lipo.

[a] Chemically synthesized peptide.
[b] Vaccine against EBV induced lymphomas in cottontop tamarins.
[c] N-acetyl-muramyl-L-alanyl-D-isoglutamine, a water-soluble component of the mycobacterial cell wall.
lipo., Liposomes; sub., Subunit

Neither pox nor rabies vaccine was to serve as model for live vaccines and the yellow fever vaccine was chosen. In the pox vaccine model, the pathogen of a mild disease, *Vaccinia virus* was used to confer immunity against a related but severe disease, namely variola or alastrim. The rabies model involved the use of an attenuated virus which was very gradually inactivated. The yellow fever model using attenuated live viruses, however, could guarantee long-term humoral and cell-mediated immunity. Efforts to produce a live, oral attenuated polio vaccine ran parallel to the development of inactivated polio vaccine in the USA as of 1950. In the main, this was the work of Sabin, Melnick, Cox and Koprowski. The work on attenuated measles vaccine began after Enders and his coworkers Peebles and Dorothy Katz had isolated the Edmonston-strain in February of 1954. In discussions on these vaccines, the desirability of long-term IgG antibodies and cell-mediated immunity was emphasized, which seemed inducable only by replicating viruses.

All possible in vivo and in vitro techniques for virus multiplication were suitable to obtain attenuation and were actually used.

In 1959, however, it was apparent that candidate viruses for a live polio vaccine differed considerably in their neurovirulence as assessed after intracerebral and intraspinal inoculation of monkeys. As a result, provisional criteria for the choice of a virus strain were set up:

- minimal neurovirulence tested in monkeys- absence of viremia in man
- maximal genetic stability
- free of any harmful effect when administered to susceptible populations of approx. 100000.

In order to control genetic stability, especially with regard to neurovirulence, the use of markers was recommended for safety tests. The usefulness of the first, the "rct marker", which had been characterized by A. and Marguerite Lwoff (1959), was demonstrated by Melnick and Matilda Benyesh-Melnick in 1960. This marker indicated to what extent the replication of the attenuated strain differed from the virulent strain at elevated temperatures, and how these correlated to neurovirulence. In the following years at least five additional in vitro and in vivo markers proved useful for developing live vaccines against poliomyelitis, measles, mumps and rubella. In time, numerous additional markers were added for other viral vaccines.

The measures which were taken to ensure utmost safety, efficacy and tolerability during production and administration served to diminish the vaccination risk but gave rise to criticism of both vaccine types. The use of inactivated vaccine raised questions of efficacy, allergy and adventitious agents. Live, attenuated strains, on the other hand, aroused questions of genetic stability, the persistence of live virus in immuno-deficient children leading to disease long after vaccination, as well as mutagenic and teratogenic effects of attenuated viruses that might be transmitted to pregnant women. These critical points led scientists later to seek other solutions for the production of viral vaccines, when they could approach the problem with far better knowledge in virology and immunology as well as with more advanced laboratory techniques. The optimal goal was an attenuated intra-nasally administered life vaccine obtained from a replicable virus with all its immunogenetic potential; the development of inactivated complete vaccines, i.e. containing complete viruses, or of "splitted vaccines" was considered an unsatisfactory alternative. As examples of the new developments which ensued, the vaccines for measles, influenza and hepatitis as well as synthetic vaccines are described.

Vaccines Produced with Viral Material

Measles. The live vaccines available in 1960 and those which were developed later brought about side effects which varied not only from country to country but also from trial to trial. Attempts were made to alleviate these unwanted effects by pre-vaccination with inactivated vaccine or by use of a passive vaccination. The efficacy of the inactivated vaccine proved unsatisfactory and side effects were also observed. The reservations about and the contra-indications of live vaccine could not be dispelled meant that new possibilities should be found. These seemed to lie in splitting the virus into its infective and its immunogenic components.

Splitting the virus under high temperature with ether and Tween 80 was accomplished by Norrby in 1962. As a result, he was able to develop the Tween-ether split vaccine with Gard. Waterson and colleagues (1962) began to develop a subunit vaccine consisting of surface and core antigens. Earliest indications of possible untoward side effects began to appear in publications in 1965, observed in children vaccinated with three doses of inactivated vaccine in field trials. In a measles epidemic 2 years later, cases of atypical measles were reported. Thereafter, serious local reactions were observed in children who had received live measles virus vaccine, although they had been immunized with inactivated vaccine several years previously. One of the hypotheses submitted to interpret these reactions was induced, delayed dermal hypersensitivity. Another, based on experimental results, was a striking similarity between SSPE viral and wild-type measles proteins. Although nothing was conclusive, the killed measles vaccines disappeared from the market.

Influenza. Inactivated influenza split and subunit vaccines were developed, as had been done for measles. Subunit vaccines contained the surface components hemagglutinin and neuraminidase, which had been isolated from the type A and B strains most relevant in the epidemiology of influenza, as recommended by WHO. Davenport and colleagues (1960) showed that highly purified HA stimulated the production of HI antibodies in animals; Lange (1966) demonstrated the antibody-inducing capacity of HA subunits. Thus, HA and neuraminidase subunits were assumed to be the most important immunogenic parts of the virion. Though the side effects of these vaccines administered subcutaneously or intramuscularly were diminished because they did not contain lipids, a fundamental enhancement of vaccine efficacy was not achieved.

Hepatitis B. Since *Hepatitis B virus* could not be propagated in suitable in vitro systems, Krugman and coworkers (1971) initiated trials by immunizing children with heated serum derived from *HBV* infected patients. Subsequent investigations showed that it was possible to divide the complete *HBV* antigen (HB Ag) into the surface antigen (HB_sAg) and the core antigen (HB_cAg). Experience with other strains prompted the assumption that HBsAG, which is synthesized in large quantities in hepatocyte cytoplasma and then released into the serum, was the more important immunogenic factor in the virus. Thus, Hilleman and colleagues undertook development of a highly purified, inactivated vaccine which had been prepared from antigenically-potent HB_sAg; they received a license for it in 1981. The inactivating and purifying process guaranteed that the vaccine was free of the known types of active adventitious agents. It was recommended for infants, children, and adults who were considered to have an increased risk of *HBV* infection.

Within the three decades following the licensing of the first oral, live polio vaccine, numerous inactivated and live vaccines for human and non-human use were also licensed. The model of choice for these vaccines was the oral, live polio vaccine. It met all the requirements set up during years of investigation and fulfilled virtually all immunological expectations: it activated the common mucosal immune system and provided the organism with long-term local, humoral and cell-mediated immunity. Interferon as well as secretory IgA induction secured primary defence mechanisms. This was reliable not only for poliomyelitis but also for all viral infections involving the mucosal surfaces.

As a result of the availability of so many vaccines, a high degree of control of serious viral diseases of humans was achieved: e.g. of poliomyelitis, yellow fever, mumps, rubella. Great success was reflected in the message wired around the world by WHO on October 26, 1979, "smallpox has now been eradicated throughout world!" Perhaps a better choice of verb would have been "brought under control". Reservoir hosts are known for only a few viruses; but we do know, however, that viruses outlive in water, plants, animals and humans. Thus, it would be unduly presumptious to assume a

Table 5.20. Viral mucosal vaccines for non-human use

Animal	Vaccine
Birds	Newcastle disease Infectious bronchitis Fowl pox Infectious laryngotracheitis Encephalomyelitis Marek's disease Infectious bursal disease
Catlle	Foot-and-mouth disease Parainfluenza Infectious rhinotracheitis Adeno Cattle plague Rota Papillomatosis Infectious vulvovaginitis
Mink	Distemper
Swine	Aujeszky disease Transmissible gastroenteritis
Duck	Hepatitis
Fox, dog, and bat	Rabies
Horse	Rhinopneumonitis African horse sickness Infectious arteritis
Fish	Haemorrhagic septicemia Infectious pancreas necrosis
Sheep	Blue tongue

successful vaccination programme could achieve complete eradication of an infective virus.

Veterinary science has been successful in administering attenuated live vaccines to protect single animals as well as populations in stalls or outdoors, in water and while being transported. Besides injection, and oral or nasal administration, these vaccines are spread via drinking water and feed, by inhalation, instilling in the eyes, on the anus, by scarification and application on the mucous membranes of chick cloaca. It was conceded that mucous membranes, including the highly active intestinal tract, were important organs in immunity. As can be seen in Table 5.20, which contains a few examples, many veterinary diseases have been brought under control.

Although licensed vaccines admit to having low but calculable risks, and are without serious side effects when used as recommended, their efficiencies differ markedly and are not always satisfactory. This pertains to all the types of vaccines described above as well as to the administration of a single or multiple-component or polyvalent vaccine, given alone or in combination with other bacterial or viral inactivated or live antigens or toxoids. For this reason, scientists re-examined the improvement of immunization methods, so that the efficacy, safety and tolerability of the vaccines might more closely approach the optimum. This should preferably be without "crutches" like pre-vaccinating with homologous hyper-immune serum or preceding IF induction or administration. To achieve this goal, chemically defined and synthetic immunogens were produced for use in antiviral vaccines. As a second measure, cloning and sequencing of genes coding for the proteins responsible for immunogenic potency were undertaken. These clones were used, in turn, for the biosynthesis of antigenic viral proteins.

Chemical and Molecular Genetically Defined Synthesis of Antigens: The fundamental chemical approach was published by Goebel (1938–1939) and Anderer in 1963. After familiarizing himself with the factors which govern the immunological specifity of carbohydrates and artificial antigens, Goebel initiated experiments in this field. He combined the diazonium derivative of the p-amino benzyl glycoside of cellobiuronic acid, the aldobionic acid from which type III pneumococcus polysaccharide is constituted, with horse serum globulin. The animals receiving this artificial antigen acquired active immunity against type III virulent pneumococci and in turn conferred passive immunity to mice against virulent types I, II, + III. Comparing his results with control groups, he proved that antibacterial immunity could be evoked by an artificial antigen which contained the pattern unit of the immune-specific polysaccharide rather than the intact carbohydrate itself.

Anderer proved that only 15–20% of the amino acid sequence of *TMV* possessed serological determinant activity and, furthermore, that *TMV*-RNA was not involved in immunity. Following Goebel's conjugate model, he used well-defined short hexapeptides from the C-terminus of the *TMV* protein. After N-acylating, he converted this to the p-aminobenzoyl derivative, and finally combined it with bovine serum albumin. Rabbits immunized with this synthetic conjugate induced antibodies which not only precipitated intact virus but neutralized its infectivity.

Since it is possible to identify antigenic determinants of proteins by using monoclonal antibodies, the peptide fragments can be synthesized and used to prepare immunogenic conjugates. Meanwhile, successful experiments with artificial conjugates have been carried out by Langbeheim and colleagues using *Coliphage MS-2n* in 1976. In order to conjugate an *FMDV* vaccine, Strohmaier and coworkers (1982) isolated peptides corresponding to the antigenic sites. Bittle and colleagues (1982) synthesized a *FMDV* peptide predicted by the viral nucleotide sequence, and this was later accomplished by Emini and coworkers (1983) for poliovirus. Because most synthetic vaccines are water soluble and tend to be less immunogenic than particulate antigens, adjuvants or immuno-potentiators had to be used to enhance the immunogenicity.

Freund's complete adjuvant (FCA) which had been formerly used, was found to induce undesirable local reactions, and a replacement was sought. The methods investigated since 1958 include the use of cell extracts and of natural or synthetic liposomes as adjuvants and/or carriers. In 1974, Anne Haywood fused *Sendai virus* with various liposomal types and found the viral glycoprotein spikes in patches at the surface of the liposomes (Table 5.19).

There are two distinct ways of synthesizing anti-viral vaccines when using the molecular-genetic approach. First, a protein is expressed in in vitro cultures of prokaryotic or eukaryotic cells, and second, the DNA sequence coding the protein is inserted into a virus which in turn replicates in vitro or in vivo. Applying the first method, Valenzuela and colleagues, and Burrell and colleagues reported the HB_sAG expression in E.coli in 1979. One year later, in vitro expression experiments were carried out by Entage and coworkers using the RNA segment of influenzavirus coding for HA. The sequence of *HBsAG* was reported by Patricia MacKay. During the following 3 years and using the same method, Küpper and colleagues (1982) synthesized immunogenic proteins of *FMDV*, R.Watson and colleagues (1982) those of *Vesicular stomatitis virus* and *HSV*, and Elizabeth Yelverton and colleagues (1983) those of *Rabies virus*. Since 1980, eukaryotic cells like yeast cells, T cells, monkey cells, cells of a human hepatoma cell line, and cultured mouse L cells have been used. HB_sAG was expressed in yeast cells by Valenzuela in 1982. The immunogenic efficacy of this antigen produced in S.cerevisiae was tested by McAleer and coworkers (1984) in various animals, and in a protection test in chimpanzees which had been challenged with the identical virus and a heterologous subtype. This was the first vaccine to be produced employing recombinant DNA technology which proved effective against a human virus infection.

In order for a protein, the gene for which had been inserted into a viral genome, to be successfully expressed, various-sized genomes had to be used. Experiments with plant viruses utilizing this technology were carried out with the *Cauliflower mosaic virus* genome by Howell and coworkers (1981). In the same year, *SV40* was used as a vector for influenzavirus glycoproteins by Sveda and C.Lai, for influenzavirus HA by Mary-Jane Gething and Sambrook, for a *Harvey murine sarcoma virus* protein by Gruss and colleagues, and for HBsAg by Ann Moriarty and colleagues. In 1981 the latter used *SV40*-infected L cells for the expression of HB_sAG. Presently, the most commonly used vector for the production of antigens to be utilized as vaccines is *Vaccinia virus*. Recombination of different strains of *Vaccinia virus* was demonstrated by Fenner (1959) and with orthopoxviruses by Fenner and Woodroofe (1960). Sam and Dumbell (1981) proved that the activity of intact orthopoxvirus DNA could be maintained, if transfected cells were co-infected with active orthopoxvirus. This was to serve as a model for Mackett and colleagues and Panicalli and Paoletti (1982), when they inserted the foreign herpesvirus thymidine kinase gene into a recombinant *Vaccinia virus* gene to get expression.

Meanwhile, the list has lengthened to more than 20 different viral nucleic acid segments inserted and expressed in recombinant *Vaccinia virus* to synthesize viral proteins for immunization with single or multiple antigens. In most of these recombinant vaccines, the foreign genetic material is inserted into the thymidine kinase (Tk) gene of *Vaccinia virus* and it should be mentioned that the DNA sequence of the Tk gene of the closely related *Monkeypox virus* shares about 98% homology with the *Vaccinia virus* Tk gene, as was demonstrated by Jerry Weir and Moss (1983) and by Esposito and Janice Knight (1984). This high level of homology makes possible the recombination of these two poxvirus types, if recombinant *Vaccinia virus* vaccines should be applied in regions with endemic *Monkeypox viruses*, e.g. in Western and Central Africa.

5.5 Viral Pathogenicity

All steps in viral replication represent more or less serious consequences for the host cell, the infected tissue and the whole organism. Viral pathogenicity entails multiple factors. Virus virulence-dependent events may range from the death of the cell, over disturbed control of cell growth and disturbance of biochemical and genetical functions, to various diseases, including also heritable forms. Recognizable, virus-induced, untoward effects in the host fall within the definition of pathogenicity. The list of organ-related virus diseases drawn up in 1957, would presently have to be augmented to include diseases of the liver, kidney, the endocrine and lymphatic system, and germ cells.

As more highly sophisticated laboratory technology was developed, specialized aspects of virulence became more conspicuous and assessable both qualitatively and quantitatively. Qualitatively recognizable aspects would include, for example, organ and tissue tropism, whilst quantitative recognition covered rate and speed of reduplication, adaptability to altered multiplication conditions judged by the mutation rate, induction of genome damage in the host organism, and resistance to physical-chemical factors, etc. Refined molecular-genetic methods have permitted scientists to investigate if and to what extent a viral genome determined the pathogenic characteristics of a virus. Problems receiving priority were cell and tissue tropism, virulence and its attenuation, and latency.

5.5.1 Histological and Anatomical-Pathological Effects

From the very beginning, the emphasis placed on pathological effects in tissues and organs meant that virus classification would be medically oriented. This situation also pertained long after the 5th International Congress for Microbiology in Rio de Janeiro in 1950, where the criteria for virus classification had been laid down. Even today, this medically oriented thinking serves as the basis for many clinical virology textbooks. The consideration of this subject that follows has been limited to results on new recognized pathological effects of viruses.

5.5.1.1 Teratogenic Virus Effects

Attention has been paid to virus-induced congenital malformations since 1934, but interest became firmly established in 1941 when Gregg described chronic infections and teratogenic effects caused by *Rubella virus*. In 1957, 10 viruses were assumed or shown to have teratogenic properties, and the list has since lengthened.

Cytomegalo virus was recognized as the cause of birth defects by Weller and colleagues (1956), and Nelson and G.Wyatt (1959) reported that such congenital infections accounted for 1–2% of all infant deaths. Hanshaw in 1966 described the symptoms of congenital and acquired *Cytomegalo virus* infections, noting that a high percentage of infants who survived often exhibited neurological sequelae including blindness and mental retardation. Even now, *Cytomegalo virus* together with *Rubella virus* are the subject of thorough investigations related to their ability to induce congenital malformations and severe symptoms following perinatally and early postnatally acquired infections.

HSV 2 responsible for genital infections, was shown to be the cause of congenital malformations of the CNS by Mary Ann South and colleagues (1969), after Sieber and coworkers (1966) had described fetal infection by the same *HSV* type. That *Smallpox virus* infection and vaccination may lead to fetal infections, still-birth, abortion, premature birth, or death was reported by many groups, including G.Brown, Eichenwald, Monif and Dorothy Horstman and their coleagues between 1966 and 1969. In 1974, R.Ackermann and coworkers confirmed the 1955 report of Komrower on the teratogenic effects of *LCM*. These malformations in the newborn resulted from an infection acquired during the late stages of intra-uterine life.

Teratogenic viruses in animals were also reported: Griner and colleagues (1964) confirmed the 1955 results of Shultz and Delay for *Blue togue virus* in sheep. Young and coworkers (1955) observed fetal abnormalities after *Hog cholera virus* infection, and Hashimi and colleagues (1966) disclosed that in congenital and neonatal infected mice *Reo virus 2* caused interstitial pneumonia and subcortical renal tubular necrosis, as well as retarded growth and rough fur. Experimental teratogenicity in general has made us more aware of the importance of choosing the right animals for viral investigation. Thus, only *Rubella virus*, *HSV* and *Coxsackie virus B* were able to pass the placenta barrier in a small number of the animal species studied. In summary, viruses can be teratogens, although this may not always be simple to prove, and animal experiments may be helpful but such results cannot necessarily be extrapolated to man.

5.5.1.2 Mutagenic Virus Effects

A virus may be considered to be a genetic macromolecule, and we can also regard it as a chromosome. Just like a chromosome, a virus can interact spontaneously with the chromosomes of other types of viruses or, either spontaneously or induced, with the replicating chromosomes of somatic or germ cells of its host. The consequences of such interactions with host cell genomes include all types of gene and chromosome mutations and a wide range of chromosomal alterations like "sister chromatid exchange" (SCE), breaks, deletions, pulverizations, formation of micronuclei, endoreduplication or the formation of syncytia or polykaryonts by effects on the spindle.

The development of several suitable cytogentic in vitro and in vivo methods has increased the possibilities of investigating the mutagenicity of viruses, i.e. the influence of the viral genome on that of the host. Table 5.21 should give an impression of how rapidly our knowledge of viral mutagenicity has expanded. It is noteworthy that not only active viruses but also inactivated and attenuated virus particles can be mutagenic. Meanwhile, it is no longer a matter of debate that carcinogens are also mutagens. This means that for every confirmed oncogenic virus, there is a need for concrete evidence of its mutagenicity. The discovery of v- and c-oncogenes led to conjecture whether they might be breaking points for spontaneous and induced changes in a chromosome, and thus might serve as the basis for various forms of recombination.

In their transfection experiments, Shih and colleagues and Cooper and colleagues (1979, 1980) showed that transforming properties can be transferred from tumour cell DNA to non-transformed cells, and they disclosed the presence of oncogenic factors in tumour cell DNA and its absence in normal cell DNA. Hayward and coworkers (1981) reported that retroviruses lacking oncogenic sequences can integrate in the vicinity of proto-oncogenes and thereby disrupt their regulating elements. In 1982, proto-oncogenic related chromosomal translocations were observed in chronic myelocytic leukemia by Annelies de Klein, in human Burkitt lymphoma and murine plasmocytoma cells by Taub and colleagues and J.Erikson and coworkers. In the same year, work on the activation of proto-oncogenes to transforming oncogenes by point mutations was undertaken by the groups of

Table 5.21. Viruses known to possess mutagenic properties

Virus	Year	Test system[a]				Authors
		soma	germ	vitro	vivo	
HSV	1961	+		+		Hampar, Ellison
Measles activ, inactivated	1962	+			+	Fjelde et al.; Holterman W.Nichols et al.
SV40	1962	+		+		Yerganian et al.
Varicella	1963	+		+	+	Pertti Aula
RSV	1963	+			+	W.Nichols et al.
HSV attenuated	1964	+		+		Stich et al.
Herpes zoster	1964	+		+		Matilda Benesh-Melnick et al.
Adeno 12	1964	+		+		Stich et a.
Sendai	1965	+		+		Saksela et al.
Rubella	1966	+		+		Chang et al.
Vaccinia	1966	+		+		Zur Hausen, Lanz
HSV 2	1969	+		+		O'Neill, Miles
HBV	1978	+		+		Kurvink et al.
Influenza A,B	1978	+	+	+		Thadani
Infl.A, inactivated	1979		+		+	Thadani, Polasa
Polyoma	1982	+		+		E.Brown, Basilico

a soma, somatic cells; germ, germ cell; vitro, in vitro tested; vivo, in vivo induced and in vitro tested.

E.Reddi, Elizabeth Taparowsky and Tabin. Meanwhile, numerous publications have dealt with the role of proto-oncogenes and oncogenes and their products in malignancy, but the basic mechanisms involved in this phenomenon are essentially just as obscure as they were prior to detection of these genes. We also do not understand the part they play in mutagenicity and especially in the interaction of the virus with chromosomes.

5.5.1.3 Virus Diseases Dependent on Virus-Induced Defects of the Immune System

Disturbances in the host organism can include virus-related derangement of the immune system involving the cells and organs responsible for genetically controlled humoral and cell-mediated immunity. This type of derangement does not include failure of the immune response which may result in insufficient elimination of viral antigens and thereby cause persistent viral infection, but cover both immune complex disease and immunodeficiency disease.

An immune complex is an aggregate of a specific antigen and its corresponding antibody, which may form locally in tissues or in blood vessels. The complexes may circulate in a harmless manner, but they may also be deposited in precipitated form in target organs, where they can activate complement as well as the blood clotting system and thus bring about pathogenic effects. The following viruses are known to cause immune complex derangements in humans:

- Adeno virus
- Cytomegalo virus
- Dengue virus
- EBV
- Echo virus
- Hepatitis B virus
- LCM virus
- Measles virus in SSPE
- Mumps virus
- Varicella virus
- Variola virus

Detection of the acquired immune deficiency syndrome (AIDS) in 1981 brought to light a new type of viral pathogenicity in humans. Viruses responsible for this disease have meanwhile been isolated from different species and called *Human immunodeficiency virus* (HIV) or *Simian immunodeficiency virus* (SIV). Previously, the term immunological deficiency was used in virology to describe certain virus-related, transient, secondary immunosuppressive states which were recognized as defects in cell-mediated immunity rather than in IG production. In 1981, immune deficiency caused by *HIV* was understood not as a transient but as a permanent suppression of the immune system caused by an attack on the T4-cells (helper lymphocytes) which normally help to produce antibodies and support the NK cells, monocytes and macrophages. In HIV disease, transformed and infected T4-cells stimulate the natural killer cell activity and thereby cytolyse infected helper cells, all of which impairs the defence mechanism in the diseased organism, and lowers resistance to subsequent infections. Opportunistic infections, which usually run a harmless course in an intact, functioning immune system, e.g.

pneumocystis carinii pneumonia, fungal infections with Candida, Aspergillus or Cryptococcus, *Cytomegalo virus*, *HSV* or *Herpes zoster virus* and Kaposi's sarcoma, present a severe threat to AIDS patients and are the main cause of death. During *HIV* infection, lymphocytes can invade the CNS, infect macrophages and cause serious neurological disfunctions; this is now observed in approximately 60% of patients. The most common neurological problems in AIDS, however, are subacute encephalitis with vacuolar myelopathy, "aseptic" meningitis and peripheral neuropathy.

5.5.2 Molecular Genetic Assessment of Virus Pathogenicity

Each stage of the virus-host interaction has a characteristic immunological response to the pathological consequences of virus multiplication. Because the entire multiplication cycle is genetically controlled, the question has arisen as to whether relevant virus properties can be altered or influenced. This is of interest not only for fundamental biological reasons, but also for practical purposes, i.e. the development of antiviral chemotherapy, or to produce live attenuated or gene technological virus vaccines without any untoward side-effects.

5.5.2.1 Cell and Tissue Tropism

A prerequisite for virus multiplication is the adherence of the virus to the surface of the host cell. In this initial step, virus polypeptides are bound to biochemical components (receptors) of the host-cell plasma membrane. Only a few receptors are known and have been analyzed with respect to specific carbohydrate moieties. The viral attachment sites for these receptors are located on the virus surface. Orthomyxovirus receptors were determined to be at the top of the HA-trimer globules, which also contain the antibody binding sites involved in evoking antigenic variation, by Wiley and colleagues (1981). The ability of a virus to adsorb to distinct molecules on the host cell surface, or to specific sites on cells in a particular tissue or organ, is termed tropism. Thus, the ease or otherwise with which a virus finds a suitable cell in vitro or in an animal experiment can be explained in terms of tropism.

The receptors or receptor components known for some viruses are listed below:

- Sialic acid-containing glycoproteins or glycolipids in epithelial cells (Influenza A)
- Acetylcholine in peripheral nerves (Rabies)
- C_3d[1] receptors on B cells (*EBV*)
- CD_4 protein with immunoglobulin-like domains on helper lymphocytes (*HIV*)
- Ia antigen of the major histocompatibility complex (MHC) on macrophages (*Lactic dehydrogenase virus*)

Using a chromosome-banding method, Dorothy Miller and colleagues (1974) discovered that human chromosome 19 carries the poliovirus receptor gene. Cathy Mendelsohn and coworkers (1986) showed that HeLa cells contained about 3000 receptors as integrated membrane proteins and that, furthermore, this receptor could be transferred via DNA from polio-susceptible HeLa cells to polio-resistant mouse L cells. In recombination experiments, Ysaline Centifanto-Fitzgerald and colleagues (1982) detected that an alteration in the *HSV* sequence region between 0.7–0.83 *HSV* was linked to a change in herpes cornea infectivity for rabbits. This might be attributed to an altered tropism.

5.5.2.2 Virulence and Attenuation

The degree of virulence is indicated by the extent of virus-induced cell, tissue and organ damage. In addition to various viral properties, virulence is dependent on the physiologically and genetically controlled immunological state of the host. One aspect of the virulence phenomenon is the cessation of host-cell protein synthesis by the inactivation of a cellular cap-binding protein responsible for host-cell RNA translation. This was reported for influenzavirus by Plotch and colleagues (1979), and for poliovirus by Trachsel and colleagues (1980). The same mechanism was assumed for reovirus by Sharpe and coworkers (1982). The results of their genetic studies in 1981 permitted Sharpe and Fields to conclude that the S1 double-stranded RNA segment triggers this mechanism. Another virulence gene for *Reo virus 1* was recognized by D.Rubin and Fields (1980) in their studies with mice. They determined the ability of the M2 double-stranded RNA genome segment to initiate local and systemic infections in newborn mice after oral administration of the virus.

---

[1] a binding protein of the human C3 molecule, part of the complement system

Significant factors in influenzavirus virulence are the envelope glycoproteins hemagglutinin and neuraminidase. These regulate the infectivity of progeny virions by altering virus immunogenicity; this is called antigen shift as a major change or antigen drift as a minor one. Shifts as well as drifts can be induced by different numbers of point mutations. A key to understanding influenzavirus pathogenicity is HA. Khatchikian and colleagues (1989) demonstrated increased pathogenicity of *Influenza virus A* by inserting a 28 S rRNA sequence into the HA gene.

The development of a live vaccine against poliomyelitis demanded diminishing the characteristic neurovirulence of the poliovirus. The first attenuation was undertaken by M.Theiler (1941); after passaging poliovirus through mouse brain tissue, he obtained a variant which no longer caused paralysis in the rhesus monkey after intracerebral inoculation. In a biological neurovirulence safety test with live vaccine strains, cynomolgus monkeys were given an intraspinal, fixed inoculum injection. Sections of the spinal cord and the brain were then scored and the results compared with an accepted reference strain. Molecular-genetic tests allowed the definition of the molecular basis for neurovirulence as part of the pathogenicity of a virus. Nomoto and colleagues (1981) investigated the attenuated polio Sabin type 1 strain using oligonucleotide fingerprinting and found about 35 mutations. In the complete nucleotide sequence of 7441 bases, 57 differed by point mutations, 21 of which resulted in a change of amino acid. Tests undertaken by Stanway and colleagues (1983–1984) with Sabin's type 3 strain resulted in the detection of 10 point mutations within 7432 nucleotides.

D.Evans and coworkers (1985) published results indicating only a single nucleotide change in the Sabin type 3 poliovirus genome increased its neurovirulence. As the tested virus was isolated from a case of vaccine-associated poliomyelitis, the change in neurovirulence was explained by selection in the human gastro-intestinal tract. Although results of molecular-genetic neurovirulence tests cannot completely explain neurovirulent effects in a host, they do demonstrate that this virus property is fixed on the genome and that it can be reduced significantly by directed point mutation. Amanda Fisher and colleagues (1986) attenuated *HTLV-III* by excising a 200-base-pair segment in the 3'region of the virus, and Taha and coworkers (1989) located the genetic information for the neurovirulence of *HSV 2* within a distinct repetitive DNA sequence in the mouse.

5.5.2.3 Latency

Several research groups considered *HSV* a particularly suitable virus for testing virus latency. Persistent and latent herpes infections have been familar since 1923, and through the efforts of Plummer and colleagues they have been confirmed and reproduced since 1970. *HSV* in vivo remains latent predominantly in the genome of sensory ganglia. During latency, a virus does not undergo synthesis and only few genes are transcribed.

Studies of Marcialies and colleagues (1975) indicated that the thymidine kinase (TK) gene is involved in the infection process of nerve tissue. In 1978, H.Field and Wildy concluded that mutation or deletion of the thymidine kinase gene decreased virulence in their test unit mice. Ysaline Centifanto-Fitzgerald and colleague (1982) located the pathogenic factor of *HSV 1* between 0.7 and 0.8 map units when applied to the rabbit cornea. In the same year, Y.Becker and colleagues showed the dependence on TK of the virus capacity to invade various types of tissue, thereby establishing latency in mice. However, Coen and coworkers (1989) showed that TK-negative *HSV* could establish latency but could not be reactivated in co-cultivation assays.

R.Thompson and coworkers (1983) determined the region between 0.7 and 0.8 to be the site of *HSV 1* neurovirulence in mice, and Angela Rösen and coworkers (1985) confirmed these results for the *HSV 1* virulence gene after intraperitoneal administration in tree shrews, an animal introduced into *HSV* latency tests by the group of Darai in 1980. In 1989, Angela Rösen-Wolff et al. were able to show that organotropism for latent *HSV 1* correlated with the presence of a 1.5 kb RNA mapped between the genome coordinates 0.760 to 0.762. One year later they correlated this 1.5 kb RNA to a single specific *HSV 1* gene. This gene was altered in all *HSV 1* strains which had lost the ability to persist as latent viruses in the ganglia of latently infected animals.

5.6 Virus Chemotherapy

Although successful results in experimental anti-viral chemotherapeutic research coincided with the beginning of this Chapter, namely in the 1950's, the necessary groundwork in methods and test substances had already been laid two decades earlier. Chick embryos which provided various organ cultures, tissue and cell cultures which offered definable test conditions replaced or complemented animal experiments. Domagk (1935) ushered in a trend away from disinfectants in anti-bacterial chemotherapy by discovering the sulfonamides. Furthermore, methods developed during World War II led to mass production of the first antibiotic, penicillin, discovered by Fleming in 1928, which also influenced and broadened the testing of anti-viral substances. The other important link in the chain was the transfiguration of the "crystal" virus via a biochemical nucleoproteid to a multiplying, genetically active DNA or RNA macromolecule.

Anti-viral chemotherapeutic research based on the above experience did not lead to an effective anti-viral compound until 1957. It did, nevertheless, produce substances which enabled scientists to influence and control virus multiplication in laboratory experiments. Moreover, experimental work was no longer limited to animal viruses; economic reasons had prompted the testing of chemical substances and protein extracts against *TMV*. The reduced virus content which was attained through such efforts was solely the result of physical reactions including aggregation and precipitation. Glancing back over the past 50 years, experimental anti-viral chemotherapy can be seen historically to have developed from two activities: First, in relation to test substances, the disinfectants were followed by antibiotics, anti-metabolites, and a variety of synthetics and substances to manipulate virus multiplication. Second, test methods progressed from serial tests with animals and plants, then tissue and cell cultures, to molecular-genetic experiments.

5.6.1 The Antibiotic Phase

This phase was initiated two years after the detection of the sulfonamides. Kelson began to test them in monkeys infected with poliovirus, whilst the Rosenthal group experimented with influenza-viruses and *LMC virus*. Up to the beginning of WorldWar II, about 1300 sulfonamides had been synthesized, and in the following 2 years many experiments in animals with *FMD*, encephalitis, yellow fever, canine distemper, rabies, herpes, smallpox as well as rabbit fibroma and myxoma were done and failed. Negative results were also reported in tests with penicillin, streptomycin, gliotoxin, aureomycin, tramycin and chloramphenicol. These had been initiated by Heilmann and

Herrell (1944) in penicillin trials for ornithosis and psittacosis, which belonged to the then group of large viruses. These causative agents were also used for testing other substances. Early unsuccessful results with penicillin, streptomycin and aureomycin with the plant pathogens *TMV* and *Potato yellow dwarf virus* were reported by Helen Beale and C.Jones (1951).

R.Thompson in 1947 published results on metabolites, metabolite-antagonists and enzyme-inhibitors on the multiplication of *Vaccinia virus* in a Maitland culture using chick embryo tissue. Reduced or inhibited viral multiplication was reported with dinitrophenol, azide, cyanide, Atabrine, proflavine, and iodoacetic acid. The inhibition of multiplication resulted from substituted amino acids, e.g. amino methane sulfonic acid, methoxinine, α-amino-isobutane sulfonic acid, and α-amino-phenylmethane sulfonic acid. Using the amino-acid analog 5-methyltryptophan in their virus-host interaction experiments, S.Cohen and Catherine Fowler showed that *Coliphage T2* required protein synthesis for growth and that substrate-dependent inhibition could be specifically reversed by tryptophan.

5.6.2 The Anti-metabolite Phase

A number of interesting anti-metabolite studies on plant and animal viruses began to appear in the early 1950's. Nickels and colleagues investigating the effects of various nucleic acid components and related compounds on viral tumours, reported that the inhibiting action of guanozolo, a pyrimidine derivate, could be reversed by guanine. Between 1951 and 1953, Matthews published the results of his earlier investigations. He had tested purine analogues and related compounds, as well as analogues of imidazole, against *Lucerne mosaic virus* and *TMV*. Water containing the substances was sprinkled onto the soil or sprayed onto the leaves of plants grown in pots in the greenhouse. Results, though interesting, did not suggest that virus-infected plants could be treated systemically with any measurable success. Commoner and Mercer (1951) demonstrated inhibition of *TMV* biosynthesis in isolated leaf tissue using the pyrimidine analogue 2-thiouracil. The inhibitory action of this only effective substance in the series was partially reversed by uracil and by pyrimidines isolated from *TMV* nucleic acid.

Synthetic regulators of plant growth, e.g. naphthalene-acetic acid (Kutsky and Rawlins,1950), indolebutyric acid (Kutsky,1952), and gibberellic acid (Raychandhuri and Mishra, 1964) were tested without any therapeutic effect. Other investigations of a similar nature were reported with little or no positive results, for example Pine (1967) injected peach trees with dimethyl sulfoxide to suppress *Peach mosaic* and *Necrotic ringspot viruses*. He was successful in suppressing the symptoms but, alas, for only one season.

As for the field of animal virology, the first hopeful results were reported with thiosemicarbazones – substances which had been used earlier with interesting results only for tuberculosis – and benzimidazoles. Hamre and colleagues (1950) reported on the activity of p-amino-benzaldehyde-3-thiosemicarbazone on *Vaccinia virus* in chick embryos and mice. G.Brown and W.Ackermann studied the influence of certain analogues of amino acids, purines or pyrimidines, and other metabolites on the propagation of poliovirus cultured in human embryo tissue in roller tube. They found the activity of some analogues of amino acids, and also noted inhibition by benzimidazoles, 2,6-diaminopurine, proflavine and, in 1952, ethionine. They were, furthermore, able to demonstrate that the inhibitory activity was reversible by appropriate metabolites, for example, of ethionine by its analogue methionine.

A.Rassmussen and colleagues (1951) used 6-methyltryptophan in experiments with *Poliomyelitis virus* strain Lansing in mice, and Mercer and coworkers (1953) inhibited *TMV* biosynthesis using 2-thiopyrimidines. Heterocyclic and other thiosemicarbazones had been synthesized and tested in *Vaccinia virus* infected mice by R.Thompson and coworkers in 1953. They recognized that thiosemicarbazones contained benzene, thiophene, pyridine, quinoline or isatin groups exhibited the greatest anti-vaccinial activity when administered cerebrally as a prophylactic or supplemented in the diet before the onset of symptoms. D.Bauer (1955) reported the chemotherapeutical activity of isatin-β-thiosemicarbazone in mice that had been infected intracerebrally with neurovaccinia virus, and D.Bauer and Sadler (1960) proved that anti-viral activity could be enhanced by alkylation in the 1-position in intracerebrally infected mice. They suggested that selected N-isatin-β-thiosemicarbazone (ITSC, Marboran^{R}, methisazone) could be of value in specific treatment of smallpox in humans. The effectiveness of ITSC was limited to poxviruses as demonstrated in cell cultures, in animals, and in several clinical trials, including a smallpox epidemic in Madras in 1963. ITSC was recommended for smallpox prophylaxis in persons who had been exposed to the infection. However, because it tended to cause severe vomiting, its use was restricted to serious cases. In 1962 some groups concluded that ITSC was effective in the late phase of replication because it blocked the synthesis of late proteins.

One year after G.Brown and W.Ackermann had established that benzimidazoles could inhibit poliovirus multiplication, two additional studies appeared concerning the action of benzimidazoles in tissue culture. G.Brown found them active against poliovirus, and Tamm and colleagues observed that 2.5-dimethyl-benzimidazole inhibited influenzavirus multiplication. During the following 10 years, several benzimidazoles were synthesized and they proved effective in vitro selectively against some RNA viruses (Tamm and colleagues,1953, 1954, 1961; Hollinshead and P.Smith, 1958; Eggers and Tamm, 1961). 2-(α hydroxybenzyl)-benzimidazole (HBB) proved to be most active of this group, especially against enteroviruses in which it selectively inhibited the synthesis of virus-specific RNA polymerase.

When the plaque method for testing anti-viral substances was introduced by Porterfield (1959), by Herrmann and Rosselett, and by Herrmann and colleagues (1960), the possibility arose of screening hundreds of substances within a short period of time. Soon thereafter, three substances were found with anti-viral activity in vitro as well as in animals and humans: 2′-deoxy-5-iodouridine (IUDR), cytosine arabinoside (CA), and 1-aminoadamantane (Amantadine, AM). IUDR was synthesized by Prusoff in 1959, and was thought to be useful for cancer in humans because it was incorporated into mammalian DNA instead of thymidine. Its anti-viral action against several viruses was demonstrated by many scientists, of which the most encouraging were against *HSV* in vitro by Herrmann in 1961, and as a treatment for herpetic keratitis in the rabbit and in humans by Kaufman and colleagues in 1962. IUDR became the first chemotherapeutic anti-viral drug licensed for use in humans. From 1962–1964, anti-herpetic activity was also observed for CA by Underwood in vitro, by Kaufman and Maloney in clinical herpes keratitis, and by Buthala in the rabbit eye.

The 1964 investigations of AM found it to be selectively active against influenza infection in vitro, in animals and in humans, according to the published results of the groups W.Davies, Jackson and Wendel. AM success in human trials was restricted to *Influenza virus A* strains, and was particularly prophylactic against A2 strains. Similar activity has been demonstrated for α-methyl-1-adamantane-methylamine-hydrochloride since 1966. Kato and Eggers (1969) concluded that AM was effective in the uncoating phase as a result of their experiments with *Fowl plague virus*.

Several substances which exhibited anti-viral activity in vitro but proved ineffective in therapeutic experiments were subsequently used as control substances in the laboratory for influencing steps in the virus multiplication cycle. W.Ackermann and Maassab (1954) and W.Ackermann and colleagues (1955) used the amino acid analogue, p-fluorophenylalanine (PFA) to inhibit the growth of poliovirus and influenzavirus. Halvorson and Spiegelman (1952) had already demonstrated that PFA replaced phenylalanine during protein synthesis; there was no inhibition of protein synthesis, but functionally-impaired proteins were synthesized through the decay of mRNA and the dissolution of polysomes. Puromycin, isolated from Streptomyces alboniger, was shown to interrupt poliovirus maturation, apparently by inhibiting synthesis of the virus-directed protein. The substance 5-fluorodeoxyuridine blocked DNA synthesis by inhibiting the conversion of deoxyuridylic acid to thymidylic acid. Actinomycin D had the same effect by binding specifically to double-stranded-DNA. Phosphonoacetic acid was found to be an anti-*HSV* agent in mice and rabbit trials by Shipkowitz and colleagues (1973). It inhibited virus-induced nucleic acid polymerases by specific and direct interactions.

5.6.3 Will the Molecular Genetic Approach Help?

With enhanced knowledge of the molecular genetic mechanisms of virus-host interaction during virus multiplication, the screening of thousands of substances became redundant and has been replaced by specific syntheses with has a direct bearing upon targeted replication stages. This was already generally accepted at the "Third Conference on Anti-viral Substances" in 1977, when antivirals were presented which were activated by cellular or viral enzymes. Specifically, this meant adenine arabinoside and ribavirin, with acyclovir, ganciclovir and zidovudine being added later.

Adenine arabinoside (9-β-D-arabinofuranosyladenine, vidarabine, ara-A) was synthesized as an anti-cancer agent. In 1968, Schabel reported its anti-viral activity. More attention was paid to the substance when Shannon and colleagues (1974) found that ara-A was active against *Gross murine leukemia virus* and, furthermore, influenced the replication of *Rauscher murine leukemia virus* and tumorigenesis by inhibiting viral reverse transcriptase. Subsequent investiga-

Table 5.22. Drugs useful in virus infections

Agent	Virus treated	Application
Marboran	Vaccinia, variola	prophylactically
IDU	HSV keratitis	therapeutically, topically
Aminoadamantan	Influenza A2	prophylactically therapeutically
Vidarabine	HSV, Varicella-zoster	therapeutically
Ribavirin	Resp.sync.virus	therapeutically, (aerosol)
Acyclovir	HSV, Varicella-zoster	therapeutically (topically and systemically
Ganciclovir	Cytomegalo virus, HSV	systemically, therapeutically
Zidovudine	HIV assoc. viruses	therapeutically

tions by Pavan-Langstone and colleagues (1975) proved its effectiveness against *HSV*, *Cytomegalo virus*, and *Varicella-zoster virus* in humans. The same year, Bodey and coworkers determined that ara-A was a useful anticancer drug for the acute blast crisis of chronic myelogenous leukemia. The drug was activated by cellular kinases which convert to ara-ATP and, thereby, selectively inhibit viral DNA synthesis. Ara-A was the first anti-viral drug licensed for systemic use in humans. For its use see Table 5.22.

Ribavirin (1-β-D-ribofuranosyl-1,2,4-triazole-3-carboxymide, virazole), a broad spectrum, nucleoside for DNA and RNA viruses in vitro and experimental animals, was synthesized and tested by the Sidwell and Huffman groups in 1972. The substance is phosphorylated to 5′-monophosphate, -diphosphate and -triphosphate metabolites by cellular adenosine kinase. There are several models for its mode of action on various viruses, but there is no evidence of its efficacy in humans. The predominant side effects, anemia and immunosuppression, are both reversible.

Acycloguanosine (Acyclovir, Zovirax, (9-(2-hydroxyethoxymethyl guanine, ACV) was the first anti-viral agent known to be activated by a virus-specific enzyme. It was synthesized by Schaeffer and colleagues (1978), who proved it anti-herpetic in herpes encephalitis-infected mice after treating them systemically. In addition, it was effective in herpes eye infections in the rabbit as well as in herpes skin lesions in the guinea-pig when applied topically. The selectivity of the agent was detected by Gertrude Elion and colleagues in 1977. The required enzyme was deoxythymidine-deoxycytidine-kinase which phosphorylates ACV intracellularly to acycloguanosine-monophosphate. A cellular guanosine monophosphate kinase phosphorylates this intermediate to the active acycloguanosine triphosphate, which inhibits the herpesvirus-specific DNA polymerase.

Ganciclovir (9-(1,3-dihydroxy-2-propoxy) methylguanine) is similar in structure to acyclovir. It is also activated in the same manner and has been shown to inhibit *HSV* and *Cytomegalo virus* replication. Because of its very low bioavailability, intravenous administration is required to obtain therapeutic levels.

Zidovudine (Azidothymidine, Retrovir, 3-azido-3′-deoxythymidine, AZT) was synthesized by Horwitz and coworkers (1964). Ten years later, Ostertag and colleagues showed that AZT inhibited murine retrovirus replication in vitro. AZT is converted to AZT-5′-triphosphate by cellular enzymes and this in turn, interacts with viral reverse transcriptase. Mitsuya and colleagues (1985) demonstrated that AZT inhibited the infectivity and CPE of HTLV III in vitro. One year later, Yarchoan and colleagues published the first results on clinical administration of AZT, which was given intravenously for 2 weeks, then orally for 4 weeks in a double intravenous dose. In 15 of the 19 patients with AIDS or AIDS-related complex, the number of circulating helper-inducer T lymphocytes increased. Results of further individual cases has shown the clinical benefit of AZT in certain HIV-associated infections.

After 40 years of intensive research, progress in anti-viral chemotherapy has been meagre, if not sorely disappointing, especially when compared with the abundance of substances developed for anti-bacterial chemotherapy. Without doubt, a considerable number of the many thousands of substances tested exhibited anti-viral activity in vitro and in animals. The number of those left for human testing has dwindled even further because of their toxicity, and only pitifully few have remained as encouraging prophylactics or therapeutics. The success in the search for effective anti-virals for use in humans is characterized by a ratio of <1 to 1000 substances tested.

It must be conceded that a therapy which interferes with the viral multiplication cycle, and is thus intimately involved with the living host cell, from the start is bound to be limited in its practical application. Moreover, potential immuno-suppressive, mutagenic, teratogenic and carcinogenic side effects demand a cautious assessment of the therapeutic benefit versus the potential damage to the organism. Most likely, anti-viral therapeutic substances will continue to gain greater specifity for more narrowly-defined indications and, like their anti-bacterial counterparts will induce an ever-increasing number of substance-resistant virus mutants along the way.

Epilogue

During the 100-year history of experimental virology, what was assumed to be a very small bacteria has turned out to be a definable macromolecule. We may confront it as a demonic parasite which surreptitiously gains access to the innermost parts of the host cell in order to exploit it for its own purposes. Apart from this gloomy aspect, however, we see yet another: the fascinating phenomenon of a replicating, genetic particle which enables us to solve molecular-genetic and molecular-biological mysteries whilst turning up new questions.

As a parasite, it threatens and worries us. This will, without a doubt, also be the case in the future, whether in the familiar forms or in forms yet to be discovered. Based on the experience gained thus far, Matthews has projected over the next 10–20 years a fairly constant rate of 4.6 newly detected viruses per year for higher plants, and an average of one new virus per year capable of infecting humans. Our scientific resourcefulness has thus been challenged to improve methods for identifying immunological hallmarks and viral fingerprints; only in this way can we avert repetition of the burden we are now encountering with HIV, which has been devastating its victims' immune networks for more than a decade.

The genetic macromolecule has given us profound insight into the questions of life itself, and it has disclosed new possibilities for diagnosis, prophylaxis and therapy. The RNA viruses, viruses as naked pathogens, and the ribozymes which are enzymatically active RNAs have been enriching the hypotheses on the origin of the virus while providing us with new ideas on the subject of evolution itself.

R. Green's (1935) and Laidlaw's (1938) hypothesis of parasitic degeneration considered viruses to be the end product of a degenerative process which emanated from higher organisms via microorganisms. In an extreme process of regressive evolution, they developed into intracellular parasites which were nourished on vital components of the host cell. They thus lost more and more of their ability to synthesize, and eventually became a simple, small fraction of the host cell's genetic system.

At the same time, phage scientists have discussed the possibility of alternative viruses being created endogenously or exogenously and released as cellular genes. Even for Andrewes (1966) and Luria and Darnell (1967), viruses were mere modifications of cellular components. Greater or lesser conformity and the relationship between viral and cellular DNA sequences led Subak-Sharpe and colleagues (1966) and Bellett (1967) to regard the evolution of different virus families as happening in the host cell as well as externally.

In the future, our knowledge and imagination will continue to mould the history of virology theoretically as well as experimentally. In this book we remember with respect to some of deceased scientists who contributed greatly to the progress of virology in its 100-year history.

Andrewes, Sir Christover Howard
June 6, 1896 – December 31, 1988

Beijerinck, Martinus Willem
March 16, 1851 Amsterdam – January 1, 1931 Gorssel

Burnet, Sir Macfarlane
September 3, 1899 Traralgon/Australia – August 31, 1985 Melbourne

Carrel, Alexis
June 28, 1873 Lyon – November 5, 1944 Paris

Delbrück, Max
September 4, 1906 Berlin – March 10, 1981 Pasadena

D'Herelle, Felix
April 25, 1873 Montreal/Quebec – February 22, 1949 Paris

Doerr, Robert
November 1, 1871 Técso – January 6, 1952 Basel

Elford, William Joseph
January 4, 1900 Malmesbury/Wiltshire – February 14, 1952 London

Enders, John Franklin
February 10, 1897 West Hartford/Connecticut - September 8, 1985 Waterford/Connecticut

Flexner, Simon
March 25, 1863 Louisville/ Kentucky - May 2, 1946 New York

Frosch, Paul
August 15, 1860 Berlin - June 2, 1928 Berlin

Haagen, Nils Eugen
1898, Berlin - August 3, 1972 Berlin

Henle, Werner
August 27, 1910 Dortmund - July 6, 1987 Philadelphia

Herzberg, Kurt
April 29, 1896 Berlin - November 15, 1976 FrankfurtMain

Isaacs, Alick
July 17, 1921 Glasgow - January 26, 1967 London

Ivanowski, Dimitri Josiforich
November 9, 1864 Gdov - June 20, 1920 UdSSR

Landsteiner, Karl
June 14, 1868 Wien - June 26, 1943 New York City

Löffler, Friedrich August
June 24, 1852 Frankfurt/Oder - April 9, 1915 Berlin

Mayer, Adolf
August 9, 1843 Oldenburg - December 25, 1942 Heidelberg

Monod, Jacques
February 9, 1910, Paris - May 31, 1976, Cannes

Noguchi, Hideo (Seisaka)
Nov. 24, 1876 Okinashimamura/Japan - May 21, 1928 Accra/Ghana

Reed, Walter
September 13, 1851 Harrisonburg/Virg. - November 22, 1902

Rivers, Thomas Milton
September 3, 1888 Jonesboro/ Georgia - May 12, 1962 New York

Rous, Francis Peyton
October 5, 1879 Baltimore - February 16, New York

Roux, Pierre Paul Emile
December 17, Confolens/France - November 3, 1933 Paris

Sanarelli, Guiseppe
April 24, 1864 Monte San Savino/Italy - April 6, 1940 Rome

Schlesinger, Max
1907, Ungarn - February 4, 1937, London

Schramm, Gerhard
June 27, 1910, Yokohama - February 3, 1969, Tübingen

Smith, Erwin Frink
January 21, 1854 Gilbert's Mills/USA - April 6, 1927 Washington

Smith, Theobald
July 31, 1859 Albany - December 10, 1934 New York

Stanley, Wendell Meredith
August 16, Ridgeville/USA - June 15, 1971 Salamanca

Theiler, Max
January 30, 1899 Pretoria/ South Africa - August 11, 1972 New Haven/Connecticut

Uhlenhuth, Paul
January 7, 1870 Hannover - December 13, 1957 Freiburg

Wollman, Elisabeth, nee Michelis
August 15, 1888 Minsk - December 22, 1943 Auschwitz

Wollman Eugéne
May 1, 1883 Minsk - December 22, 1943 Auschwitz

Glossary

Accessory cells: Lymphoid cells which cooperate with T and B Lymphocytes in immune reactions.

Acronym: A word created from the initial letters of words.

Adjuvant: A substance which enhances the immune response when given together with an antigen.

Allergy: An overshouting hypersensitivity reaction.

Annealing: The formation of a double-stranded nucleic acid from single-stranded molecules with complementary base sequences.

Antibody: Antigen-induced specific part of the humoral defence system in invertebrates.

Antigen: Organic substances, mostly proteins or glycoproteins, which induce the formation of specific antibodies in vertebrates and react with these antibodies in vitro as well as in vivo.

Attenuation: Treating microbes to lower their pathogenicity.

Autoradiography: A photographic technique for the detection of radiolabelled biological cell elements.

Clon: A group of genetically identical descendants from a single cell.

Concanavalin A (Con A): A lectin which stimulates predominantly T lymphocytes.

Core: The central part of a virus. It consists of the nucleocapsid and associated binding proteins.

Cytoskeleton: The structural framework of the cell consisting of proteins which form microfilaments and microtubuli.

Dalton: Mass unit related to the hydrogen atom.

Density gradient centrifugation: Separating macromolecules of different density in a gradient.

Disseminated infection: A multi-step event following a localized infection. It includes virus multiplication in regional lymph nodes, the spread of progeny virus through the blood stream and the lymphatics to reach target organs.

Endogeneous virus: Virus integrated into the host cell genom.

Evolutionary conservation: The conservation of defined DNA sequences through evolution.

Early (late) enzymes: Virus-coded enzymes synthesized before (after) replication.

Exogenes Virus: Free virus.

Exon: A coding DNA region.

Focus: A group of morphologically transformed cells within normal cells.

Hybridization: The formation of double-stranded nucleic acid by base-pairing between single-stranded nucleic acids of different origin.

Icosahedron: A regular crystal built of 20 equilateral triangulars.

Immune tolerance: A specific reduction in immune responsiveness to a given antigen.

Interferenz: The ability of an microbe to inhibit a following infection by another agent.

Intron: A non-coding DNA region.

Lattice: A formation of cyclic structures (like a crystal)

Lipmann system: Processes involved in intermediary metabolism.

Liposomes: Natural or synthetic spheres of phospholipid bilayers separated by aqueous compartments. They were used as adjuvants or antigen-carriers in vaccine production. Antigens were attached to their surface or incorporated.

Localized infection: The situation when only tissues near the site of entry are infected.

Lymphoma: A solid tumour of lymphatic tissue.

Mitogen: A substance which causes DNA synthesis, blast transformation, and ultimately the division of lymphocytes.

Multiplicity of infection (m.o.i.): The number of virions infecting a cell simultaneously.

Opportunistic infections: Infections with microorganisms normally harmless but cause of illness in immunodeficient patients.

Passive agglutination: When a hemagglutinating agent does not react directly with cell components but with soluble particles attached to the surface of carriers.

Permissive culture: A cell culture in which suitable conditions exist for a virus to multiply.

Phagocytosis: The engulfment of particles by leukocytes.

Pinocytosis: The ingestion of soluble materials by cells.

Plaque forming unit (PFU): An undefinite number of virions inducing one plaque.

Plasmid: A small, extrachromosomal, self-replicating DNA molecule.

Prodromal period: The time between infection and the onset of symptoms.

Pyrogen: A substance inducing fever.

Redundancy: A sequence present in more than one copy per genome.

Replication: The process whereby an identical nucleic acid molecule is synthesized.

Resolving power: The ability of a microscope to produce separate images of closely spaced objects.

Sedimentation rate: The rate at which a macromolecule moves in a standard gravitational field. Expressed in units of $cm \times sec^{-2}$.

Sigla: Identical to acronym.

Splicing: The mechanism of processing mRNA.

Svedberg-unit S: Sedimentation coefficient, the unit of sedimentation within a defined gravitational field.

Synthetic adjuvants: Non-toxic, well defined organic immunoregulators.

Systemic infection: An infection that spreads throughout the body.

Terminal redundancy: Identical nucleotide sequences at both ends of a linear nucleic acid molecule.

Terminal repetition: Identical, or identical but inverted nucleotide sequences at both ends of a nucleic acid molecule.

Titre: A measure of antibodies or of infecting particles per defined volume. In serology it is usually quoted as the highest serum dilution causing an antibody-mediated reaction.

Transfection: The transfer of infectious nucleic acid.

ts mutants: Temperature-sensitive mutants.

Vector: Carrier. In virology, a carrier of a virion; in molecular genetics, a carrier of a nucleic acid sequence.

Viremia: After a localized infection and virus multiplication in regional lymph nodes progeny virus can be transported via the blood stream to other organs (primary viremia). In a secondary viremia, newly synthesized viruses are spread to target organs.

Viroid: Helper-indipendent, uncoated RNA plant pathogen. Unusual covalently closed circles, unable to code for proteins.

Virosome: Purified HA and neuraminidase spikes relocated on the surface of unilamellar liposomes.

Virusoid: Helper-dependent, encapsidated satellites of plant viruses with a linear or circular genome.

Viroplasm: The accumulation of synthesized virions or virus components within the cytoplasm.

Bibliography

A Books and Reviews
B Special papers

Chapter 1

A

Bayon HP (1947) William Harvey. J Hist Med 2:51–96

Brock TD (1988) Robert Koch. A life in medicine and bacteriology. Science Tech Madison, WI and Springer, Berlin Heidelberg New York Tokyo

DeBary A (1853) Untersuchungen über die Brandpilze. Müller, Berlin

Duclaux E (1882) Ferments et Maladies. Masson, Paris

Ehrenberg C (1838) Die Infusionsthierchen als vollkommene Organismen. Leipzig

Flügge C (1886) Die Mikroorganismen, 2nd edn. Vogel, Leipzig

Geison GL (1974) Louis Pasteur. Dict Sci Biogr 10:350–416

Haden RL (1939) The origin of the microscope. Ann Med Hist 1:30–44

Haeser H (1882) Lehrbuch der Geschichte der Medizin und der epidemischen Krankheiten. Fischer, Jena

Hueppe F (1891) Die Methoden der Bakterien-Forschung, 5th edn. Kreidel, Wiesbaden

Lechevalier HA, Solotorovsky M (1965) Three centuries of microbiology. McGraw-Hill, New York

Koch R (1912) Gesammelte Werke. Thieme, Leipzig

Mayor RH (1939) Athanasius Kircher. Ann Med Hist 1:105–120

Medical Research Council (ed) (1930) A system of bacteriology in relation to medicine, vol 1. H M Stationery Office, London

Oppenheimer C (1900) Die Fermente und ihre Wirkungen. Vogel, Leipzig

Pouchet FA (1859) Héterogénie ou traité de la génération spontanée, basé sur de nouvelles experiences. Paris

von Nägeli C (1877) Die niederen Pilze in ihren Beziehungen zu den Infektionskrankheiten und der Gesundheitspflege. Oldenbourg, München

Weed LA (1942) John Tyndall and his contribution to the theory of spontaneous generation. Ann Med Hist 4:55–62

B

Arthur JC (1885) Proof that bacteria are the direct cause of the disease in trees known as pear blight. Bot Gaz 10:343–345

Behring E (1892) Ueber die Immunisierung und Heilung von Versuchsthieren beim Tetanus. Z Hyg Infektionskr 12:45–57

Behring, Wernicke (1892) Ueber Immunisierung und Heilung von Versuchsthieren bei Diphtherie. Z Hyg Infektionskr 12:10–44

Burrill TJ (1880) Anthrax of fruit trees; or the so-called fire blight of pear, and twig of apple trees. Proc Am Assoc Adv Sci 29:583–597

Davaine C (1864) Nouvelles recherches sur la nature de la maladie charbonneuse connue sous le nom de 'sang de rate. C R Seances Acad Sci 59:393–396

Ehrlich P (1897) Die Wertbemessung des Diphtherieheilserums und deren theoretische Grundlagen. Klin Jb 6:299–328

Fischer A (1899) Die Bakterienkrankheiten der Pflanzen. Zbl Bakt II/Orig 5:279–287

Kartulis (1893) Ueber pathogene Protozoen bei den Menschen. Z Hyg Infektionskr 13:1–14

Kitasato S (1891) Experimentelle Untersuchungen über das Tetanusgift. Z Hyg Infektionskr 10:267–305

Klebs AC (1913) The historic evolution of variolation. Bull J Hopk Hosp 24:69–83

Metschnikoff E (1884) Ueber die Sporenpilzkrankheit der Daphnien. Beitrag zur Lehre über den Kampf der Phagozyten gegen Krankheitserreger. Virchows Archiv 96:177–195

Pasteur L (1861) Mémoire sur les corpuscules organisés qui existent dans l'atmosphére, examen de la doctrine des géneérations spontanées. Ann Sci nat Zool, 4th ser 16:5–98

Remak R (1852) Ueber extrazellulare Entstehung thierischer Zellen und über Vermehrung derselben durch Theilung. Müllers Archiv 47–57

Salmon, Smith (1886) On a new method of producing immunity from contagious diseases. Proc Biol Soc, Washington 3:29–33

Sédillot C (1878) De l'influence des découvertes de M.Pasteur sur le progrés de la chirurgie. C R Seances Acad Sci 86:634–640

Chapter 2

A

Doerr R, Hallauer C (eds) (1938/1939) Handbuch der Virusforschung. Erste und zweite Hälfte. Springer, Wien

Hughes SM (1977) The virus, a history of the concept. Heinemann, London

Joest E (1902) Unbekannte Infektionsstoffe. Zbl Bakt I/Orig 31:361–384, 410–422

Koch R (1912) Gesammelte Werke. Thieme, Leipzig

Kolle W, Kraus R, Uhlenhuth P (eds) (1929) Handbuch der pathogenen Mikroorganismen, vol 9. Fischer, Jena; Urban & Schwarzenberg, Wien

Loeffler F (1907) Zum 25 jährigen Gedenktage der Entdeckung des Tuberkelbacillus. Dtsch med Wochenschr 33:489–495

Waterson AP, Wilkinson L (1978) An introduction to the history of virology. Cambridge Univ Press, Cambridge

Wolbach SB (1912) The filterable viruses, a summary. J med Res 27:1–25

B

Baur E (1904) Zur Aetiologie der infektiösen Panachierung. Ber dtsch Bot Ges 22:453–460

Beijerinck MW (1899) Ueber ein Contagium vivum fluidum als Ursache der Fleckenkrankheit der Tabakblätter. Zbl Bakt Orig/II 5:27–33

Borrel A (1903) Etude expérimentale de la clavelée. Ann Inst Pasteur 17:123–137

Centanni E (1902) Die Vogelpest. Zbl Bakt I/Orig 3:145–152, 182–201

Evans AS (1976) Causation and disease. The Henle-Koch Postulates revisted. Yale J Biol Med 175–195

Goodpasture EW (1929) Cellular inclusions and the etiology of virus disease. Arch Pathol 7:114–132

Grafe A (1988) Die sogenannten Koch'schen Postulate. Gesnerus 45:411–418

Hitchens AP, Leikind MC (1939) The introduction of agar-agar into bacteriology. J Bateriol 37:485–493

Iwanowski D (1903) Über die Mosaikkrankheit der Tabakpflanze. Z Pflanzenkr 13:1–41

Lode A, Gruber J (1901) Bakteriologische Studien über die Aetiologie einer epidemischen Erkrankung der Hühner in Tirol. Zbl Bakt I/Orig 30:593–604

Loeffler, Frosch (1897, 1898) Berichte der Kommission zur Erforschung der Maul- und Klauenseuche bei dem Institut für Infektionskrankheiten in Berlin. Zbl Bakt I/Orig 22:257–259; 23:371–391

Maggiora, Valenti (1901) Su una epizooia di tifo essudativo dei gallinacei. Accad med Modena

Martin L (1898) Production de la toxine dipthérie. Ann Inst Pasteur 12:26–48

Mayer A (1886) Ueber die Mosaikkrankheit des Tabaks. Landwirtsch Vers Stant 32:451–467, Paray, Berlin

M'Fadyean J (1900) African horse-sickness. J comp Pathol Ther 13:1–20

Nicolle M, Adil-Bey (1902) Expériences sur la filtration du virus. Ann Inst Pasteur 16:56–64

Nocard, Roux (1898) Le microbe de la péripneumonie. Ann Inst Pasteur 12:240–262

Reed W (1902) The etiology of yellow fever. Am Med 3:301–305

Reed W, Carroll J, Agramonte A (1901) The etiology of yellow fever. J Am Med Assoc 36:431–440

Remlinger P (1906) Les microbes filtrants. Bull Inst Pasteur 4:337–345, 385–392

Rivers TM (1937) Viruses and Koch's postulates. J Bacteriol 33:1–12

Roux E (1903) Sur le microbes dits "invisibles". Bull Inst Pasteur 1:7–12, 49–56

Chapter 3

A

Adams MA (1959) Bacteriophages. Interscience, New York

Bechhold H (1929) Die Kolloide in Biologie und Medizin, 5th edn. Steinkopf, Dresden-Leipzig

Burnet FM, Andrewes CH (1933) Ueber die Natur der filtrierbaren Vira. Zbl Bakt Orig/I 130:161–183

Doerr R und Hallauer C (eds)(1938, 1939, 1944) Handbuch der Virusforschung. Erste und zweite Häfte; 1. Ergänzungsband. Springer, Wien

Friedberger E und Pfeiffer R (eds)(1919) Lehrbuch der Mikrobiologie. Fischer, Jena

Gildemeister E, Haagen E, Waldmann O (eds)(1939) Handbuch der Viruskrankheiten. Fischer, Jena

Goodpasture EW, Teague OT (1923) Transmission of the virus of herpes febrilis along nerves in experimentally infected rabbits. J med Res XLIV:139–184

Gross L (1971) Oncogenic viruses, 2nd edn. Pergamon Press, Oxford

Kolle W, Kraus R, Uhlenhuth P (eds) (1929) Handbuch der pathogenen Mikroorganismen. vol 1, 3rd edn. Fischer, Jena; Urban & Schwarzenberg, Wien

Medical Research Council (ed) (1930) A system of bacteriology in relation to medicine. vol VII, H M Stationery Office, London

Paul J (1970) Cell and tissue culture, 4th edn. Livingstone, Edinburgh

B

Bawden FC, Pirie NW (1937) The isolation and some properties of liquid crystalline substances from solanaceous plants infected with three strains of tobacco mosaic virus. Proc R Soc Lond, Ser B Biol Sci 123:274–320

Bawden FC, Pirie, Bernal, Fankuchen (1936) Liquid crystalline substances from virus-infected plants. Nature (London) 138:1051–1052

Bechhold H, Schlesinger M (1931) Die Grössenbestimmung von subvisiblem Virus durch Zentrifugieren. Biochem Z 236:387–414

Bechhold H, Villa L (1926) Die Sichtbarmachung subvisibler Gebilde. Z Hyg Infektionskr 105:601–613

Bland JOW (1928) Filter and centrifuge experiments with guinea-pig vaccinia virus. Brit J Exp Pathol 9:282–290

Borrel A (1904) Sur les inclusions de l'épithelioma contagieux des oiseaux. C r Soc Biol 57:642–643

Burnet FM (1929) A method for the study of bacteriophage multiplication in broth. Brit J exp Pathol 10:109–115

Craigie J (1932) The nature of vaccinia flocculation, and observations on the elementary bodies of vaccinia. Brit J Exp Pathol 13:259–268

d'Herelle F (1917) Sur un microbe invisible antagonists des bacilles dysenteriques. C R Seances Acad Sci 165:373–375

Elford WJ (1931) A new series of graded collodion membranes suitable for general bacteriolgical use, especially in filterable virus studies. J Pathol Bacteriol 34:505–521

Gildemeister E, Herzberg K (1925) Experimentelle Untersuchungen über Herpes. Dtsch med Wochenschr 51:97–98, 1647–51

Goodpasture EW (1929) Cellular inclusions and the etiology of virus diseases. Arch Pathol 7:114–132

Goodpasture EW, Woodruff A, Buddingh GJ (1931) The cultivation of vaccine and other viruses in the chorio-allantoic membrane of chick embryos. Science 74:371–372

Grüter W (1924) Das Herpesvirus, seine ätiologische und klinische Bedeutung. Münch med Wochenschr 1058–1060

Hallauer C (1932) Über das Verhalten von Hühnerpestvirus in der Gewebekultur. Z Hyg Infektionskr 113:61–74
Herzberg K (1933) Mikrophotographische Darstellung einer intrazellulären Virusentwicklung. Zbl Bakt I/Orig 130:326–329
John-Brooks (1937) Second international congress for microbiology, London 1936. Rep Proc, London 1937
Kolle W, Turner G (1898) Über Schutzimpfungen und Heilserum bei Rinderpest. Z Hyg Infektionskr 29:309–375
Kraus R, von Eisler, Fukuhara (1909) Über Adsorption des filtrierbaren Virus. Z Immunitätsforsch 1:307–315
Ledingham JCG (1931) The aetiological importance of the elementary bodies in vaccinia and fowl-pox. Lancet II:525–526
Levaditi C, Nicolau S (1923) Propriétés physiques des ultravirus neurothopes. C r Soc Biol 88:66–69
Lewis MR (1931) Production of tumors by means of purified (protein removed) tumor extracts. Am J Canc (Suppl) 15:2248–2251
MacCallum WG, Oppenheimer EH (1922) Differential centrifugalization. J Am Med Assoc 78:410–411
Maisin J (1921) Adaption du bactériophage. C r Soc Biol 84:468–472
Malfitano G (1904) Sur la conductibilité électrique des solutionscolloidales. C r Acad Sci 139:1221–1223
Martin CJA (1896) A rapid method of separating colloids from crystalloids in solutions containing both. J Phys 20:364–371
Murphy JB, Rous P (1912) The behavior of chicken sarcoma implanted in the developing embryo. J exp med 15:119–132
Rivers TM, Haagen E, Muckenfuss RS (1928) A method of studying virus infection and virus immunity in tissue cultures.Proc Soc Exp Biol Med 26:494–496
Rous P, Johns FS (1916) A method for obtaining suspensions of living cells from the fixed tissues, and for the plating out of individual cells. J exp med 23:549–554
Roux, Yersin (1889) Contribution à l'étude de la diphthérie. Ann Inst Pasteur, Paris 3:273–288
Siedentopf H, Zsigmondy R (1903) Über die Sichtbarmachung und Grössenbestimmung ultramikroskopischer Teilchen mit besonderer Anwendung von Goldrubingläsern. Ann Phys 10:1–39
Svedberg T (1934) Sedimentationsmessungen mit der Ultrazentrifuge. Naturwissenschaften 22:225–231
Svedberg, Nichols JB (1923) Determination of size and distribution of particles by centrifugal methods. J Am Chem Soc 45:2910–2917
Schlesinger M (1933) Die Bestimmung der Teilchengrösse und spezifischem Gewicht des Bakteriophagen durch Zentrifugierversuche. Z Hyg Infektionskr 114:161–176
Schlesinger M (1934) Zur Frage der chemischen Zusammensetzung des Bakteriophagen. Biochem Z 273:306–311
Schulz EW, Hoyt J (1928) Studies on the antigenic properties of the ultraviruses. IV. The antigenic propertiesof herpes virus. J Immunol 15:411–419
Schulz EW, Bullock LT, Lawrence F (1928) Studies on the antigenic properties of the ultraviruses. II. The antigenic properties of vaccinia virus. J Immunol 15:243–263

Stanley WM (1935) Isolation of a crystalline protein possessing the properties of tobacco-mosaic virus. Science 81:644–645

Steinhardt E, Israeli C, Lambert RA (1913) Studies on the cultivation of the virus of vaccinia. J Infect Dis 13:294–300

Sternberg GM (1896) Wissenschaftliche Untersuchungen über das spezifische Infektionsagens der Blattern und die Erzeugung künstlicher Immunität gegen diese Krankheit. Zbl Bakt I/Orig 19:805–815, 857–869

Takahashi WN, Rawlins TE (1932/1933) Method for determining shape of colloidal particles; application in study of tobacco mosaic virus. Proc Soc Exp Biol Med 30:155–157

Tizzoni G, Centanni E (1892) Ueber die Art, bei Thieren die schon ausgebrochene Rabies zu heilen. Dtsch med Wochenschr 18:624–626

Twort FW (1915) An investigation on the nature of ultramicroscopic viruses. Lancet II:1241–1243

Vallée, Carré (1921) Sur l'adsortion du virus apthteux. C R Acad Sci 172:185–187

Vinson CG (1927) Precipitation of the virus of tobacco mosaic. Science 66:357–358

Vinson CG, Petre AW (1931) Mosaic disease of tobacco. II. Activity of the virus precipitated by lead acetate. Contrib Boyce Thompson Inst Plant Res 3:131–145

Weyl T (1877) Beiträge zur Kenntnis thierischer und pflanzlicher Eiweisskörper. Z Physiol Chem 1:72–100

Woodruff CE, Goodpasture EW (1929) The infectivity of isolated inclusion bodies of fowl-pox. Am j pathol 5:1–9

Woodruff AM, Goodpasture EW (1931) The susceptibility of the chorioallantoic membrane of chick embryos to infection with fowl-pox virus. Am j pathol 7:210–222

Chapter 4

A

Burnet FM (1960) Principles of animal virology. 2nd edn. Academic Press, New York

Burnet FM, Stanley WM (eds) (1959) The viruses, vols 1–3. Academic Press, New York

Chargaff E, Davidson JN (ed) (1955) The nucleic acids, vol II. Academic Press, New York

Demerec M (1933) What is a gene? J Hered 24:369–378

Doerr R, Hallauer C (eds) (1950) Handbuch der Virusforschung. II.Ergänzungsband, Springer, Wien

Francis T Jr (1945) The development of the 1943 vaccination study of the commission on influenza. Am J Hyg 42:1–11

Haas R, Vivell O (eds) (1965) Virus- und Rickettsieninfektionen des Menschen. Lehmanns, München

Hallauer C, Mayer KF (eds) (1958) Handbuch der Virusforschung, vol 4. Springer, Wien

Herrlich A (ed) (1965) Handbuch der Schutzimpfungen. Springer, Berlin Heidelberg New York

Horsfall FL, Tamm I (eds)(1965) Viral and rickettsial infections of man. 4th edn. Pitman Medical Publ, London

Jordan P (1939) Die Stellung der Quantenphysik zu den aktuellen Problemen der Biologie. Arch ges Virusforsch 1:1–20

Judson HF (1979) The eighth day of creation. Simon andSchuster, New York

Luria SE, Darnell JE Jr (1968) General virology, 2nd edn. Wiley, New York

Mayr A, Eißner G, Mayr-Bibrack B (1984) Handbuch der Schutzimpfungen in der Tiermedizin. Parey, Berlin

Paul JR (1971) A history of poliomyelitis. Yale Univ Press, New Haven London

Rivers TM (ed) (1948) Viral and rikettsial infections of man. Lippincott, Philadelphia

Rhodes AJ, van Rooyen CE (1958) Textbook of virology. Williams and Wilkins, Baltimore

Röhrer H (1960) 50 Jahre Forschung auf dem Riems. Arch exp Vet Med 9:713–731

Stent GS (1963) Molecular biology of bacterial viruses. Freeman, San Francisco

Weiss R, Teich N, Varmus H, Coffin J (eds) (1982) RNA tumor viruses, 2nd edn. Cold Spring Harbor Lab

B

Astbury WT (1961) Molecular biology or ultrastructural biology. Nature (London) 190:1124

Burnet FM, Lush D (1936) Induced lysogenicity and mutation of bacteriophage within lysogenic bacteria. Aust J exp Biol Med Sci 14:27–38

den Dooren de Jong LE (1931) Studien über Bakteriophagie. I. Ueber Bac.megatherium und den darin anwesenden Bakteriophagen. II. Fortsetzung der Untersuchungen über den Megatherium-Bakteriophagen. Zbl Bakt I/Orig 120:1–15, 15–24

Dulbecco R (1952) Production of plaques in monolayer tissue cultures by single particles of an animal virus. Proc Natl Acad Sci USA 38:747–752

Dulbecco R, Vogt M (1954) Plaque formation and isolation of pure lines with poliomyelitis virus. J exp Med 99:167–182

Ellis EL, Delbrück M (1939) The growth of bacteriophage. J gen Physiol 22:365–384

Enders JF, Weller TH, Robbins FC (1949) Cultivation of the Lansing strain of poliomyelitis virus in cultures of various human embryonic tissues. Science 109:85–87

Evans CA, Chambers VC (1948) Growth of neurotropic viruses in extraneural tissues. I. MM virus in the feet of hamsters. Proc Soc Exp Biol Med 68:436–442

Flexner S, Amoss HL (1914) Penetration of the virus of poliomyelitis from the blood into the cerebrospinal fluid. J exp Med 19:411–416

Fraenkel-Conrat H (1956) The role of the nucleic acid in the reconstitution of active tobacco mosaic virus. J Am chemSoc 78:882–883

Gallo R et al (1983) Association of the human type C retrovirus with a subset of adult T-cell cancers. Cancer Res 43:3892–3899

Gey GO, Coffman WD, Kubicek MT (1952) Tissue culture studies of proliferative capacity of cervical carcinoma and normal epithelium. Cancer Res 12:264 - 265

Gierer A, Schramm G (1956) Infectivity of ribonucleic acid from tobacco mosaic virus. Nature (London) 177:702–703

Green RG (1935) On the nature of filterable viruses. Science 82:443–445

Horstmann DM (1952) Poliomyelitis in the blood of orally infected monkeys and chimpanzees. Proc Soc Exp Biol Med 79:417–419

Horstmann DM, McCollum RW, Mascola AD (1954) Viremia in human poliomyelitis. J exp Med 99:355–369

Kellenberger G, Kellenberger E (1957) Electron microscopical studies of phage multiplication. III. Observation of single cell burst. Virology 3:275–285

Luria SE (1951) The frequency distribution of spontaneous bacteriophage mutants as evidence for the exponential rate of phage reproduction. Cold Spring Harbor Symp Quant Biol 18:463–470

Lwoff A (1957) The concept of virus. J gen Microbiol 17:239–253

Markham R, Smith JD (1950) Chromatographic studies on nucleic acids. 3. The nucleic acids of five strains of tobacco mosaic virus. Biochem J 46:513–517

Morgan IM (1949) Level of serum antibody associated with intracerebral immunity in monkeys vaccinated with Lansing poliomyelitis virus. J Immunol 62:301–310

Moscona A (1952) Cell suspensions from organ rudiments of chick embryos. Exp Cell Res 3:535–539

Rivers T (1941) The infinitely small in biology. Diplomate13:184–86

Robbins FC, Enders JF, Weller TH (1950) Cytopathogenic effect of poliomyelitis viruses in vitro on human embryonic tissues. Proc Soc Exp Biol Med 75:370–374

Salk JS (1955) Considerations in the preparation and use of poliomyelitis virus vaccine. J Am Med Assoc 158:1239–1248

Schrödinger E (1945) What is Life? Cambridge Univ Press, New York

Sertic V (1929) Procédé d'obtention de variants de bactériophage adaptées a lyser des formes bactériennes secondaires. C R Soc Biol 100:612–615

Steinhaus EA (1953) Taxonomy of insect viruses. N Y Acad Sci 56:517–537

Weißenberg R (1938) Studies on virus disease of fish. Am J Hyg 28:455–462

Williams RC, Fraser D (1956) Structural and functional differentiation in T2 bacteriophage. Virology 2:289–307

Wyckoff RWG, Corey RB (1936) X-ray diffraction pattern of crystalline tobacco mosaic proteins. J Biol Chem 116:51–55

Chapter 5

Periodic book series

Advances in virus research. Academic Press, New York London. Vol 1, 1953

Comprehensive virology. Plenum Press, New York London. Vol 1, 1974

Current topics in microbiology and immunology. Springer, Berlin Heidelberg New York. Vol 40, 1967

Monographs in virology. Karger, Basel New York. Vol 1, 1968

Progress in medical virology. Karger, Basel München Paris London New York Sydney. Vol 1, 1958
Virology monographs. Springer, Wien New York. Vol 1, 1968

A
Ahne W (1985) Virusinfektionen bei Fischen: Ätiologie, Diagnose und Bekämpfung. Zbl Vet Med B 32:237–264
Ahne W, Kurstak E (eds)(1989) Viruses of lower vertebrates. Springer, Berlin Heidelberg New York Tokyo
Almond JW (1987) The attenuation of poliovirus neurovirulence. Ann Rev Microbiol 41:153–180
Andrewes C (1964) Viruses of vertebrates. Baillière, Tindall and Cox, London
Baltimore D (1976) Viruses, polymerases, and cancer. Science 192:632–636
Bible DJ (ed) (1988) Milestones in immunology. Sci Tech Publ, Madison, Wi, Springer, Berlin Heidelberg New York Tokyo
Brown F (1984) Synthetic viral vaccines. Ann Rev Microbiol 3:221–235
Burnet FM (1962) The immunological significance of the thymus: an extension of the clonal selection theory of immunity. Aust Ann Med 11:79–91
Casjens S, King J (1975) Virus assembly. Ann Rev Biochem 44:555–611
Dales SD (1973) Early events in cell-animal virus interactions. Bacteriol Rev 37:103–135
Diener TO (ed) (1987) The viroids. Plenum Press, New York
Diener TO, McKinley MP, Prusiner SB (1982) Viroids and prions. Proc Natl Acad Sci USA 79:5220–5224.
Dimmock NJ (1982) Initial stages in infection with animal viruses. J gen Virol 59:1–22
Duesberg PH (1987) Cancer genes: rare recombinants instead of activated oncogenes. Proc Natl Acad Sci USA 84:2117–2124
Dulbecco R (1976) From the molecular biology of oncogenic DNA viruses to cancer. Science 192:437–440
Edelman R (1980) Vaccine adjuvants. Rev infect Dis 2:370–383
Eiserling FA, Dickson RC (1972) Assembly of viruses. Ann Rev Biochem 41:467–502
Elizan TS, Fabiyi A (1970) Congenital and neonatal anomalies linked with viral infections in experimental animals. Am J Obstet Gynecol 106:147–165
Fenner F (1968) The biology of animal viruses. Academic Press, New York
Fenner F, Gibbs A (eds) (1988) Portraits of viruses. A history of virology. Karger, Basel
Fields BN (ed) (1990) Virology, 2nd edn, vol 1 and 2. Raven, New York
Fraenkel-Conrat H (1969) The chemistry and biology of viruses. Academic Press, New York
Fraenkel-Conrat H, Kimbell PC (1982) Virology. Prentice-Hall, Englewood Cliffs, New York
Fucillo DA, Sever JL (1973) Viral teratology. Bacteriol Rev 37:19–31
Gajdusek DC (1977) Unconventional viruses and the origin and disappearance of Kuru. Science 197:943–959
Gordon M, Buchanan RL, Crenshaw RR, Kerridge KA, MacNintch JE, Siminoff P (1981) A review of approaches to viral chemotherapy. J Med 12:289–382

Gregoriadis G (1976) The carrier potential of liposomes in biology and medicine. New Engl J Med 295:704–710, 765–770
Habel K, Salzman NP (eds) (1969) Fundamental techniques in virology. Academic Press, New York
Harris KF (1981) Arthropod and nematode vectors of plant viruses. Ann Rev Phytopathol 19:391–426
Harris RJC (ed) (1964) Techniques in experimental virology. Academic Press, London
Heidelberger M (1967) Some contributions of immunochemistry to biochemistry and biology. Ann Rev Biochem 36:1–12
Herrmann EC (ed) (1977) Third conference on antiviral substances. Ann N Y Acad Sci 284
Herrmann EC, Stinebring WR (eds) (1970) Second conference on antiviral substances. Ann N Y Acad Sci 173
Horsfall FL, Tamm I (eds) (1965) Viral and rickettsial infections of man, 4th edn. Pitman Med Publ, London
Joklik WK (ed.)(1985) Virology, 2nd edn. Appleton-Century-Crofts/ Norwalk, Connecticut
Kabat EA (1971) Einführung in die Immunchemie und Immunologie. Springer, Berlin Heidelberg New York
Kaper JM (1975) The chemical basis of virus structure, dissociation and reassembly. North-Holland, Amsterdam
Kleber W (1983) Einführung in die Kristallographie. Bearbeitet von Bautsch H, Bohm J, Kleber I. VEBVerlag Technik, Berlin
Klein G, Klein E (1986) Conditioned tumorigenicity of activated oncogenes. Cancer Res 46:3211–3224
Knight CA (1975) Chemistry of viruses, 2nd edn. Springer, Wien New York
Krauss H (1968) Inapparente Virusinfektionen beim Geflügel und ihre Bedeutung unter Berücksichtigung eines erstmals isolierten Agens. Zbl Vet Med 15:286–314,317–352
Matthews REF (1982) Classification and nomenclature of viruses. Fourth report of the international committee on taxonomy of viruses. Intervirology 17:1–199
Matthews REF (1985) Viral taxonomy for the nonvirologist. Ann Rev Microbiol 39:451–474
Mayr A, Bachmann PA, Bibrack B, Wittmann G (1974, 1977) Virologische Arbeitsmethoden. Vol 1 and 2. Fischer, Stuttgart
McAuslan BR (1974) Virus-associated enzymes. Life Sci 14:2085–2097
Merigan TC, Stevens DA (1971) Viral infections in man associated with acquired immunological deficiency states. Fed Proc 30:1858–1864
Meselson M, Yuan R, Heywood J (1972) Restriction and modification of DNA. Ann Rev Biochem 41:447–466
Miller G (1971) Human lymphoblastoid cell lines and Epstein-Barr virus: a review of their interrelationships and their relevance to the etiology of leukoproliferative states in man.Yale J Biol Med 43:358–384
Mims CA (1986) Virus receptors and cell tropisms. J infect Dis 12:199–203
Müller-Eberhard HJ (1988) Molecular organization and function of the complement system. Ann Rev Biochem 57:321–347

Nathans D, Smith HO (1975) Restriction endonucleases in the analysis and restructuring of DNA molecules. Ann Rev Biochem 44:273–293

Nossal GJV (1987) Current concepts: immunology. The basic components of the immune system. New Engl J Med 316:1320–1325

Ogra PL, Fishaut M, Gallagher MR (1980) Viral vaccination via the mucosal route. Rev Infect Dis 2:352–369

Pollock RR, Teillaud J, Scharff MD (1984) Monoclonal antibodies: a powerful tool for selecting and analyzing mutations in antigens and antibodies. Ann Rev Microbiol 38:389–417

Rima BK, Martin SJ (1976) Persistent infection of tissue culture cells by RNA viruses. Med Microbiol Immunol 162:89–118

Roizman B, Sears AE (1987) An inquiry into the mechanisms of herpes simplex virus latency. Ann Rev Microbiol 41:543–571

Rossmann MG, Johnson JE (1989) Icosohedral RNA virus structure. Ann Rev Biochem 58:533–73

Sager R (1989) Tumor suppressor genes: the puzzle and the promise. Science 246: 1406–1412

Spiegelman S, Haruna I (1966) A rationale for an analysis of RNA replication. Proc Natl Acad Sci USA 55:1539–1554

Stent GS, Calendar R (1978) Molecular genetics. 2nd edn. Freeman, San Francisco

Watson JD et al (1987) Molecular biology of the gene, 4nd edn. Benjamin/ Cummings, Menlo Park

White JM, Littman DR (1989) Viral receptors of the immunoglobin superfamily. Cell 56:725–728

Temin HM (1971) Mechanism of cell transformation by RNA tumor viruses. Ann Rev Microbiol 25:609–648

Ter Meulen V, Hall WW (1978) Slow virus infections of the nervous system: virological, immunological and pathogenic considerations. J gen Virol 41:1–25

Tyrrell J (1987) Interferons and their clinical value. Rev Infect Dis 9:243–249

Wetter C (1965) Serology in virus-disease diagnosis. Ann Rev Phytopathol 3:19–42

Whipple HE (ed) (1965) Antiviral substances. Ann N Y Acad Sci 130

Wittmann G, Gaskell R, Rziha H-J (eds) (1984) Latent herpes virus infections in veterinary medicine. Martinus Nijhoff, Boston

Yewdell JW, Gerhard W (1981) Antigenic characterization of viruses by monoclonal antibodies. Ann Rev Microbiol 35:185–206

B

Adler SP, Nathans D (1970) Cleavage of simian virus 40 DNA by bacterial restricting enzymes. Fed Proc(Abstr) 29:725

Agrawal HO, Kent JW, MacKay DM (1965) Rotation technique in electron microscopy of viruses. Science 148:638–640

Bär F (1940) Erfahrungen und Ergebnisse über die chemotherapeutische Beeinflußbarkeit des Lymphogranuloma inguinale und anderer Viruskrankheiten. Z Immunitätsforsch 97:344–363

Baltimore D (1970) Viral RNA-dependent RNA polymerase. Nature (London) 226:1209–1211

Baltimore D (1971) Expression of animal virus genomes. Bacteriol Rev 35:235–241

Bernice E, Borman GS, Grubbs GE, Young RD (1962) Identification of the oncogenic substance in rhesus monkey kidney cell cultures as simian virus 40. Virology 17:65–75

Blumberg BS (1977) Australia antigen and the biology of hepatitis B. Science 197:17–25

Brenner S, Horne RW (1959) A negative staining method for high resolution electron microscopy of viruses. Biochem Biophys Acta 34:103–110

Brenner S, Jacob F, Meselson M (1961). An unstable intermediate carrying information from genes to ribosomes for protein synthesis. Nature (London) 190:576–581

Burnet FM (1962) The immunological significance of the thymus: An extension of the clonal selection theory of immunity. Aust Ann Med 11:79–91

Caspar DLD, Klug A (1962) Physical principles in the construction of regular viruses. Cold Spring Harbor Symp Quant Biol 27:1–24

Chargaff E (1950) Chemical specificity of the nucleic acids and mechanism of their enzymatic degradation. Experientia 6:201–209

Cohen SN, Chang ACY, Boyer HW, Helling RB (1973). Construction of biologically functional bacterial plasmids in vitro. Proc Natl Acad Sci USA 70:3240–3244

Cooper GM, Okenquist S, Silverman L (1980) Transforming activity of DNA of chemically transformed and normal cells. Nature (London) 284:418–421

Crick F (1968) The origin of the genetic code. J Mol Biol 38:367–379

Crick F, Watson J (1956) Structure of small viruses. Nature (London) 177:473–475

de-The G et al (1978) Epidemiological evidence for causal relationship between Epstein-Barr virus and Burkitt's lymphoma from Ugandan prospective study. Nature (London) 274:756–761

Dulbecco R, Vogt M, Strickland AGR (1956) A study of the basic aspects of neutralization of two animal viruses, western equine encephalitis virus and poliomyelitis virus. Virology 2:162–205

Edgar RS, Wood WB (1966) Morphogenesis of bacteriophage T4 in extracts of mutant-infected cells. Proc Natl Acad Sci USA 55:498–505

Epstein MA, Achong BG (1973) Various forms of Epstein-Barr virus infection in man: established facts and a general concept. Lancet II 836–839

Evans DMA (1985) Increased neurovirulence associated with a single nucleotide change in a noncoding region of the Sabin type 3 poliovaccine genome. Nature (London) 314:548–550

Fazekas de St.Groth S (1948) Regeneration of virus receptors. Nature (London) 162:294–295

Fiers W, Contreras R, Duerinck F, Hageman G, Inserentant D, Merregaert J, Min Jou W, Molemans F, Raemoekers A, Van den Berghe A, Volkaert G, Ysebaert M (1976) Complete nucleotide sequence of bacteriophage MS2 RNA: primary and secondarystructure of the replicase gene. Nature (London) 260:500–507

Franchini G et al (1987) Sequence of simian immunodeficiency virus and its relationship to the human immunodeficiency viruses. Nature (London) 328:539–547

Gallo RC et al (1983) Association of the human type C retrovirus with a subset of adult T-cell cancers. Cancer Res 43:3892–3899
Gillespie D, Spiegelman S (1965) A quantitative assay for DNA-RNA hybrids with DNA immobilized on a membrane. J Mol Biol 12:829–842
Goebel WF (1939) Immunity to experimental pneumococcus infection with an artificial antigen. Nature (London) 143:77–78
Groß HJ et al (1978) Nucleotide sequence and secondary structure of potato spindle tuber viroid. Nature (London) 273:203–208
Habel K (1965) Specific complement-fixing antigens in polyoma tumors and transformed cells. Virology 25:55–61
Hahn BH et al (1984) Molecular cloning and characterization of the HTLV-III virus assiciated with AIDS. Nature (London) 312:166–169
Hampar B, Ellison SA (1961) Chromosomal aberrations induced by an animal virus. Nature (London) 192:145–147
Harris H, Watkins JF (1965) Hybrid cells derived from mouse and man: artificial heterokaryons of mammalian cells from different species. Nature (London) 205:640–646
Henle W (1972) Role of Epstein-Barr virus infectious mononucleosis and malignant lymphomas in man. Fed Proc 31:1674
Huebner R, Todaro GJ (1969) Oncogenes of RNA tumor viruses as determinants of cancer. Proc Natl Acad Sci USA 64:1087–1094
Huebner RJ, Rowe WP, Turner HC, Lane WT (1963) Specific adenovirus complement-fixing antigens in virus-free hamster and rat tumors. Proc Natl Acad Sci USA 50:379–389
Huse WD, Sastry L, Iverson SA, Kang AS, Alting-Mees M, Burton DR, Benkovic SJ, Lerner RA (1989) Generation of a large combinatorial library of the immunoglobulin repertoire in phage lambda. Science 246:1275–1281
Köhler G, Milstein C (1975) Continuous cultures of fused cells secreting antibody of predefined specificity. Nature (London) 256:495–497
Koltin Y, Day PR (1976) Suppression of the killer phenotype in Ustilago maydis. Genetics 82:629–637
Krugman S, Giles JP, Hammond J (1967) Infectious hepatitis. J Am Med Assoc 200:365–373
Küpper H (1981) Cloning of cDNA of major antigen of foot-and-mouth disease virus and expression in E.coli. Nature (London) 289:555–559
Lenette EH, Horsfall FL (1940) Studies on epidemic influenza virus. The nature and properties of the complement-fixing antigen. J exp Med 72:233–246
Manaker RA, Groupé V (1956) Discrete foci of altered chicken embryo cells associated with Rous sarcoma virus in tissue culture. Virology 2:838–840
Meselson M, Yuan R (1968) DNA restriction enzyme from E.coli. Nature (London) 217:1110–1114
Muesing MA et al (1985) Nucleic acid structure and expression of the human AIDS/lymphadenopathy retrovirus. Nature (London) 313:450–458
Murant AF, Mayo MA (1982) Satellites of plant viruses. Ann Rev Phytopathol 20:49–70
Murray NE, Murray K (1974) Manipulation of restrictiontargets in phage λ to form receptor chromosomes for DNA fragments. Nature (London) 251:476–481

Nirenberg MW, Mathhaei JH (1961) The dependence of cell-free protein synthesis in E.coli upon naturally occurring or synthetic polyribonucleotides. Proc Natl Acad Sci USA 47:1588–1602

Nonoyama M, Pagano JS (1973) Homology between Epstein-Barr virus DNA and viral DNA from Burkitt's lymphoma and nasopharyngeal carcinoma determined by DNA-DNA reassociation kinetics. Nature (London) 242:44–47

Norkin LC (1982) Papovaviral persistent infections. Microbiol Rev 46:384–425

Noyes WF (1965) Studies on the human wart virus. II. Changes in primary human cell cultures. Virology 25:358–363

Plummer G, Hollingsworth DC, Plungseb A (1970) Chronic infections by herpes simplex virus and by the horse and cat herpesviruses. Infect Immun 1:351–355

Reddy VB, Thimmappaya B, Dhar R, Subramanian KN, Zain BS, Pan J, Ghosh PK, Celma ML, Weissman SM (1978) The genome of simian virus 40. Science 200:494–502

Rubin H, Franklin RM (1957) On the mechanism of Newcastle disease virus neutralization by immune serum. Virology 3:84–95

Schwartz DE, Tizard R, Gilbert W (1983) Nucleotide sequence of Rous sarcoma virus. Cell 32:853–869

Scolnick EM, Rands E, Williams D, Parks WP (1973) Studies on the nucleic acid sequences of Kirsten sarcoma virus: a model for formation of a mammalian RNA-containing sarcoma virus. J Virol 12:458–463

Seiki M, Hattori S, Hirayama Y, Yoshida M (1983) Human adult T-cell leukemia virus: complete nucleotide sequence of the provirus genome integrated in leukemia cell DNA. Proc Natl Acad Sci USA 80:3618–3622

Stehelin D, Varmus HE, Bishop, Vogt PK (1976) DNA related to the transforming gene(s) of Avian sarcoma viruses is present in normal avian DNA.Nature 260:170–173

Stehelin D, Guntaka RV, Varmus HE, Bishop JM (1976) Purification of DNA complementary to nucleotide sequences required for neoplastic transformation of fibroblasts by Avian sarcoma viruses. J Mol Biol 101:349–365

Steinschneider A, Fraenkel-Conrat H (1966) Studies of nucleotide sequences in tobacco mosaic virus ribonucleic acid. Biochemistry 5:2729–2734, 2735–2743

Stent G (1968) That was the molecular biology that was. Science 160: 390–395

Temin HM (1962) Separation of morphological conversion and virus production in Rous sarcoma virus infection. Cold Spring Harbor Symp Quant Biol 27:407–414

Temin HM (1963) Further evidence for a converted, non-producing state of Rous sarcoma virus-infected cells. Virology 20:235–245

Temin HM (1964) The participation of DNA in Sarcoma virus production. Virology 23:486–494

Temin HM (1964) Homology between RNA from Rous sarcoma virus and DNA from Rous sarcoma virus-infected cells. Proc Natl Acad Sci USA 52:323–329

Temin HM (1976) The DNA provirus hypothesis. Science 192:1075–1080

Temin HM, Mizutani S (1970) DNA-dependent A polymerase in virion of Rous sarcoma virus. Nature (London) 226:1211–1213

Thomas M (1974) Viable molecular hybrids of bacteriophage λ and eukaryotic DNA. Proc Natl Acad Sci USA 71:4579–4583

Thompson RL (1947) The effect of metabolites, metabolite antagonists and enzyme-inhibitors on the growth of the vaccinia virus in Maitland type of tissue culture. J Immunol 55:345–352

Uchiyama T, Yodoi J, Sagawa K, Takatsuki K, Uchino H (1977) Adult T-cell leukemia: clinical and hematologic features of 16 cases. Blood 50:481–492

Volkin E, Astrachan L (1956) Intracellular distribution of labeled ribonucleic acid after phage infection of Escherichia coli. Virology 2:433–437

Wickner RB (1986) Double-stranded RNA replication in yeast: the killer system. Ann Rev Biochem 55:373–395

Williams RC, Smith KM (1958) The polyhedral form of the Tipula iridescent virus. Biochim Biophys Acta 28:464–469

Wolf K, Snieszko SF, Dunbar CE, Pyle E (1960) Virus nature of infectious pancreatic necrosis in trout. Proc Soc Exp Biol Med 104:105–108

Wood HA (1973) Viruses with double-stranded RNA genomes. J gen Virol 20:61–85

Yarchoan R et al (1986) Administration of 3′-acido-3′-deoxythymidine, an inhibitor of HTLV-III/LAV replication, to patients with AIDS or DS-related complex. Lancet I:575–580

zur Hausen H, Schulte-Holthausen H, Wolf H, Dörries K, Egger H (1974) Attempts to detect virus-specific DNA in human tumors. II. Nucleic acid hybridizations with complementary RNA of human herpes group viruses. Int J Cancer 13:657–664

Name Index

Abbe, Ernst 28
Abercrombie, Michael 231
Abraham, Gordon 199
Achong, BG 208, 240
Ackermann, R 270
Ackermann, W Wilbur 220, 280, 281, 282
Adams, Jerry M 105
Adams, Mark H 183, 184
Adil-Bey 42
Adler, SP 168
Agramonte y Simoni, Aristide 41
Ahne, Winfried 219
Aizawa, K 128
Akiyama, Y 232, 240
Albrecht, Thomas B 232
Alexander, Hattie E 188, 206
Alexander, RA 64
Allard, HA 58
Allison, Anthony Clifford 198, 201
Almeida, June D 261
Aloni, Yosef 234
Alper, Tikvah 228
Alter, Harvey J 215, 216
Alwine, JC 256
Alzheimer, Alois 226
Ambronn, Hermann 91
Amies, CR 75
Amoss, Harold L 73
Anderer, Friedrich Alfred 182, 266
Anderson, HK 249
Anderson, JF 105
Anderson, Katherine 115
Anderson, Stewart G 118, 245
Anderson, Thomas Foxen 135, 136, 144, 145
Andervont, Howard B 54, 113
Andrewes, Christover H, Sir 69, 73, 77, 88, 105, 114, 158, 185, 288
Andry, 10
Angerer, Karl von 53, 55
Anson, Mortimer L 59
Aoki, Kaoru 75, 85, 91
Aposhian, H Vasken 184, 204
Appelmans, R 147
Appert, Nicolas Francois 9
Arber, Werner 145, 148, 168
Ardenne, Manfred von 136
Aristotl 4
Armbruster, Otto 185
Armstrong, Charles H 105, 112
Ashkenazi, A 221
Askonas, BA 252
Astbury, William Thomas 95, 164, 172
Astrachan, Lazarus 166
Atchison, Robert W 187
Audouin, Jean Victor 11, 13
August, J Thomas 184
Aula, Pertti 272
Avery, Oswald Theodore 94, 133, 138
Aycock, William Lloyd 73

Babes, Victor 70, 71, 81
Babott, FL Jr 253
Bachmann, Wilhelm Eduard 50
Bachrach, Howard L 143
Bachthold, John G 198
Bader, John P 234
Baer, Karl Ernst von 13
Bailey, WT 149, 157
Baker, R 135
Balch, RE 128
Balsamo-Crivelli, Riccardo 11

Baltimore, David 168, 183, 184, 188, 207, 210
Baluda, Marcel A 232, 234
Banerjee, Sunanda 202
Bang, Oluf 87
Bang, Frederik B 220
Barlow, James L 139
Barnard, Joseph Edwin 56
Barr, M 119
Barr, YM 240
Barré-Senoussi, F 243, 244
Barrington, LF 147
Bary, Heinrich Anton de 15
Barry, Richard D 233
Basilico, Claudio 272
Baskin, JL 112
Bassi, Agostino Maria 10, 11, 15
Bastian, Henry Charlton 17, 18
Bastian, FO 222
Bather, R 233
Bauer, D John 280
Bauer, Heinz 186
Bauer, WR 205
Baur, Erwin 44
Bawden, Frederick Charles 60, 133, 158
Beale, Helen 73, 75, 279
Beard, Joseph W 139
Beaurepaire Aragão, Henrique de 76, 82
Bechhold, Heinrich 49, 53, 55, 57
Becker, A 184
Becker, Yechiel 205, 277
Bedson, SP 75, 119
Behring, Emil von 26
Beijerinck, Martinus W 35, 36, 37 45, 58, 90, 288
Beintker, E 76
Beiß, U 185
Bellett, Alan JD 288
Bendinger, 62
Benjamin, B 69
Benjamin, Thomas L 234, 236
Bennet, 17
Benyesh-Melnick, Matilda 262, 272
Bercks, R 256
Berg, FT 15
Berget, Susan M 166
Bergold, Gernot H 127, 129, 136, 213
Berkefeld, Wilhelm 43
Bernal, John Desmond 60, 142, 160, 173, 177
Bernhard, Wilhelm 229
Bernkopf, H 102
Berrks, R 256
Berry, George P 69, 96, 230
Berson, Salomon A 256
Bertani, G 168
Berzelius, Jöns Jakob von 11, 12, 28
Best, Rupert J 180
Bevan, EA 220
Bielka, Heinz 114
Biggs, PM 239
Billroth, Christian Albert Theodor 20
Billitzer, J 55
Biquard, JM 236
Bird, FT 128, 129
Bishop, RF 217
Bittle, James L 261, 267
Bittner, John Joseph 113
Black, Francis L 253
Black, L 131
Black, Paul H 232
Blair, Phillis B 181
Bland, JOW 53
Blanden, Robert V 254
Blix, Gunnar Fritiof 141
Bloomer, AC 210
Blumberg, Baruch S 215
Bodey, GP 283
Bodian, David 110
Boez, Louis J 55
Bohr, Niels Henrik 94
Bolle, J 61
Bollinger, Otto von 37
Bolognesi, Daniel P 186
Bommer, Sigwald 98
Boncquet, PA 131
Bordet, Jules Jean 29, 74, 76, 147
Borel, Pierre 7
Borrel, Amédée 42, 45, 61, 62
Borries, Bodo von 98
Borsa, J 183
Borst, P 184
Bosc, FJ 61
Bouteille, M 224
Bozarth, R 187
Brachet, Jean Louis 138, 161

Bragg, William 174
Brandes, Jürgen 136
Bréchot, Christian 241
Brefeld, Oskar 28
Breindl, Václav 127, 133
Brenner, Sidney 165, 166, 173, 197
Brodie, Maurice 99, 106
Brody, B 141
Bronfenbrenner, JJ 53, 91
Brown, Eric H 272
Brown, JAH 245
Brown, Mckay 183
Brown, Gordon C 270, 280, 281
Brown, Robert 13
Brugge, JS 168
Bruton, OC 245
Bruynoghe, R147
Brzosko, WJ 216
Buffon, Compte Georges, Louis Leclerc, de 8, 9
Bull, DR 156
Burgess, WL 75
Burke, JM 167
Burkitt, Denis 240
Burnet, F Macfarlane Sir 69, 74, 75, 79, 91, 96, 100, 102, 105, 121, 134, 141, 147, 148, 153, 156, 157, 158, 160, 162, 170, 185, 245, 253, 288
Burrell, JC 267
Burril, Thomas J 23
Burrows, Edwin 66
Burton, K 152
Butenandt, Adolf 134
Buthala, DA 281
Butler, P Jonathan 210

Cagniard de La Tour, Charles 11, 12, 16
Caldwell, John 59
Calmette, Albert Leon Charles 101
Calnek, BW 222
Cameron, Gladys 231
Campbell, E 112
Cardan, Geronimo 7
Carrée, H 54, 72
Carroll, James 41
Carrel, Alexis 66, 69, 88, 288
Carsner, E 154
Carus, Lucretius 4
Carusi, EA 182
Casagrandi, Oddo 64
Casals, Jordi 116, 121
Casazza, AR 249
Caspar, Donald L 142, 173, 174, 175, 177, 209
Caspersson, Torbjörn 137, 138, 161
Cavanagh, JB 224
Cech, Thomas R 167
Celsus, Aulus Cornelius 4, 5
Centanni, Eugenio 40 41 42, 70, 71, 78
Centifianto-Fitzgerald, Ysoline M 275, 277
Chamberland, Charles-Edouard 18, 25 43, 80
Chambers, Leslie A 116
Chambers, Velma C 103, 220
Chang, TH 272
Chardonnet, Yvette203
Chargaff, Erwin 164, 165
Charrin, Albert 26
Chase, Martha 138, 145
Chase, Merill W 245
Chauveau, JBA 62
Chelle, PL 223
Chen, Sheng-Tu 228
Chester, Kenneth S 73
Choppin, Purnell W 200
Chow, Louise T 166
Churchill, AE 239
Ciuca, A 74, 76
Ciuca, M 147
Cieciura, Steven J 220
Clavel, Francois 244
Clark, MC 228
Coen, Donald M 277
Cohen, Seymour Stanley 139, 141, 152, 184, 279
Cohen, Stanley N 164, 204, 207
Cohn, Ferdinand Julius 17, 18, 20, 23, 27
Cole, FE 221
Colonno, Richard J 199
Colter, John S 161, 206
Commoner, Barry 279
Compans, Richard W 202
Confer, AW 224
Conolly, J 224

Conseil, E 65
Convent, A 72
Cook, Margary L 222
Coons, Albert Hewett 98, 111, 256, 257
Cooper, Geoffrey M 271
Corey, Robert B 98, 142
Cork, James M 174
Cornalia, E 61
Correns, Carl 94
Cox, Herald R 158, 261
Craigie, James 53, 117
Crawford, EM 187
Crawford, Lionel V 187, 201, 235, 236
Cremer, NE 222, 225
Creutzfeld, Hans Gerhard 226
Crick, Francis Harry Compton 160, 161, 164, 165, 166, 167, 173, 209
Cuillé, J 223
Curasson, G 105

Dales, Samuel 200, 202, 203
Dalldorf, Gilbert 105
Dalrymple, Joel M 186
Dane, DS 216
Daniel, MD 244
Daniels, Joan B 224
Danner, Kurt 222
Darai, Gholamreza 232, 277
Darnell, James E Jr 206, 207, 288
Davaine, Casimir Joseph 19, 20, 23
Davenport, Fred M 263
Davidson, Norman 187
Davies, WL 281
Davis, Larry E 222
Davies, Patricia 201
Davies, Sheila P 182
Dawson, IM 135, 136, 144
Dawson, JR 105
Dedrick, Helen M 96, 230
Deidier, Antoine 14
Deinhardt, Friedrich 215, 220
Delay, PD 270
Delbrück, Max 94, 95, 148, 149, 157, 164, 288
Demerec, Milislav 94
Den Dooren de Jong, EL 148
Desjardins, Paul R 228
DeTray, DE 222
Deutsch, HF 244
DeVries, Hugo 94
D'Herelle, Felix-Hubert 48, 76, 91, 96, 101, 146, 147, 288
Democritus 4
Diel, V 241
Diener, Theodor O 167, 181, 227, 228
Dinter, Z 105
Dochez, Alphonse Raymond 69
Doermann, AH 149, 150
Doerr, Robert 54, 96, 288
Dolin, Raphael 217
Domagk, Gerhard 278
Dougherty, RM 235
Douglas, SR 55
Dourmashkin, RR 202
Downie, Allan Watt 69
Drescher, Joachim 260
Dryer, I 112
Duclaux, Pierre Émile 19
Duesberg, Peter H 168, 181, 182, 186, 234, 236
Duff, Ronald G 232
Dujardin, Felix 9
Dulbecco, Renato 104, 116, 149, 232, 253
Dumanski, Anton Vlamidirovic 52
Dumbell, KR 268
Dun, Finley 37
Duran-Reynals, Francisco 113
Durham, ACH 210
Durham, Herbert E 26, 28
Dusch, Theodor von 17
Dussoix, Daisy 168

Earle, Wilton R 104, 230
Eckert, Edward A 141
Eddy, Bernice Elanine 114, 231, 232
Edgar, RS 204, 209
Edney, M 157
Eggers, Hans J 207, 281
Ehrenberg, Christian Gottfried 9, 12
Ehrlich, Paul 26, 28, 76, 92, 170, 245
Eichenwald, HF 270
Eichstedt, Karl F von 15
Eklund, Curtis 183, 184
Elford, William Joseph 50, 88, 135, 136, 288

Elion, Gertrude B 284
Ellermann, Vilhelm 87
Ellis, Emory 148, 149
Ellison, Solon A 272
Emini, Emilio A 267
Entage, John Spencer 267
Enders, John Franklin 102, 103, 105, 116, 232, 261, 289
Engvall, Eva 256
Epicurus 4
Epstein, Michael Anthony 208, 232, 233, 240, 261
Epstein, Richard H 197
Erikson, Raymond L 168
Erikson, Jan 271
Erlich, Henry 170
Esche, H 182
Escherich, K 85
Espejo, T 187
Esposito, Joseph J 268
Essex, Max 243
Estes, Mary K 182
Euler, Hans von 119
Evans, Charles A 103
Evans, DMA 276
Everitt, Einar 182
Ewing, James 62

Faber, James 8
Fabricius de Aquapendente, Hyeronimus 7
Fagraeus, Astrid E 244
Fairbrother, RW 120, 155
Fallopio, Gabriel 7
Fankuchen, Isidore 60, 142, 160, 173, 177
Farnham, Ann E 198
Fastier, Lyle B 112
Faulkner, GH 69
Fazekas de St.Groth, S 199, 200
Feix, Günter 188
Fekadu, Makommen 222
Fenner, Frank 170, 245, 255, 268
Fermi, Claudio 81
Fernandes, MV 221
Ferris, RD 105
Ferry, JD 69
Field, Anne Margaret 225
Field, Hugh F 276
Fields, Bernhard N 275
Fiers, Walter F 169, 187, 205
Finch, John Thomas 177, 210
Finkelstein, Martin S 247
Findlay, G Marshall 55, 69
Fischer, E 62
Fischer, H 151
Fisher, Amanda G 276
Fisher, H 104
Flaks, Joel G 184
Fleming, Alexander Sir 278
Flewett, TH 217
Flexner, Simon 65, 66, 69, 73, 103, 289
Flodin, P 256
Flügel, Rolf 208
Foley, GE 240
Foley, EJ 235
Fowler, Catherine B 279
Fracastoro, Girolamo 6, 37
Fraenkel-Conrat, Heinz 138, 142, 161, 169, 209
Francis, Thomas Jr 116, 117, 120, 121, 123, 124, 155, 156, 252
Franklin, Rosalind 142, 164, 177, 185
Franklin, Richard M 207, 253
Fraser, Dean 136, 140, 144, 151
Freund, Jules T 121
Freundlich, Herbert Max 91
Frey, S 91
Freyer, M 75
Friedberger, Ernst 92
Friedenthal, H 52
Friedewald, WF 121, 252
Friedrich, Walter 98
Friend, Charlotte 114
Frobisher, M Jr 76
Frosch, Paul 38 41 45 47, 64, 72, 82, 90, 289
Fujii-Kawata, I 187
Fujinami, A 87
Fukushi, T 131
Fultz, PN 244
Furth, Jacob 88

Gajdusek, Daniel Carleton 226
Galindo, Annibal 234
Galindo, Jorge A 228
Galenus, Claudius 5, 86

Galibert, F 181
Gallo, Robert C 243, 244
Galloway, IA 69
Galtier, Pierre Victor 80
Gamov, George 165
Garcia, A 127
Gard, Sven 143, 263
Gardner, Sylvia D 225
Garen, Alan 165
Gautheret, RJ 67
Garros, F 43
Gay-Lussac, Joseph-Louis 9, 11, 12
Gear, James HS 112
Geder, Laszlo 232
Gengou, Octave 29, 76
Gerloff, RK 256
Gesteland, Raymond F 201
Gething, Mary-Jane 268
Gey, George Otto 66, 104
Gibbs, Clarence J Jr 226
Gierer, Alfred 138, 161
Gildemeister, Eugen 72, 73, 147
Gillespie, David H 165
Ginsberg, Harold S 220
Girard, Marc 188
Glade, PR 241
Glenny, A 119
Glusmann, MP 73
Godeau, A 241
Goebel, Walther Frederick 153, 266
Goetz, IE 232
Gold, E 54
Gold, M 184
Gold, P 183
Goldberger, J 105
Goldblatt, Harry 231
Goldschmidt, Richard 67
Goldstein, B 62
Gollub, H 57
Gomatos, Peter John 181, 233
Gönnert, Rudolf 112
Good, Robert A 117, 245
Goodpasture, Ernest William 56, 57, 62, 67, 68, 69, 85, 105, 222
Goodsir, John 15
Gordon, MH 76
Gordon, JE 253
Gotlieb, Tamar 157
Gottlieb, Michael S 243
Gottschalk, Alfred 141
Graaf, Regnier de 8
Grabar, Pierre 256
Grabowski, Paula J 167
Grace, JT Jr 240
Grafe, Alfred 260
Graffi, Arnold 114
Graham, Angus F 183
Graham, DM 200
Granados, Robert R 202
Granata, S 65, 105
Green, RG 96, 103, 287
Green, Maurice 183
Greenfield, C 112
Greenman, RL 209
Gregg, Norman McAlister, Sir 112, 270
Greve, JM 200
Griffith, Frederick 91, 96, 133, 230
Griner, LA 270
Gross, Ludwik 113, 114, 239
Groß, Hans J 227
Groupé, Vincent 231, 232
Gruber, J 40 41
Gruber, Max von 26, 28
Gruby, David 12, 15
Grunert, S 222
Gruss, Peter 268
Grüter, Wilhelm 64, 105
Guarnieri, Guiseppe 34, 61, 64, 105
Guérin, Camille 101
Guérin, M 229
Gutmann, A 153

Haagen, Nils Eugen 68, 69, 289
Haas, Richard 105
Haase, Ashley T 224
Habel, Karl 235
Hagemann, Paul K 98
Hahn, Beatrice H 243, 254
Haitinger, Max 98,
Halberstaedter, Ludwig 230
Hall, Benjamin D 165
Hallauer, Curt 68, 69, 105
Hallis, WD 215
Halvorson, Harlyn O 282
Hamilton, John 245
Hampar, Berge 272
Hamre, D 280

Hanafusa, Hidesaburo 188, 202, 236
Hanafusa, Teruko 232
Hanshaw, James Barry 270
Hare, Ronald 100
Haring, CM 76
Harris, J Ieuan 171, 173
Harris, Kerry F 214
Harrison, Ross Granville 65
Harrison, W 103
Hartig, Robert 45
Harvey, William A 7
Haseltine, William 208,
Hashimi, A 270
Hausmann, Rudolf 184
Havens, W Paul Jr 118
Hayashi, Masaki 181
Hayflick, Leonhard 215
Hayward, William S 271
Haywood, Anne 267
Heaysman, JEM 231
Hebra, Ferdinand von 14
Hecke, Friedrich 69
Heilmann, FR 278
Heine, Jochen W 202
Heinmetz, F 135
Helm, K von der 183
Hempt, A 81
Henderson, W 34, 61
Henle, Gertrude 105, 112, 121, 162, 221, 240, 241, 250
Henle, Jakob 12, 13, 14
Henle, Werner 112, 116, 121, 157, 162, 197, 240, 250, 252, 289
Henriot, E 53
Henseval, M 72
Hercik, Ferdinand 151
Herrmann, Ernest C Jr 281
Herrell, WE 279
Herriott, Rouger M 59, 139
Hershey, Alfred Day 138, 145, 148, 152, 187, 188
Herzberg, Kurt 57, 68, 69, 72, 73, 101, 289
Hess, Alfred F 82
Hesse, Fanny Angelika 22
Hesse, Walther 43
Hetler, DM 53
Hetrich, FM 221
Higashi, Noboru 200
Higbie, E 69
Hilleman, Maurice Ralph 116, 215, 264
Hindle, Edward 55, 82
Hinze, HC 221
Hippocrates of Kos, 3 4, 5
Hirst, George Keble 100, 140, 157, 181, 183, 198, 252
Hoagland, CL 139, 141
Hogan, Thomas F 225
Hoggan, M David 202
Högyes, Andreas 81
Holden, Henry Francis 107
Holland, John J 199
Hollinger, F Blaine 216
Hollings, M 228
Hollinshead, AC 281
Holmes, Francis O 101, 155
Holmes, Oliver Wendell 15
Holt, SJ 233
Holterman, OA 272
Home, Francis 14
Honess, Robert W 182
Horne, Robert W 173, 174
Horsfall, Frank Lappin 120
Horst, Jürgen 236, 256
Horst, R Kenneth 228
Horstmann, Dorothy Millicent 110, 185, 270
Horwitz, Jack P 284
Hoskins, Meredith 114
Hotchin, John E 222
Hotchkiss, Rollin D 164, 165
Howe, Howard A 123, 253
Howell, Stephen H 268
Howitt, Beatrice F 69, 72
Hoyle, Leslie 120, 155, 182, 201, 211
Hoyt, Anson 105
Huang, Alice S 186
Huang, CH 116
Huebner, Robert Joseph 168, 235, 237
Hueppe, Ferdinand 33
Huffman, John H 283
Hughes, Thomas PA 75, 117
Huguenard, E 53
Hull, Roger N 113
Hultin, JV 253
Human, SL 168

Humboldt, Alexander von 9
Huse, William D 171
Hutchison, Clyde Allen 182

Imagawa, David T 105
Imperial, JS 227
Inamoto, K 87
Innes, RJ 224
Isaacs, Alick 157, 250, 289
Issaeff, Vasily I 26
Ishida, Nakao 221
Ishimori, N 61, 129
Itakawa, S 240
Ivanovski, Dimitri Josifovich 35, 36, 37 45, 61, 289

Jablot, Louis 8
Jackson, GG 281
Jacob, Francois 166
Jacobson, M 210
Jakob, Alphons 226
Jamuni, Abhaya 107
Janssen, L 1, 133
Jenner, Edward 19, 70, 79
Jensen, EM 240
Jensen, Keith E 116
Jerne, Niels K 170, 249
Jírovec, O 133
Jobling, James Wesley 76
Johnson, James 135
Johnson, John E 179
Joho, 236
Joklik, Wolfgang K 181, 186, 205
Jones, CR 279
Jones, FS 66, 104
Josse, John 169
Joubert, Jules 18 43
Jungeblut, Claus W 105

Kaesberg, Paul J 136
Kalyanaraman, VS 243, 244
Kapikian, Albert Z 217
Kaplan, Paul M 208
Kaplan, MH 98, 256, 257
Kärber, G 117
Karstad, L 227
Kassanis, Basil 186, 221
Kates, M 183
Katekar, GF 180
Kato, N 281
Katz, Dorothy D 261
Kaufman, Herbert E 281
Kausche, Gustav Adolf 99, 135
Kawai, Sadaaki 236
Kawase, S 181
Keck, Konrad 184
Kellenberger, Eduard 145, 148, 151
Kelly, Thomas J Jr 168
Kelson, Saul R 278
Kew, Michael C 241
Khatchikian, Detlev 276
Khorana, Har Gobind 165
Kidd, John G 102
Kikuth, Walter 57, 112
Kilborne, FL 27
Kilham, Lawrence 105, 231
Kingsbury, David W 207
Kircher, Athanasius 7, 8, 10
Kit, Saul 184
Kitamura, Naomi 206
Kitasato, Shibasaburo, Baron 25, 26 43
Klatzo, I 226
Klebs, Theodor Albrecht Edwin 20, 26, 28 43
Klein, Anneliese de 271
Klein, George 235
Klement, V 232
Klenk, Ernst 141, 199
Klenk, Hans-Dieter 182, 186
Klug, Aaaron 174, 177, 209, 210
Knapp, E 137
Knight, Claude Arthur 140, 141, 154, 169, 173
Knight, Janice 268
Knipping, Paul 98
Koch, Karl 55
Koch, Meinrad A 234
Koch, Robert 20, 21, 22, 23, 25, 31, 32, 33, 38 47, 63, 81, 93
Koegh, EV 230
Köhler, George 170, 171, 256
Kofoid, Charles Atwood 76
Kölliker, Albert von 12, 13
Kolle, Wilhelm 71, 78, 82
Kolmer, John Albert 99, 158
Komarek, J 127
Komrover, GM 270

Koprowski, Hilary 123, 158, 232, 261
Kopytowski, W 61
Korn, David 184
Kornberg, Arthur 165, 184, 204
Kotte, Walter 67
Koziorowska, Jadwiga 232
Kozloff, Lloyd M 147, 152
Kramer, SD 73
Kraus, Rudolf 26, 29, 54, 76
Krieg, Aloysius 212
Krontiris, Theodore G 199
Kruger, Kelly 167
Krugman, Saul 215, 264
Küpper, Hans 267
Kützing, Friedrich Traugott 12, 16
Kuhn, P 62
Kunert, H 105
Kunitz, Moses 59
Kunkel, Louis Otto 58, 131, 155
Kurstak, Edouard 187
Kurtz, Hilda 220
Kurvink, K 272
Kutsky, RJ 279

Labarraque, AG 15
Lacassagne, A 113
Lai, Michael MC 236
Lai, Ching-Juh 268
Laidlaw, Patrick P, Sir 96, 287
Lancisi, Giovanni-Maria 10
Landsteiner, Karl 65, 72, 105, 245, 289
Landtman, Bernhard 112
Lange, PC 263
Langbeheim, Harry 267
Lanthrop, A 113
Lanz, E 272
Larsen, Austin E 227
Lauda, E 62
Laue, Max von 98
Laveran, Charles Louis Alphonse 34
Lavoisier, Antoine Laurent 11
Lawman, Michael JP 261
Lawson, RH 228
Lazear, Jesse Williams 41
Lebon, P 225
Leclerc (Count de Buffon), 8
Lederberg, Joshua 170
Ledingham, John Charles Grant, Sir 53, 75
Ledinko, Nada 105
Lee, Chong Sung 187
Leeuwenhoek, Antony van 1, 8
Lehmann, H 98
Leidy, Joseph 23
Lempfried, Heinrich 199
Lepp, M 70, 71
Levaditi, Constantin 54, 65, 66, 72
Levinthal, Cyrus 151
Levy, Jay A 240
Lewis, Margaret Reed 54, 66
Lewis, Paul A 65
Lewis, Warren Harmon 66
L'Heritier, P 130
Liebig, Justus von 11, 12
Lillie, R 112
Lind, Patricia 157
Lindenmann, Jean 250
Linné, Karl von 10
Linsbauer, L 98
Lipmann, Fritz Albert 159
Lipton, HL 222, 224
Lipschütz, Benjamin 48, 57, 62
Lister, Joseph, Lord 19, 28
Little, SP 186
Littau, Virginia C 131
Liu, Oscar C 157
Lloyd, W 125
Lode, Alois 40 41,
Loeb, Leo 113
Loeb, Tim 206
Löffler, Friedrich August 33, 38, 39 41 45 47, 64, 72, 82, 90, 118, 289
Loewe, L 73
Lonberg-Holm, K Karl 199
London, William Thomas 241
Long, PH 56
Loring, Hubert S 137
Lotmar, R 129
Lucké, Balduin 88
Luger, Alfred 62
Luria, Salvador Edward 136, 144, 148, 150, 155, 164, 168, 288
Lush, Dora 96, 147, 185
Lustgarten, Sigmund 57,
Lwoff, André M 148, 153, 158, 159, 171, 262

Lwoff, Marguerite 262
Lynfield, YL 249
Lyon, G 225

Maaløe, Ole 150, 152
Maas, D 183
Maassab, Hunein F 282
MacCallum, AF 215
MacCallum, William George 52
MacGregor, DR 129
MacKay, Patricia 267
Mackenzie, RD 69
Mackett, Michael 268
MacNamara, Joan, Dame 73, 79
MacPherson, Ian 233
Madsen, J 112
Maestri, A 61
Magill, TP 116, 120
Maggiora, Arnaldo 40
Magnus, Herdis von 106
Magnus, Preben von 100, 106, 161
Maisin, Joseph Henry 54
Maitland, Hugh Bethune 67
Maitland, MC 67
Maizel, Jacob V Jr 182, 210
Malfitano, G 49
Malmsten, Peter Henrik 15
Maloney, ED 281
Manaker, Robert A 231, 232
Manson, Patrick, Sir 34
Mandel, Benjamin 247, 253
Maramorosch, Karl 131
Marchal, J 62, 72
Marcialies, MA 277
Marie, P 78
Marion, PL 216
Markham, Roy 138, 173
Martin, GS 168, 236
Martin, AE 120
Martin, Charles JA, Sir 49
Martin, Louis 39
Marx, E 42
Masur, Henry 243
Matthaei, Johann Heinrich 165
Matthews, Richard Ellis Ford 279, 287
Maurer, Bernd 208
Mayer, Adolf 27, 35, 36 45, 289
Maynard, James E 216
Mayr, Anton 222
McAleer, William J 267
McAuslan, Brian R 183
McBride, William David 232
McKee, Albert P 253
McKinney, Harold H 101, 154
McClelland, L 100
McCormick, M 143
McFarland, VW 181
McFarlane, AS 136, 144
McGaughey, CA 69
McKie, M 148, 153
McLaren, Leroy C 199
McLorinan, H 118
Measroch, V 112
Melnick, Joseph L 124, 135, 158, 221, 261, 262
Mellors, FC 239
Mendel, Johann Gregor 94
Mendelson, Cathy L 200, 275
Mendelson, William 54
Menezes, Josef 199
Mercer, Frank L 279, 280,
Merill, MH 73
Merselis, JG 249
Meselson, Matthew Stanley 168
Metchnikoff, Ilja 25, 39
M'Fadyean, John, Sir 40
Michelson-Fiske, Susan 202
Miles, Charles Philipp 272
Miller, Dorothy A 199, 275
Miller, E 153
Miller, HC 112
Mills, RP 241
Milstein, César 170, 171, 256
Mishra, MD 280
Mitchel, AG 78
Mitsuya, Hiroaki 284
Mizutani, Satoshi 168
Miyajima, M 85
Molliard, M 67
Monif, GR 270
Mommaerts, E 141
Monod, Jaques Lucien 166, 289
Mooney, John 261
Moore, GE 240
Moorehead, Paul Sidney 215
Montagnier, Luc 181, 188
Mora, Peter T 181

Morein, B 261
Morgan, Councilman E 135, 211
Morgan, Doris Anne 242
Morgan, Isabel M 110, 253
Morgan, Thomas Hunt 94, 104
Moriarty, Ann M 268
Moses, A 76, 87
Moses, HL 241
Moskona, Aron 103
Moss, Bernhard 183, 268
Moyer, AW 105
Muckenfuß, RS 68
Muesing, Mark A 208
Mühlbock, O 113
Müller, Hans 137, 140
Müller, Johannes, Peter 12
Müller, Otto Frederik 9
Münch, H 117
Mulder, J 156
Muller, Hermann Joseph 133, 147, 154
Munk, Klaus 143, 197, 232
Munyon, W 183
Murphey-Corb, M 244
Murphy, HW 105
Murphy, James 67, 87, 91
Murray, Kenneth 170
Murray, Noreen E 170
Murray, WS 113
Mussgay, Manfred 201

Nägeli, Karl Wilhelm von 20, 24
Nagler, FP 100, 144
Nakajima, H 187
Nathans, Daniel 168
Nayak, DP 234
Naylor, Paul T 261
Nedrigailoff, 76
Needham, John Turberville 8, 9
Neefe, JR 118
Negri, Adelchi 61, 64
Nelson, JS 270
Netter, Arnold 76, 79
Nettleship, Anderson 230
Nevins, JR 205
Newman, S 135
Newton, Alison Anne 198
Nichols, JB 52
Nichols, Warren W 272
Nickels, AG 279
Nicolau, Sébastian S 54, 76
Nicolle, Charles Jules Henri 65
Nicolle, Maurice 42
Niederman, James C 241
Nieschulz, Otto 64
Nikkilä, E 116
Nirenberg, Marshall Warren 165
Nocard, Edmond Isidore 39 47
Noguchi, Hideyo Seisaka 66, 69, 289
Nomoto, Akio 276
Nordtmeier, H 43
Norrby, Erling Carl 263
North, JR 261
Northtrop, John Howard 59, 153
Noyes, Wilbur Fiske 105, 111, 232, 239
Nunn, Josua A 40
Nye, Robert N 69
Nygaard, AP 165

Obara, J 187
Okada, Yoshimi 202
Oker-Blom, Nils 116
Okochi, Kazuo 216
Oldenburg, Henry 8
Olitzky, Peter K 55, 56, 73
O'Neill, Frank J 272
Oppenheimer, Ella Hutzler 52
Orgel, Leslie E 167,
Ostertag, W 284
Ouchterlony, Örjan 116, 256
Oudin, Jaques 116
Owens, Robert A 228

Padgett, Billie L 225
Paillot, A 62, 127
Panicalli, Dennis 268
Paoletti, Enzo 183,268
Pape, J 64, 105
Pardee, Arthur Beck 152, 166, 184
Park, Bo Han 183, 184
Parker, Frederic Jr 69
Parker, Robert F 123
Parks, Wade P 237
Parsons, R 215
Paschen, Enrique 57, 62, 75
Pasteur, Louis 16, 17, 18, 19, 23, 25, 27, 31 43, 76, 79, 80, 154

Pauker, Kurt 250
Paul, John Redman 118
Pauling, Linus 164, 165
Pavan-Langstone, Deborah 283
Pease, D 135
Peebles, TC 105, 261
Peleg, J 221
Pelon, W 214
Penman, Sheldon 207
Perlmann, Peter 256
Peters, Dietrich 144
Peters, D 228
Petre, AW 58, 59, 93
Perty, Joseph A Maximilian 18, 20, 23
Pettit, Auguste 79, 83
Pfankuch, Edgar 99, 137, 160
Pfeiffer, Richard Friedrich 26
Piepenbrock, F 160
Pillemer, Louis 246
Piña, Magdalena 183
Pine, TS 280
Pirie, Norman Wingate 60, 133, 158
Plenĉiĉ, Marcus Anton von 10, 11
Plotch, Stephen J 275
Plowright, Walter 105
Plummer, G 222, 277
Pogo, Beatriz GT 183
Poiesz, Bernhard J 242, 244
Polasa, Hanamanlu 272
Pollender, Aloys 23
Polovtsev, VV 35
Pons, Marcel W 181
Ponsen, M 131
Pope, JH 241
Popovic, Miculas 244
Popper, Erwin 65, 105
Porath, Jerker 256
Porterfield, James S 105, 281
Porters, David D 227
Potts, Barbara J 222
Pouchet, FA 17
Price, Winston H 214
Pridham, JG 227
Priestley, Joseph 15
Prince, Alfred M 216, 233
Prival, J 242
Provost, P 215
Prowazek, Stanislaus von 48, 56, 57
Prusiner, Stanley B 182, 228
Prusoff, William H 281
Puck, Theodore T 104, 198
Puhalla, JE 220
Pukall, W 43
Pullen, von 220,
Pulvertaft, RJV 240
Purdy-Beale, Helen 73
Pyl, Gottfried 119,136

Quevenne, TA 12

Ragetli, HWJ 256
RajBhandary, Uttam L 167
Ralstone, DJ 183
Ramon, Gaston 119
Randles, John W 228
Rapp, Fred 232
Rappaport, Irving 132
Rasmussen, R 235
Rassmussen, AF 280
Ravenel, SF 78
Rawlins, TE 91, 279
Raychandhuri, SP 280
Raymer, WB 167, 181, 227
Reddy, V Bhaskar 205
Redi, Francesto 7
Reed, Lowell J 117
Reed, Walter 41, 290
Reedman, BM 235
Reich, Edward 206
Reissig, Magdalena 135
Reitz, Marvin S 242
Remak, Robert 12, 13
Remlinger, Paul Ambroise 43, 64, 81
Rennie, J 128
Retzius, Anders Adolf 15
Rhodes, Andrew J 123
Rich, Alexander 207
Richardson, C 184
Richter, A 207
Richter, R 222
Rinaldo, CR 249
Rinde, H 52,
Rivolta, Sebastiano 34, 61
Rivers, Thomas Milton 68, 69, 77, 93, 290
Rizetto, M 216

Robbins, Frederick Chapman 102, 103, 105, 116
Robbins, William Jacob 67
Roberts, FM 212
Robinson, Williams S 208, 209, 236
Roca-Garcia, G 158
Rodriguez, MR 228
Rösen (-Wolff), Angela 277
Roger, CA 26
Rogers, Kenneth D 253,
Rogers, S 184
Roizman, Bernhard 182, 186, 202
Rokutanda, Makoto 234
Romaine, CP 228
Roosevelt, Franklin Delano 100, 122
Rosen, Leon 253
Rosenthal, SM 278
Rosner, J 207
Ross, A Frank 154
Rosselett, J 281
Rossmann, Michael G 179
Rott, Rudolf 233
Rous, Francis Peyton 66, 67, 87, 104, 231, 290
Roux, Pierre Paul Émile 19, 24, 25, 39 42 43 47, 54, 80, 290
Rubin, Donald H 275
Rubin, Harry 186, 230, 233, 253
Ruckle, Gisela 253
Rushizky, GW 169
Ruska, Ernst August 98
Ruska, Helmut Georg 99, 144
Russ, Victor 51
Russell, JD 227
Rustigian, Robert 113, 221

Sabo, A 222
Sabin, Albert Bruce 54, 113, 124, 158, 234, 261, 276
Sachs, F 92
Sachs, Leo 232
Saddington, RS 69, 105
Sadler, PW 280
Saiki, Randall K 170
Saksela, E 272
Salimbeni, AT 39
Salk, Jonas Edward 116, 121, 123, 124
Salmon, Daniel 25, 26
Sam, 268
Sambrook, Joseph F 268
Sanarelli, Guiseppe 39 40, 86, 290
Sanchez, Yanuario 261
Sanders, F Kingsley 181, 188
Sandford, Katherine S 230
Sanger, Frederick 205
Sänger, Heinz L 187, 227, 228
Sarov, I 186
Sasaki, M 228
Saunders, LZ 224
Sawtschenko, 76
Sawyer, William 82
Schabel, Frank M Jr 282
Schaeffer, Howard J 283
Schaffer, Frederick Leland 143, 160, 173
Schäfer, Werner 143, 160, 161, 197, 206
Schapira, R 105
Scharff, MD 207
Scherer, WF 105
Schleiden, Jacob 12, 13
Schlesinger, Max 53, 59, 75, 96, 133, 134, 149, 290
Schlesinger, R Walter 221
Schluederberg, Ann 254
Schmidt-Jensen, HO 119
Schmidt, Sven 119
Schnaitman, Carl A 202
Schönbein, Christian Friedrich 49
Schönlein, Johann Lukas 12
Scholtissek, Christoph 233
Schramm, Gerhard Felix 137, 138, 140, 142, 143, 161, 173, 182, 209, 290
Schröder, H 17
Schrödinger, Erwin 95
Schroeter, J 28
Schulte-Holthausen, Heinrich 241
Schumacher, G 142
Schumacher, Wilhelm 28, 142
Schwann, Theodor Ambrose 12, 13, 16
Schwartz, Dennis E 208, 237
Schwerdt, Carlton Everett 136, 143, 160, 173
Scollosiy, E 198
Scolnick, Edward M 237

Séchaud, Janine 151
Sédillot, Charles Emmanuel 1
Sellers, RF 105
Semancik, JS 187, 228
Semmelweiss, Ignaz Philipp 15
Semple, D 81
Sertic, Vladimir 96, 141, 147
Shafritz, David A 241
Shanbrom, S 249
Shannon, William M 282
Sharp, D Gordon 136
Sharpe, AH 275
Shatkin, Aaron J 169, 183, 187
Sheffield, FML 62
Shein, Harvey M 232
Shelokov, A 104
Shepherd, Robert J 181
Sherlock, S 241
Shi, C 271
Shikata, Eishiro 228
Shigasaki, Y 75, 85, 91
Shimada, FT 112
Shipkowitz, NL 282
Shishkina, O 107
Shope, Richard Edwin 72, 87, 88
Shortridge, KF 222
Shultz, G 270
Shuman, H 112
Sickles, GM 105
Sidwell, Robert W 283
Sieber, Otto F 270
Siedentopf, Heinrich Friedrich 57
Siegal, Frederick P 243
Siegmann, B 182
Sigurdson, Björn 223
Siminovitch, L 200
Simizu, B 221
Simon, Edward H 206
Singer, SJ 256
Singh, RP 228
Sinsheimer, Robert L 169, 181, 182, 187, 205
Sipe, JD 183
Skinner, H 105
Sluyter, T 15
Smadel, Joseph Edwin 139, 141, 232
Smith, Erwin Frink 23, 24, 37 44, 290
Smith, Hamilton Othanel 168
Smith, JD 139, 173, 239
Smith, Kenneth Manley 138, 173
Smith, Paul K 281
Smith, RE 131
Smith, Richard T 254
Smith, Theobald 25, 26, 27, 290
Smith, MG 222
Smith, Wilson 55, 65, 69, 72, 100, 105, 107, 120, 157
Smorodintsev, Anatoli A 107, 121
Somers, JM 220
Sotelo, J 226
South, Mary Ann 270
Southern, Edwin M 256
Spallanzani, Lazarro 9, 10
Spear, Patricia G 186
Spearman, C 117
Spiegelman, Sol 165, 282
Splendor, A 62
Ssolowjewa, JW 73
Stahl, CF 154
Stahmann, M 136
Stanley, Wendell Meredith 59, 60, 90, 93, 94, 132, 134, 136, 137, 140, 154, 158, 290
Stanway, G 276
Starling, Ernest H 49
Staunton, Don E 200
Stavrou, D 224
Steere, Russel L 136,
Stegwee, D 131
Stehelin, Dominique 237
Steinhardt, Edna 65, 69
Steinhaus, Edward A 129, 130, 213
Steinmann, Ralph M 201
Steinschneider, A 169
Stelos, Peter 247
Stent, Gunther S 152, 153
Sternberg, George Miller 70
Stevens, Jack G 222
Stewart, Sarah E 114, 231
Stich, HF 272
Sticker, Anton 42
Stoeckenius, W 144
Stoel, G 69
Stoker, Michael 233
Stokes, Adrian 64, 105
Stokes, George Gabriel Sir 51
Stokes, Joseph Jr 72, 112, 118
Stone, JD 144, 228

Stone, OM 228
Storey, Harold, Haydon 131
Strauß, I 73
Streisinger, George 187, 188
Strickland, AGR 104
Strohl, William A 221
Strohmeier, Karl 267
Stroian, N 76
Stuart-Harris, Charles H, Sir 156
Stubbs, E 88
Stubbs, Gerald J 210
Stuhlberg, CS 105
Subak-Sharpe, John H 288
Summers, Donald F 210
Summers, Jesse 59, 187, 208, 209, 216, 241
Sumner, James Batcheller 59
Sugai, T 76
Sugiyama, T 181
Sutnick, Alton 216
Sutton, William 94
Sveda, Michael M 268
Svedberg, Teodor 52, 97
Svehag, Sven-Eric 247
Svet-Moldavsky, GJ 231
Symonds, N 150
Syverton, Jerome T 69, 123, 199
Szybalski, Waclaw 187

Tabor, Edward 216
Tabin, Clifford J 272
Taha, Mahmoud J 276
Takahashi, William N 91
Takaki, J 76
Takatsuki, K 242
Takayama, N 221
Talmage, David 170
Tamm, Igor 181, 207, 281
Taparowsky, Elizabeth 272
Taub, R 271
Taylor, A136, 139, 141, 143
Taylor, RM 155
Teague, Oscar 68, 85
Teissier, P 78, 117
Teixeira, J de Castro 69, 105
Temin, Howard M 167, 168, 233, 234
TenBroek, Carl 73
Thadani, Meneka A 272
Theiler, Arnold, Sir 40
Theiler, Max 64, 69, 72, 105, 112, 158, 276, 290
Thomas, M 170
Thompson, David 66,
Thompson, Randall L 277, 279, 280
Thormar, Halldor 224
Thullier, LF 25
Thung, SN 241
Tiegel, Ernst T 28 43
Tillet, William Smith 77
Tiselius, Arne Wilhelm 97, 143
Tizzoni, Guido 70, 71, 78
Todaro, George J 168, 237
Todd, Charles 55, 107
Tokumaru, T 182
Tolmach, LJ 198
Topacio, T 69
Torres, CM 69, 105
Towbin, H 256
Townsend, Jeanette J 225
Toyoshima, Kumao 168, 236
Trachsel, H 275
Trainin, Ze'ev 243
Trask, James D 103
Traub, Erich 69, 112, 245
Trautwein, Karl 72
Tschermak, Erich 94
Tsuchida, Nabuo 237
Tubeuf, Carl von 85, 112
Turner, Georg 71, 78, 82
Turpin, F 12
Twort, Frederick William 48, 146
Tyndall, John 17, 18
Tyrrell, David AJ 202, 214

Uchiyama, T 242
Uhlenhuth, Paul 38 45, 290
Uhr, Jonathan W 247
Ullmann, Sigrid 143
Underwood, Gerald E 249, 281
Unger, Franz 13
Unger, Ludwig J 82
Urbain, Achille 76

Vago, C 212
Vallée, H 54, 72, 119
Valenti, GL 40
Valentine, RC 198
Valenzuela, Pablo 267

Vallisnieri, Antonio 8
Van Bekkum, JG 222
Van Beveren, C 238
Van Bogaert, L 224
Van der Veen, J 156
Van Dorst, HJM 228
Van Nazerian, Key 240
Varmus, Harold E 224
Varro, Marcus Terentius 4
Verson, E 61
Verwoerd, DW 181
Vesely, V 232
Vierucci, Alberto 216
Vigier, P 236,
Villa, L 57
Villaginés, R 183
Villemin, Jean Louis 23
Vinson, Carl George 57, 58, 59, 93
Visnich, Sam 215
Vogt, Marguerite 104, 116, 232
Vogt, Peter K 168, 186, 234, 236
Volkin, Elliot 166

Wagner, Robert R 112
Wahl, Bruno 62, 85
Waldmann, Otto 64, 72, 78, 82, 105, 119
Walker, Duard L 221
Walter, B 228
Wang, K 187
Ward, SM 69
Wasser, H 130
Waterson, Antony Peter 263
Watkins, John Frederick 171
Watson, David H 182
Watson, James Dewey 139, 160, 164, 165, 166, 173, 209
Watson, MA 212
Watson, Roger J 267
Weathers, LG 228
Weaver, Warren 95
Webster, Robert G 261
Wecker, Eberhard 161, 197, 206, 207
Weech, AA 78
Weed, SD 249
Weidel, Wolfhard 145, 183
Weidel, Joseph 201
Weigert, Carl 21, 28, 61
Weigle, JJ 168
Weil, ML 225
Weintraub, M 256
Weir, Jerry 268
Weiser, J 129
Weiss, Susan Reich 236
Weissenberg, Richard 163
Weiß, Bernhard 184
Weißbach, A 184
Weißman, Charles 184, 188
Welch, FV 56
Welch, William Henry 41
Weller, Thomas Huckle 98, 102, 222, 256, 270
Wendel, HA 281
Werner, Jacqueline H 116
Wésslen, T 105
Whalen, M 129
White, PR 67
White, RG 107
White, G 129
Widal, Fernand Georges 29
Wide, Leit 256
Wiener, M 232, 252
Wildmann, Sam G 132
Wildy, Peter 158, 174, 176, 277
Wiley, DC 274
Wilkins, Maurice Hugh Frederick 164
Williams, CA 256
Williams, I 152, 184
Williams, Robley Cook 135, 136, 140, 142, 143, 151, 173, 209
Willstätter, Richard 54, 59, 119
Winocour, Ernest 232
Wishart, FO 117
Wöhler, Friedrich 11, 12
Wolbach, SB 44
Wolf, Ken 218
Wolinsky, JS 225
Wollmann, Elisabeth 148, 150, 290
Wollmann, Eugene 148, 150, 290
Wood, William Barry Jr 112, 204, 209
Woodroofe, Gwendolyn M 255, 268
Woodruff, Alice Miles 67, 69
Woodruff, C Eugene 56, 57, 62
Wright, Arthur M 113
Wulff, H 105
Wyatt, Gerard R 139, 270

Wyckoff, Ralph Walter 98, 135, 140, 142, 151
Wyss, Orville 183, 184

Yärvinen, 112
Yalov, Rosalyn S 256
Yamafuji, K 212
Yarchoan, Robert 284
Yelverton, Elizabeth 267
Yerganian, G 272
Yersin, Alexandre John 22, 25, 54
Ylinen, 112
Yokken, RH 249
Young, George A 270
Yuan, Robert 168

Zak, SK 117, 245
Zenker, Friedrich Albert von 23
Zhdanov, Victor M 121
Zigas, V 226
Zilber, LA 231, 235
Zillig, Wolfram 142, 143
Zilliken, Friedrich 199
Zimmer, Arthus J 136
Zinder, Norton David 206
Zinke, Georg Gottfried 14, 80, 105
Zsigmondy, Richard Adolf 50, 57
Zur Hausen, Harald 239, 241, 272
ZuRhein, Gabriele M 225
Zwick, Wilhelm 72

Subject Index

Abbe illuminating apparatus 21
Abiogenesis 3
Accessory cells 248, 291
Acetone-ether extraction 116
Acetone-precipitation 58
Acid agglutination 55
Acridine orange 98, 172
Acrylamide gel electrophoresis 257
Actinomycin D 206, 233, 282
Activation of satellitism 188
Active immunization 79–83
Active vaccines 126, 260–268
Acute disease (infection) 212
Acyclovir 283
Adenine arabinoside 282
Adeno-associated virus (AAV) 187
Adenosine triphosphatase 141
Adjuvants 119, 121, 122, 260, 261, 267, 291
Adsorbate vaccine 55
Adsorbents 54, 97
Adsorption 54–55, 97, 161, 198–200
- capacity 149
- mechanism 145
- rate 200
column a. 97
leucocyte a. 54
Adult T-cell leukemia 242
African horse sickness 40, 64
Agammaglobulinemia 117
Agar 22, 119
Agar plate 28
Agglutination 75, 256
Agglutinating antibodies 26
Agglutinins 70
Aggregation tests 71, 74–76, 256–257
AIDS (aquired immune deficiency syndrome) 243
AIDS-associated retroviruses (ARV) 243, 273, 274
Alcohol precipitation 36, 58
Alcoholic fermentation 11
Aleutian disease 226
Allantois sack 67
Allergens 125
Allergic reaction 122
2-(α-hydroxybenzyl)benzimidazole 281
α-methyl-1-adamantane-methylamine-HCl 281
Alum 119
Aluminium hydroxide 54, 119, 122
–, –, gel 54
–, oxide 260
–, phosphate 122
Alzheimer disease 226
Amoeboid bodies 62
Aminopterin 233
Amino acid analysis 155
1-Aminoadamantane 281
Ammonium sulphate 58
Amyl alcohol 58
Aniline dyes 21, 28
Animalcula 7
–, minima 10
–, insensibilia 10
Animal charcoal 54
Animalia minutissima 4
Animal viruses 73, 75, 155
Anthrax 19
Antibiotics 278
Antibodies 26, 70, 97, 244
agglutinines 26, 70

lysines 26, 70
precipitines 26, 70
humoral a. 109, 117
neutralizing a. 70, 71
virucidal a. 71
Anti-contagionists 22
Antigen 70–71, 291
Antigen drift and shift 276
Antigenic vaccines 81
Anti-metabolites 279–282
Anti-oncogenes 238
Anti-Pasteurians 23
Antisepsis 14, 19, 24
Anti-toxins 26, 70
Anti-vaccinators 23
Anti-viral drugs 283
Anti-viral chemotherapy 92, 277–285
Antonius pest 5
A-particles 229, 230
Apochromatic lense 28
A-protein 206
Aquired immune deficiency syndrome (AIDS) 243
–, immunity 76
Archigenesis 3
Archiv für die gesamte Virusforschung 96
Arsenic 15, 92
Asbestos 17
– filters 43
Asepsis 14, 15, 19, 24
Assembling 161, 209
Atoxyl 92
Attachment 198
Attenuated vaccines 81, 99
Attenuation 25, 79, 275, 291
Attic plague 4
Attic school of medicine 1
Australia antigen 215
Autocatalytic protein 60, 90
Autocatalytic reproduction 60
Auto-antibody 226
Auto-immune disease 227
Auto-immunity 245
Auto-interference 115
Aureomycin 278
Avian leucosis-sarcoma complex 229
Avian tumour viruses 229
5-Azacytidine 236

Bacteria 3, 10, 54
Bacterial pathogenicity 30
Bacteriolysis 26
Bacteriophages 48, 74, 75, 90, 144–145, 146
Base plate 145
Basic functions of life 1
Basis of life 94
Bentonite 116
Benzimidazoles 280, 281
Bilateral symmetry 175
Bile salts 163
Binding protein 205
–, site 198
–, time 200
Biochemical findings 57–61
Biotin 141
B-particles 229
Bittner's milk factor 113
Blind passages 156
Blood 41
–, corpuscules 54
–, poisoning 19
Boiled potatoe slices 21
Boiling 15
Borrowed life 96
Bovine pleuropneumonia 39
Budding 211
Burkitts lymphoma 240
Bursa fabricii 7
Burst-size 148

Calcium 15
Calcium chloride 15, 198
–, oxide 54, 122
–, phosphate 54, 122
Calcium phosphate adsorbate vaccine 121
Cancer 86
Canine distemper demyelinating encephalomyelitis 224
Cap-binding protein 275
Capsid 175, 176
– form 175
– symmetry 172, 175, 178
nucleo c. 176
Capsomer 175
Capsule disease 129
Carbohydrates 141, 185

Carbol-fuchsin staining 57
Carbonic acid 15
Carcinoma 88
Carmine 28, 54
Carriers 27, 41, 89, 90, 267
Catalytic principle 12
Cataphoresis 55
Cell 159
- culture 104
- division 13
- fusion 202
- inclusions 34, 61–63
- membrane 200
- nucleus 13
- reproduction 13
- theory 11, 13
- transformation 168, 236
- transformation test 231, 232
- tropism 274
Cell-free system 165
Cell-mediated immunity 248, 251
Cellular immunity 25
–, (c-) oncogenes 238
Central dogma of biology 139, 161, 166, 167
Centrifugation 51–54
Charcoal 54
Chemotherapy 92, 277–285
Chemistry of viruses 159
Chemotherapeuticals 283
Chick embryo 67, 87
–, –, influenza-canine distemper vaccine 120
Chicken tumours 87
Childbed fever 10
Chlamydozoa-strongyloplasma 48
Chloramphenicol 278
Chlorine/water solution 15
Chloroform 82
Cholesterol 141
Chorioallantois membrane (CAM) 102
Chromatographic ion exchange 97
Chromatography 97
Chromosome 94
Chronic infection 112, 222
Chronic progressive myelitis of mice 224
–, wasting disease 228
Circular permutation 188
Classification of viruses 89–90, 189–197
Clathrin 201
Clay cells 28
–, filters 43
Cleaving-point 169
–, specifity 168
Clonal selection theory of aquired imunity 170, 245
Clone 149, 291
Cloning 149, 169
Cloning vector 170
Co-cultivation 220
COFAL test 235
Cohesive ends 188
Colchicin 201
Cold stability of viruses 160
Collodion 49
Collodion-agar-membrane method 148
–, membrane 28
–, membrane sack 28, 49
–, sack culture 39
Colloid suspension method 51
Column adsorption 97
Colour test 103, 116
Complement 76
- fixation reaction (CFR) 29, 76
- system 246
- type-specifity 76
Complete virus 161
Complex vaccines 120
–, chick-embryo influenza-canine-distemper v. 120
Compound microscope 1, 8
Compression turbine 53
Concanavalin A 220, 291
Congenital immunity 245
Congenital malformations 270
Contact inhibition 231
Contagious disease 2, 13
Contagium, contagia 3, 4, 5, 10, 11, 13, 14, 15, 20, 22
–, animatum 8, 20, 21
–, fixum 36
–, vivum 7
–, vivum fluidum 36, 45, 90
Coombs' test 257

Copper 141
- sulphate 15
Core 143, 145, 160, 177, 291
Cowpox 19
C-particles 229, 230
Creutzfeldt-Jakob disease (CJD) 226, 229
Crick's sequential hypothesis 166
Critical air-pressure method 51
Cross-hemagglutination-inhibition test 156
-, immunity 88
-, immunity test 155
-, neutralization test 74
Cryptogram 190
Crystal(s) 58, 98
Crystal lattice structure 134
Crystalline protein 59
Crystallization 58
Crystallography 97
Crystal needles 60
Cultures
fractionated c. 28
insect cell c. 67
long-term c. 67
Maitland c. 67
monolayer c. 231
non-permissive c. 232
organ c. 66
permanent c. 66, 106
permissive c. 232
primary c. 106
pure c. 28, 63, 65–69
roller tube c. 66
plant cell c. 67
suspension c. 104
tissue c. 67, 69
Cytochalasin B 201
Cytopathic effect 68, 102
Cytopathogenicity 68
Cytopathogenic virus 102
Cytoplasmatic inclusions 33, 62, 68
-, polyhedra 129
Cytoplastema 13
Cytoryctes vaccinia 61
Cytosine arabinoside 281
Cytoskeleton 202, 291
Cytotoxic cells 248
Dane particle 216
Dark-field method 57
De-embryonated chick embryo 102
Defective virus 187
Deficient virus 187
Definition of the virus 42–43, 89, 96
Delayed-type hypersensitivity 245, 249
Delta antigen 216
Density gradient centrifugation 52, 292
2-Deoxy-5-iodouridine 281
Depot vaccine 55
D-form 156
Diagnostic methods 255–260
Dialysis 55
2,6-Diaminopurine 280
Diatomaceous earth 43, 54
Differential centrifugation 52
Differential staining 21
Diffusion tests 255
Diffusion trials 36
Dimethyl sulphoxide 280
Diploid cell strains 215
Diptera 128
Direct contact 84
-, fluorescence antibody technique 11
Disease control 264
Disinfectants 15, 56
Disinfection 15, 56
Disinfection centres 24
DNA 138, 164
- binding site 206
- pool 153
- double helix 164
- polymerase 165
- provirus hypothesis 238
- replication 186, 203, 293
- transfer technique 259
- viruses 191–193, 204–206, 238–241
DNA-RNA hybridization 165
Double helix 164
-, infection 115
-, shadow-casting technique 173
D-particles 230
Drift 252
Dry vaccines 125
Dust and droplet infection 84
Dyes 28

Early phage enzyme 153
–, replication 204
EBNA 235, 240
Eclipse 150, 162, 201, 203
Efficacy of vaccines 71
Egg-allergy 125
Electron microscope (EM) 98
Electrophoresis 55–56, 97
Elementary bodies 56, 62
Elution 54, 100
Embryo extract 66
End group analysis 169
Endocytosis 201
Endogenous interferon 250
Endogenous virus 234, 292
Endoplasmatic reticulum 211
Endospore 20
English vernacular name of viruses 190
Engulfment 161
Enteroviruses 143
Envelope 176
Env gene 208, 236
Enzyme 58, 140, 207
– linked immunosorbent assay (ELISA) 258
Epstein-Barr virus 240
Erythrocytes 54, 107
Essential nutrients 30
Ether 160, 163
Ethionine 280
Evolutionary conservation 238
Exhaustion theory of immunity 19
Exogenes interferon 250
Exon 167, 292
Exonuclease 166
Exotic virus diseases 217–218
Experimental animals 72, 105
Exponentially growth 150

Family (of viruses) 190
Feeder-layer system 104
Fermentation 12, 15, 17, 19
Ferments 26
Fertilization 13
Feulgen reaction 133, 142
Fever 84
– poison 30
Fibres 199
Filters 43, 49, 50
Filter candles 49
–, layers 43, 49
Filterable microbes 43
–, viruses 44, 47
Filtration 17, 49–51
– resistance 49
Fingerprinting 258
Fish viruses 218–219
Flacherie 19
Flavin 141
Flemming solution 61
Flocculation 75
Fluorescence antibody technique 98
5-Fluorodeoxyuridine 282
Fluorosceine-isothiocyanate 257
Fluorescein solution 101
Fluorescence microscope 97
–, antibody technique 257
Fluorochromes 98
Foci 104, 231, 233
Fomites 7
Food conservation 9
Foot-and-mouth disease 37–39, 64
Formaldehyde 82
Formalin 82, 119, 120, 122
Founder of virology 45–46
Fowl plague 40
Fractionated culture 28
Free nucleic acid 227
Freund's (complete) adjuvant 267
Fucose 141
Fundamental criteria of classification 163, 177
Fungi 3
Fusion technique 171

Gag gene 208, 236
Galactose 141
Galactosamin 185
Globulins 244
Ganciclovir 284
Gastroenteritis viruses 217
Galenus pest 5
Gelatine 21
Gel diffusion technique 116
Gene 94, 235
Gene-enzyme hypothesis 197
General symptoms 27

Generatio aequivoca 3
–, spontanea 3, 5–10, 17, 18
Genetic code 165
–, engineering 164
–, map 169
–, markers 156
–, structure of viruses 181, 187
Generative substance 96
Gene technology 164
Genus (of viruses) 190
Germ doctrine 16–29
–, theory 10–16, 19
Gerstmann-Sträußler syndrome 229
Gibberellic acid 280
Giemsa staining 57
Gliotoxin 278
Globulin 54
Glucose 185
Glucoseamine 141
Glycerin 82
Glycolipids 185, 211
Glycoproteids 183, 185, 211
Government regulations 125
Gradient electrophoresis 97
Gradocol membranes 50
Granula 62
Granulations élémentaires 62
Granular cytoplasmatic inclusions 85
Granulosis 62, 127, 129
Group (of viruses) 190
Guanine 279
Guidelines for classification and nomenclature 189–190
Gum arabic 125
Guncotton 17, 49
Glycerin 82

Hanging droplet method 57
Healthy carriers 221
HeLa cell strain 104
Helper cells 248
–, virus 186
Hem-adsorption test 104
Hemagglutination (HA) 100
– inhibition (HI) test 116, 252
Hemagglutinin 100, 179
Hematoxyllin 28
Hepadnaviruses 208
Hepatitis B vaccine 264
–, B virus (HBV) 215–216
–, δ virus 187
Hepatitisviruses 215–217, 241
Hepatocellular carcinoma 217
Herpes simplex 64
Herpesviruses 239–241
Heterogenesis 3
Hexamer 176
Hexamethylenamine 92
Hexamethylenetetramine 92
High resolution electron microscopy 179
HI test 116
Hospitals 6
Host-specifity of bacteriophages 168
Human immunodeficiency virus (HIV) 243, 244, 254
Human wart virus 233, 239
Humoral antibodies 109, 117
–, immunity 25, 28, 83, 251
Hybridization 220, 292
– techniques 259
Hybridoma technique 164, 170, 250
Hydroxyl apatite 97
Hygienic institutes 24
Hymenoptera 128
Hypogammaglobulinemia 245

Icosahedron 292
Imidazole 279
Immortalization 240
Immune complexes 258
–, complex disease 273
–, defects 272–274
–, electrophoresis 256
–, precipitation 257
Immunity 76, 77–78, 244–255
aquired i. 76
auto-i. 245
cellular i. 25, 246
congenital i. 245
cross i. 88
humoral i. 25, 28, 83, 246
natural i. 76
non-specific i. 246
specific i. 246
tissue i. 83
tuberculin-type i. 245

Immunization 76, 78–83, 118–119
active i. 79–83, 118
passive i. 26, 78–79, 118
simultaneous i. 82
Immunodeficiency disease 273
–, viruses 254, 273–274
Immune electron microscopy (IEM) 217
Immunofluorescence 172
Immunoglobulins 200, 244, 246–248
Immunologic tolerance 245
Immuno-potentiators 267
Immunoprophylaxis 260
Inactivated vaccines 81, 99
Inactivating substances 120
Inactivation by heat 120
Inapparent infection 109, 212, 213, 220–221
Inclusion bodies 56, 61, 66, 67, 204
Inclusion diseases 127–129
Inclusions 34
cytoplasmatic i. 34, 61, 62, 68
granular cytoplasmatic i. 85
nuclear i. 61, 62, 68
polyhedral cytoplasmatic i. 85
Incomplete viruses 100, 114, 161, 187
Incubation period 110, 111
–, time 84, 111
Indolebutyric acid 279
Induction 222
Infection
– capacity 149
generalized i. 110
acute i. 212, 214
inapparent i. 109, 212, 213, 214, 220–221
latent i. 112, 212, 213, 214, 222–223
lysogenic i. 148
lytic i. 148
mixed i. 147
multiple i. 147, 149
non-persistent i. 212
occult i. 213
persistent i. 112, 212, 214, 221–222
persistent-latent i. 213
secondary i. 110
slow virus i. 214
subclinical i. 213
super-i. 150
symptomatic i. 110
tumour virus i. 113–114
Infectious diseases 2, 84–86
–, mononucleosis 241
Infectivity 83
Inflammation 117
Inflammatory reaction 119, 249
Influenza 65, 100
– centres 122
– vaccines 120–122, 263–264
Infusoria 9
Inhibitors 100, 116
Initial bodies 62
–, lesions 109
Insect 35, 42
– cell culture 67
– viruses 75, 85
– for biological pest control 162
In situ hybridization 259
Institute Pasteur 19, 80
Interference (phenomenon) 114–115, 155, 250
Interferon 250
– synthesis 220
endogenous i. 250
exogenous i. 250
Interferon-inducing substances 250
Interleukin-1 249
Interleukin-2 242
International Association of Microbiological Societies 189
International Committee on Nomenclature of Viruses 189
International Committee on Taxonomy of Viruses 189
Intervening sequences 167
Intracellular virus 234
– multiplication 57, 68, 132
– reproduction 45
Intrachromosomal defect virus 234
Intravital staining 57
Introns 167, 292
Invisible microbes 42
–, organisms 43
In vitro hybridization 220
Iron acetate 58
Ios 1, 4, 5
Iso-electric point 55
Isomorphic replacement method 174

Jennerization 80
Jerne's network theory 170, 249

Kaolin 54
Killer system (viruses) 219
Koch's postulates 30–33
Kuru 225

Labelling techniques 257
Laboratory centrifuges 51
Lag phase 92, 162
Large viruses 48, 279
Late replication 204
Latent disease (infection) 112, 212, 213
–, membrane protein (LMP) 235
–, period 148
–, viruses 85
Latency 213, 277
Laue diagram 98
Lazaretti 6
L-cells 104
Lecithin 119
Lentogenic strains 213
Leprosaria 6
Leucocytes 107
- adsorption 54
Leukemia in chickens 87
–, in mice 113
–, viruses 229
Leukosis 87
Levende dierkens 8
Lipids 141, 184–185
neutral l. 185
peripheral structure l. 185
phospho l. 185
glyco l. 185, 211
Lipman-system 159, 292
Lipoproteins 183
Liposomes 260, 267, 292
Liquid substrate 27
Live vaccines 81, 82, 262
Living cells 60
Localists 22
Local lesion test 146
–, reactions 83
–, symptoms 27
Locke nutrient solution 52
Löffler corrosion 57
–, hyperimmune serum 120
Long-term cultures 67
Low temperature method 136
Luminescence microscope 97
Lymphadenopathy-associated virus (LAV) 243
Lymphocyte-detected membrane antigen (lydma) 235
Lysin 26, 29, 70
Lysogenic bacteria 154
–, cultures 148
–, phage 147
–, response (lysogeny) 148
Lysogenization 153
Lysozyme 141
Lytic enzyme 147
–, phage 154
–, response (lysis) 148

Macrophage immigration inhibition factor (MIF) 249
Macropinocytosis 202
Maedi disease 223
Magnesium chloride 198
–, sulphate 58
Maitland culture 67
Mala aria 4
Mal calcino 11
–, de segno 11
Mannose 141
Markers 156
- for vaccines 262
- of influenza viruses 156
genetic m. 156
Maturation 150, 153, 162
Measles vaccines 263
–, virus 109
Medium 199, 104
Membrane filters 50
Memory B lymphocytes 247
Mercury 15
Meristem 133
Merthiolate 122
Message 161
Mesogenic strains 213
Metabolic inhibition test 116
Metallic oxide 54
–, shadowing 135

Methylcholanthrene 235
Miasma 1, 4, 5, 7, 10, 13, 14
Microbiological institutes 24
Microfilament 201
Microorganism 1, 158
Micropinocytosis 202
Microscopes 8
compound m. 8
electron m. 98
fluorescence m. 97
luminescence m. 97
ultra m. 57
UV m. 56
Microspectrophotometry 142
Microvilli 201
Milk 17
Mineral oil Falba vaccine 121
Minus strand 188
Mixed infection 147
Modification enzymes 168
Molecular biology 95
-, genetic tests 258
-, oncology 229
-, sieve chromatography 257
Monoclonal antibodies 164, 170, 171, 220, 260
Monolayer cultures 231
Morbilli 5
Morbillization 14
Morbus Gallicus 6
Morbus Neapolitanus 6
Morphological unit 174, 175
Morphology of the virion 172
Mosquitoes 10, 41
Mouse protection test 122, 125
mRNA 165, 204
early -, 204, 206
late -, 204, 206
Mucinase 141
Mucopolysaccharides 21
Multiple infection 147, 149
Multiplication cycle 161–162, 197–212
Multiplicity reactivation 158
Mumps 65
Muscardine 11
Mutagenic virus effects 271–272
Mutants 154
Mutation 83, 133, 146
- frequency 149
- index 149
Mystical period 2
Mycoplasma 47

Naphthaleneacetic acid 279
Nasopharyngeal carcinoma 241
National Foundation for Infantile Paralysis (NFIP) 100, 122
Natural immunity 76
-, killer cells (NK cells) 248, 250
Nearest neighbour analysis 169
Neoplastic transformation 231
-, tumours 40
Negative staining 173
N-acetylneuraminic acid 141
Network theory of the immune system 170, 249
Neuraminic acid 141, 185
Neuraminidase 141, 179, 202
Neutral fat 141
Neutralization test 70, 71–74, 116
Neutralizing antibodies 70
N-glycolyl-neuraminic acid 141
N-isatin-β-thiosemicarbazone 280
Nitrocellulose 49
Non-culturable microbes 43
Non-inclusion diseases 129–130
Non-permissive cell culture 232
Non-persistent infection 212
Nonsense codon 165
Non-structural proteins 182
Normal bacteria nutritive solution 27
-, solution 27
Northern blotting 259
NP antigen 232, 255
Nuclear inclusions 61, 62, 68
-, polyhedrosis 127, 129
Nucleic acid 138, 140, 180
- content 140, 155
- molecular weights 140
- structures 181, 187
Nucleocapsid 160
Nucleoproteid 59, 60, 96
Nucleoprotein (NP) antigen 232, 255
Nucleotide rates 155
Nutrient agar plate 28

Occult infection 213
O-form 156

Oil turbine ultracentrifuge 52
Omne vivum ex ovo 7
Onc gene 168, 208, 236
Oncogene hypothesis 168, 237
Oncogenes 168, 229, 238
Oncogenic transformation 231
ON-diacetyl-neuraminic acid 141
One-step-growth experiment 148
Opportunistic infections 273, 293
Optical methods 56–57
Optic immersion systems 28
Organ affinity 85
–, culture 66
–, manifestation 84
Organism 159
Organized pathogenic agents 33
Origin of assembly 210
–, of replication 206

Palindromic sequences 168
Paper electrophoresis 97
Papilloma 88
Papillomaviruses 239
Papovaviridae 229
Paraffin oil 121
Parasitic degeneration hypothesis 287
–, germs 30
Particle weights 140
Passive immunization 26, 78–79
Pasteurization 17
Pathogenic factor 26
Pathogenicity 83
–, of viruses 83–84
Pathological changes 85
Pebrine disease 18
Penetration 200
Penicillin 92, 278
Penny Parade Program 100, 122
Pentamer 176
Peplomer 177
Peplos 177
Pepsin 136
Peptone medium 27
Peripheral structure lipids 185
Permanent immune (antigen) stimulator 122
Permanent cell strain 104
Permanent (monolayer) cell culture 66, 106
–, secondary immunodeficiency 243
Permissive culture 232, 293
Persistent infection 112, 212, 221–222
–, latent infection 213
–, virus infection in vitro 221
–, virus infection in vivo 222
Persisting virus 78
p-fluorophenylalanine 282
Pest 5, 14
- control by insect viruses 162
Phage
- lysozyme 141
incomplete p. 151
lytic p. 154
lysogenic p. 154
temperate p. 148, 154
vegetative p. 152
pro-p. 153, 154
Phagocytes theory 25
Phagocytosis 26, 107, 109, 161, 201, 293
Phenol 81, 82, 120, 122
- glycerin mixture 82
- red 103
pH (range) 160, 198
Phospholipids 141
Phosphonoacetic acid 282
Photographing bacteria 21
Physical map 169
Phytohemagglutinin 220, 242
Phytopathogenic insect viruses 131
Pinocytosis 109, 161, 202
Pits 201
Plague 1, 4
- of Thucydides 4
Plant cell culture 67
–, diseases 89, 130–133, 219
–, viruses 73, 75, 85, 89, 219–220
Plaque-forming units (PFU) 104, 157, 293
Plaque test 104
Plasma embedding 66
Plasmid 169, 293
Plasmodium-like masses 62
Pleuropneumonia-like organisms (PPLO) 47
PLP (progressive rubella panencephalitis) 225
Plus strand 188

PML (progressive multifocal leucoencephalopathy) 224
Pock counting method 102
Poison 5
Poisonous matter 30
Pol gene 236
Poliomyelitis 65
- vaccines 122–125
Polyhedra disease 62
Polyhedral cytoplasmatic inclusiuons 85
Polyhedrosis 62, 127
nuclear p. 127, 129
cytoplasmatic p. 129
Polymerase chain reaction (PCR) 170
Polyoma virus 239
Polysome 207
Porcelain filters 43
Potassium hydrate 80
-, hydroxide 15
-, iodide 15
-, nitrate 15
-, phosphotungstate 173
Potato slices 21
Precipitation 29, 58, 75
Precipitating antibodies 26, 70
Precipitines 26, 70
Precursor pool 153
Presence of living cells 60
Primary cell culture 106
Primary hepatocellular carcinoma 241
Primary monolayer cell culture 106
Primulin 98
Prion 172, 182, 225, 228
Prodromal period 110, 111, 293
Proflavin 280
Progressive rubella panencephalitis 225
Progressive multifocal leucoencephalopathy (PML) 224
Promoter 206
Properdin system 246
Prophage 153, 154
Protection test 117
Protective effect of antitoxin 26
Protein 93, 140, 181–184
- disc 210
- pool 152
A-p. 206
autocatalytical p. 60, 90
binding p.,205
crystalline p. 59
glyco-p. 211
lipo-p. 183
non-structural p. 182
structural p. 182
virus-specified p. 235
Proto-oncogene 168, 238
Protozoa 3, 8
Provirus 208, 233
Provirus hypothesis 167, 233, 234
Provisional criteria for attenuated vaccines 262
Provisorial Committee for the Nomenclature of Viruses 175
Psittacose-lymphogranuloma-trachoma group 48
Pure culture 21, 28, 63, 65–69
functional p.c. 25
morphological p.c. 25
physiological p.c. 25
Purifying substances 58
Purine analogues 279
Puromycin 282
Putrefaction 10, 11, 15
Pyrimidine derivatives 279
Pyrogen 112, 293

Quarz lenses 56
-, prisms 56
Quasi-equivalence theory 174
Quinine 92

Rabies 64
- vaccines 80–81
Rabbit myxomatosis 39
Radioimmunoassay (RIA) 258
Rapid ether method 81
Rate of water flow method 51
Receptor 107, 179, 198, 274, 275
- gene 199
- map 200
- destroying enzyme (RDE) 141
Recognition site 175
Recombinant DNA methods 258
Recombination experiment 157
Reduplication 203

Reinfection 251
Repeats 208
Replication 186, 203, 293
Replicative RNA 188
Resistance 155, 244
Resistance-inducing factor test (RIF) 230
Restriction endonuclease 166
Restriction enzymes 168
–, map 169
Retroviruses 207–208, 229, 242
Retrovirus integration 234
Reverse transcriptase 168, 207, 208, 229, 234
Rhinoviruses 214–215
Rhodamin 257
Ribavirin 283
Ribose 141
Ribozyme 167, 172, 228
Rickettsiae 48
RNA-dependent DNA polymerase 168
RNA-cDNA hybridization 259
RNA processing 167
RNA tumour viruses 167, 242–244
RNA viruses 194–196, 206–208
Roller tube culture 66
Room disinfection 15
Rota virus 217
Rotation symmetry 175
Rous associated virus (RAV) 186

Saccharomyces 12
Safranine 58
Sandwich technique 258
Saponin 119, 135
Saprophytes 26, 30
Satellite virus 186
Satellitism 186
Schizomycete 20
Schizophyte 20
Schiff's reagent 142
Schroeter's boiled potato slices 28
Schultz-Peet prophylaxis 103
Scrapie 182, 223, 227, 228
Secondary bacterial infection 109
–, immunodeficiency 249
–, infections 110
Sedimentation constant 52
–, equilibrium 52
–, velocity 52
Seeds of disease 6
Selective in vitro growth 220
Self-assembly 174, 209
–, -processing 167
Semina vitalia 4
Seminaria contagionis 6
–, morbi animata 6
Sendai virus 171
Sephadex 257
Sequencing 169
Sequential hypothesis 166
Serum hepatitis 215
Sexual organs 9
Shift 252
Sialine acid 141
Side-chain theory 26, 170
Silent rage 80
Simian immunodeficiency virus (SIV) 243, 244
Simultaneous immunization 82
–, vaccination 71, 118
Simultation method 173
Single-burst experiment 147
Single-cell dilution method 28
Slices of cooked potatoes 21
Slow virus infection 223–229
Small animals 10
Small bacterium 90
Smallest genetic entity 96
Smallpox 64
– control 264
– method 101
Small viruses 48
So-called invisible microbes 42
Sodium chloride 15
–, hypochlorite 15
Solid substrate 27
Soluble antigen 117
Southern blotting 259
Species (of viruses) 190
Species specifity 74
Specific gravity 51
Specified proteins 235
Spermatozoa 8
Spherules 62
Spikes 176, 178, 185, 199
Splitting 263

Split vaccine 263
Spontaneous generation 3, 5–10, 17, 18
Spores 18
Src-gene 168, 208
SSPE (subacute sclerosing panencephalitis) 224
Staining methods 57
Stanley's interpretation 60
Starch oil 119
Step-by-step multiplication 92
Sterilization 18
Strain specifity 74
Stressors 213
Streptomycin 278
Structural units 160, 174, 175
Structural properties of viral nucleic acids 187
Structural protein 182
Subacute sclerosing panencephalitis (SSPE) 224
Subclinical infection 213
Subfamily (of viruses) 190
Subgroup (of viruses) 190
Sulfonamides 92, 278
Substituted amino acids 279
Subunit vaccine 263
Superinfection 150
Suppressor cells 248
Suspension culture 104
Swedish filter paper 28
Symmetry 172–178
- axes 175
- faces 175
bilateral s. 175
capsid s. 175
rotation s. 175
Symptomatic infection 110, 111
Symptoms of disease 27, 109
Synthesis phase 203–209
Synthetic antigens 266
Synthetic media 102
Syphilis 6, 86
Systemic reaction 83, 84

Taches vierges 147
Tail fibres 176
–, pin 145
Tannic acid 256
T-antigen 206
Tapioca 119
Tat gene 208
Taxonomy of viruses 189–190
Temperate phage 154
Temperature-sensitive mutants 236
Teratogenic effect 112, 270–271
–, viruses 112, 270–271
Terminal redundancy 188, 293
–, sequence 169
Terminator 165
Test methods 278
–, substances 278
–, tubes 66
Theory of evolution 172
–, of phagocytes 25
Thin-layer method 57
2-Thiopyrimidines 280
Thiosemicarbazones 280
2-Thiouracil 279
Thymidin kinase 205
Thymonucleic acid 94, 137
Ticks 27
Tissue affinity 85
–, immunity 83
–, tropism 274
T-lymphocytes 248
Tobacco mosaic disease 35–37
–, –, virus (TMV) 137
Toepler'e Schlieren method 97
TO strain 106
Topoisomerase 205
Toxic effects (reactions) 122
Toxins 26, 27, 30, 112
Tramycin 278
Transactivator gene 208
Transcription 203
Transfection 236, 239, 294
Transformation 91, 94, 96, 138, 236
oncogenic t. 231
neoplastic t. 231
Translation 204
Transmissible mink encephalopathy 228
Transmissible mutagen 88
Transmission 106, 120
Triangulation 174
Trigger mechanism 213
Tropism 274

True viruses 48
Trypsin 66, 104
Tuberculin-type immunity 245
Tuberculosis (bacilli) 22, 121
Tumour (T) antigen 235
–, development 84, 86–88, 128
–, -like masses 231
–, suppressor genes 238
–, viruses 113–114, 229–244
Turbine ultracentrifuge 52, 53
Tween-ether split vaccine 263
Tyndall effect 57
–, method 91
–, phenomenon 17
Tyndallization 18
Type (of virus) 190
Type-specifity of the CFR 76

Ultracentrifugation 97
Ultracentrifuge 52–54
Ultrafilters 49
Ultrafiltration 49
– devices 50
Ultramicroscope 57
Ultramicroscopic organisms 43
Ultramicrotomy 135
Ultraviolet dark-field illumination 56
–, microscope 56
–, photography 56
Uncoating 161, 203
– protein 203
Unconventional virus (agent) 223, 225, 227, 228
Unstable form of virus 188
Uracil 279
Urease crystallization 59
Urine 15, 17
U-shaped tube 55, 97

Vaccination 14, 80, 270
  simultaneous v. 71, 118
Vaccines 19, 24, 80, 126, 263, 265
– for non-human use 265
– to protect fishes 265
  active v. 126, 260–268
  asorbate v. 55
  antigenic v. 81
  attenuated v. 81, 99
  efficacy of v. 71
  calcium phophate adsorbate v. 122
  cattle plague v. 81–82
  complex v. 120
  complex chick-embryo influenza-canine-distemper v. 120
  depot v. 55
  dry v. 125
  foot-and-mouth v. 82, 119–120
  formol-adsorbat v. 119
  hepatitis B v. 264
  inactivated v. 81, 99
  infectious v. 81
  influenza v. 120–122, 263–264
  live v. 81, 82, 262
  measles v. 263
  mineral oil Falba v. 121
  passive v. 78
  poliomyelitis v. 122–125
  rabies v. 80–81
  smallpox v. 126, 264
  split v. 263
  subunit v. 263
  tween-ether split v. 263
  yellow fever v. 82–83, 125–126
Vacuolate bodies 62
Variation 83, 146
Variolation 14
Vectors 84, 214
Vegetative phage 151
Velogenic strain 213
Victoria blue 57
Viral glycoproteids 183, 185
Viremia 84, 103, 110, 294
Virion 172, 176, 223
– assembling 161, 209
– capsid symmetry 172, 175
– model 176
– maturation 150, 153, 162
– morphology 178
– size 178
– structure 179–189
Viroid 167, 172, 197, 219, 225, 294
– diseases 228
Virological-serological diagnosis 70
Viropexis 161, 199
Viroplasm 204, 294
Virosome 204, 294
Virulence 25, 83, 275–276

Virus 1, 2, 4, 5, 30, 44, 47, 91, 158
- adsorption 54–55, 97, 161, 198–200
- allergens 112
- antibody complex 253
- associated enzymes 184
- attachment 198
- attenuation 25, 79, 275, 291
- carriers 89
- chemotherapy 92, 277–285
- classification 89–90, 163, 191–197
- contacts 107, 130
- culture 63–65
- definitions 42–43, 89, 90, 158
- diagnosis 255–260
- elution 54, 100
- enzymes 140, 183, 184
- fixe 80
- inactivation 120
- induced enzymes 184
- induced immune defects 272–274
- latency 213, 277
- multiplication cycle 161, 197–212
- mutants 154
- paticle weights 140
- pathogenicity 200, 212–214, 269–277
- protein 140
- pyrogens 112
- release 211
- reproduction 89–90, 162
- reservoirs 108
- toxins 112
- transmission 106, 120
- tropism 89, 90, 200
- uptake 200
Viruses
  animal v. 73, 75, 90, 214–219
  complete v. 161
  complex v. 143
  cryptic v. 221
  defective v. 187
  deficient v. 187
  endogenous v. 234
  fish v. 218–219
  helper v. 186
  incomplete v. 100, 114, 187
  insect v. 75, 85, 90.
  intracellular v. 234
  intrachromosomally v. 234
  large v. 48
  latent v. 85
  leukemia v. 229
  mutagenic v. 272
  persisting v. 78
  plant-pathogen insect v. 131
  plant v. 73, 75, 85, 89, 132, 219–220
  sattelite v. 186
  small v. 48
  teratogenic v. 270
  true v. 48
  tumour v. 113–114, 229–244
  unconventional v. 223, 224
  wart v. 239
Visna/maedi disease 223

Watson-Crick model 160
Western blotting 259
WHO 264
Worms 10
Wound infections 21

X-bodies 62, 204
X-ray crystal lattice structure analyses 142, 173
X-ray diffraction 98

Yellow fever 41–64, 100, 109
–, –, commission 41
–, –, virus 109

Zidovudine 284
Zone electrophoresis 97
Zovirax 283
Zuckerpilz 11

MIX
Papier aus verantwortungsvollen Quellen
Paper from responsible sources
FSC® C105338

If you have any concerns about our products,
you can contact us on
ProductSafety@springernature.com

In case Publisher is established outside the EU,
the EU authorized representative is:
Springer Nature Customer Service Center GmbH
Europaplatz 3, 69115 Heidelberg, Germany

Printed by Libri Plureos GmbH
in Hamburg, Germany